REGULATION OF
GLUCONEOGENESIS

Regulation of Gluconeogenesis

9th Conference of the Gesellschaft für Biologische Chemie

Edited by

Hans-Dieter Söling

and

Berend Willms

106 Illustrations

1971

Georg Thieme Verlag · Stuttgart
Academic Press · New York and London

Editors:

Professor Dr. H.D. SÖLING
Division for Clinical Biochemistry
Department for Internal Medicine
University of Göttingen
Göttingen / Germany

Priv.-Doz. Dr. B. WILLMS
Division for Clinical Biochemistry
Department for Internal Medicine
University of Göttingen
Göttingen / Germany

For Georg Thieme Verlag: ISBN 3 13 465501 2
For Academic Press: ISBN 0-12-654350-X
Library of Congress catalog card number 72-132014

Preface

This meeting was first planned as a continuation of a former Conference of the Gesellschaft für Biologische Chemie on "Stoffwechsel der isoliert perfundierten Leber".

To the organizers of the present meeting it seemed far more reasonable to restrict the subject to a special biochemical problem than to a technical procedure which had not changed very much during the last years.

Thanks to the aid of several sponsors and thanks to the excellent scientific contributions of the participants, the conference on "Regulation of Gluconeogenesis" became a real success. We feel that the extensive discussions were the most important aspects of the meeting. The problem of "cross over plot" interpretation, which has never been discussed so openly may serve as an example. We therefore are grateful to the Georg Thieme Verlag for the opportunity to publish the discussion in full.

Göttingen, January 15, 1971 H. D. SÖLING and B. WILLMS

1
2
3
4
5
6
7
8
9
10
11
12
13
14
15
16
17
18
19
20
21
22
23
24
25
26
27
28
29
30
31
32
33
34
35
36
37
38

Participants of the Meeting

1. B. Willms, Göttingen
2. B. Stumpf, Göttingen
3. W. Guder, München
4. D. H. Williamson, Oxford
5. G. Weiss, Göttingen
6. L. Weiss, München
7. F. Gabrielli, Pisa
8. B. Störmer, Düsseldorf
9. J. Fröhlich, München
10. I. Brand, Göttingen
11. U. Schacht, Frankfurt
12. P. Walter, Bern
13. J. Papenberg, Heidelberg
14. O. Wieland, München
15. F.J. Ballard, Philadelphia
16. C. Seufert, Göttingen
17. R. Haeckel, Hannover
18. K. H. Rudorff, Düsseldorf
19. H. D. Söling, Göttingen
20. H. A. Krebs, Oxford
21. M. F. Utter, Cleveland
22. J. Kleineke, Göttingen
23. L. Herberg, Düsseldorf
24. R. G. Thurman, München
25. G. Schäfer, Hannover
26. W. Schoner, Göttingen
27. U. Panten, Göttingen
28. B. Dugal, Göttingen
29. R. Scholz, München
30. H. Stork, Mannheim
31. J. R. Williamson, Philadelphia
32. G. Müllhofer, München
33. J. H. Exton, Nashville
34. R. Heerdt, Mannheim
35. R. W. Hanson, Philadelphia
36. N.B. Ruderman, Oxford
37. W. Seubert, Göttingen
38. W. Staib, Düsseldorf

Contributing Authors and Discussants

ALBRECHT, E., Physiologisch-chemisches Institut der Universität, Göttingen, Germany

ANDERSON, J. H., Johnson Research Foundation, University of Pennsylvania, Philadelphia, Pennsylvania, U.S.A.

BALLARD, F. J., C.S.I.R.O., Division of Nutritional Biochemistry, Kinlore Ave, Adelaide, Australia

BLANK, B., Johnson Research Foundation, University of Pennsylvania, Philadelphia, Pennsylvania, U.S.A.

BROD, H., Physiologisch-chemisches Institut der Universität, Göttingen, Germany

EXTON, J. H., Department of Physiology, Vanderbilt University, School of Medicine, Nashville, Tennessee, U.S.A.

FRÖHLICH, J., Institut für Klinische Chemie und Forschergruppe Diabetes, Krankenhaus München-Schwabing, Munich, Germany

FUNG, Chien-Hung, Department of Biochemistry, Case Western Reserve University, Cleveland, Ohio, U.S.A.

GABRIELLI, F., Istituto di Chimica Biologica, Universita di Pisa, Italy

GUDER, W., Institut für Klinische Chemie und Forschergruppe Diabetes, Krankenhaus München-Schwabing, Munich, Germany

HAECKEL, H., Institut für Klinische Chemie, Medizinische Hochschule Hannover, Hannover, Germany

HAECKEL, R., Institut für Klinische Chemie, Medizinische Hochschule Hannover, Hannover, Germany

HANSON, R. W., Fels Research Institute and Department of Biochemistry, Temple University Medical School, Philadelphia, Pennsylvania, U.S.A.

HASSELBLATT, A., Pharmakologisches Institut der Universität, Göttingen, Germany

HERLEMANN, E., Physiologisch-chemisches Institut der Universität, Göttingen, Germany

KLEINEKE, J., Abteilung für klinische Biochemie, Medizinische Universitätsklinik, Göttingen, Germany

KREBS, H. A., Metabolic Research Laboratory, Nuffield Department of Clinical Medicine, Radcliffe Infirmary, Oxford, England

KUNTZEN, O., Institut für Physiologische Chemie und Physikalische Biochemie der Universität München, Munich, Germany

LEWIS, S. B., Department of Physiology, Vanderbilt University, School of Medicine, Nashville, Tennessee, U.S.A.

LOWY, C., Division of Endocrinology, Hammersmith Hospital, London, England

MAYOR, F., Departamento de Bioquimica, Universidad de Granada, Granada, Spain

MÜLLHOFER, G., Institut für Physiologische Chemie und Physikalische Biochemie der Universität München, Munich, Germany

NICKLAS, W. J., Johnson Research Foundation, University of Pennsylvania, Philadelphia, Pennsylvania, U.S.A.

OHLY, B., Physiologisch-Chemisches Institut der Universität, Göttingen, Germany

PANTEN, U., Pharmakologisches Institut der Universität, Göttingen, Germany

PAPENBERG, J., Medizinische Universitätsklinik, Heidelberg, Germany
PARK, C. R., Department of Physiology, Vanderbilt University, School of Medicine, Nashville, Tennessee, U.S.A.
PATEL, M. S., Division of Nutritional Biochemistry, Commonwealth Scientific and Industrial Research Organization, Adelaide, South Australia
PHILIPPIDIS, H., Fels Research Institute and Department of Biochemistry, Temple University Medical School, Philadelphia, Pennsylvania, U.S.A.
POSER, W., Pharmakologisches Institut der Universität, Göttingen, Germany
REFINO, C., Johnson Research Foundation, University of Pennsylvania, Philadelphia, Pennsylvania, U.S.A.
RESHEF, L., Department of Biochemistry, Hadassah Medical School, Jerusalem, Israel
RUDERMAN, N. B., Joslin Research Laboratory, Department of Medicine, Harvard Medical School, Peter Bent Brigham Hospital and The Diabetes Foundation Inc., Boston, Massachusetts, U.S.A.
RUDORFF, K. H., Institut für Physiologische Chemie der Universität, Düsseldorf, Germany
SCHÄFER, G.P., Abteilung für Biochemie, Abteilung III, Medizinische Hochschule Hannover, Hannover, Germany
SCHMIDT, F. H., Forschungslaboratorien der Boehringer Mannheim GmbH, Mannheim, Germany
SCHOLZ, R., Institut für Physiologische Chemie und Physikalische Biochemie der Universität München, Munich, Germany
SCHONER, W., Physiologisch-Chemisches Institut der Universität, Göttingen, Germany
SEUBERT, W., Physiologisch-Chemisches Institut der Universität, Göttingen, Germany
SEUFERT, C. D., Physiologisch-Chemisches Institut der Universität, Göttingen, Germany
SHAFRIR, E., Department of Biochemistry, Hebrew University, Hadassah Medical School, Jerusalem, Israel
SÖLING, H. D., Abteilung für Klinische Biochemie, Medizinische Universitäts-klinik, Göttingen, Germany
STAIB, W., Institut für Physiologische Chemie der Universität, Düsseldorf, Germany
STÖRMER, B., Chirurgische Klinik der Universität, Düsseldorf, Germany
STORK, H. Forschungslaboratorien der Boehringer Mannheim GmbH, Mannheim, Germany
STUCKI, J. W., Medizinisch-Chemisches Institut der Universität, Bern, Switzerland
THURMAN, R. G., Johnson Research Foundation, University of Pennsylvania, Philadelphia, Pennsylvania, U.S.A.
TOEWS, C. J., Joslin Research Laboratory, Department of Medicine, Harvard Medical School, Peter Bent Brigham Hospital and The Diabetes Foundation Inc., Boston, Massachusetts, U.S.A.

UI, M., Department of Physiology, Vanderbilt University, School of Medicine, Nashville, Tennessee, U.S.A.

UTTER, M. F., Department of Biochemistry, Case Western Reserve University, Cleveland, Ohio, U.S.A.

VELOSO, D., National Institute of Mental Health, WAW Building, Washington, D.C., U.S.A.

WALTER, P., Medizinisch-Chemisches Institut der Universität, Bern, Switzerland

WEISS, G., Physiologisch-Chemisches Institut der Universität, Göttingen, Germany

WIELAND, O., Institut für Klinische Chemie und Forschergruppe Diabetes, Krankenhaus München-Schwabing, Munich, Germany

WILLIAMSON, D.H., Metabolic Research Laboratory, Nuffield Department of Clinical Medicine, Radcliffe Infirmary, Oxford, England

WILLIAMSON, J.R., Johnson Research Foundation, University of Pennsylvania, Philadelphia, Pennsylvania, U.S.A.

WILLMS, B., Abteilung für Klinische Biochemie, Medizinische Universitätsklinik, Göttingen, Germany

WINDECK, R., Institut für Physiologische Chemie der Universität, Düsseldorf, Germany

Contents

Possible Control Mechanisms of Liver Pyruvate Carboxylase

Merton F. Utter and Chien-hung Fung
Department of Biochemistry, Case Western Reserve University, Cleveland, Ohio, U.S.A.

Summary

As the first step in the gluconeogenic pathway from pyruvate, the pyruvate carboxylase reaction is an attractive possibility for a control site and studies from the laboratories of Williamson, Park and Söling have indicated that this reaction may be rate-limiting under some circumstances. The control of this enzyme may be complex since the rate of the reaction is affected by a variety of factors including concentrations of substrates (pyruvate, $MgATP^{2-}$), activators (acetyl-CoA, ß-hydroxybutyryl-CoA) and inhibitors (PEP, acetoacetyl-CoA). Our report will describe some of the interactions of these factors at the enzymic level. The studies were conducted mainly with the enzyme isolated from chicken liver but also to some extent with the comparable enzymes from livers of rats, rabbits, calves and turkeys.

The K_m for pyruvate is about 0.3 mM for this enzyme but the relationship between substrate and initial velocity does not follow Michaelis-Menten kinetics at pyruvate concentrations exceeding 2 mM. At the higher concentrations an apparent substrate activation is observed. It is likely that pyruvate concentrations existing in vivo are in the general range of the K_m value and thus that the reaction rate will be affected by changes in pyruvate levels but it should be emphasized that the effects of changes mediated through activators and inhibitors will be superimposed on rate limitations due to pyruvate concentration.

The absolute requirement for acetyl-CoA in this reaction has focused attention on this metabolite as a possible control factor. With the avian liver species of the enzyme, the relationship between acetyl-CoA and initial reaction velocity is highly cooperative (Hill constant ≅ 2.9) and the K_a values under optimal conditions are very low (2 μM). Species of the enzyme from mammalian livers (rat, rabbit, calf) differ markedly from the avian species in the quantitative aspects of their interactions with acetyl-CoA. The Hill constants are 2.2-2.3 and the K_a under optimal conditions are an order of magnitude higher (10-20 μM). Under mitochondrial conditions, the K_a values would be expected to increase several-fold and to approach the lower limit of the range of values reported for intramitochondrial acetyl-CoA concentration in rat liver (70-200 μM).

A number of inhibitors can increase the apparent K_a for acetyl-CoA. Acetoacetyl-CoA is the only inhibitor found thus far which destroys the cooperative relationship between acetyl-CoA and chicken liver pyruvate carboxylase. At At acetoacetyl-CoA concentrations of 0.5 mM, the Hill constant for acetyl-Coa is reduced to 1 and the apparent K_a for the activator is increased 5-fold. These properties suggest that acetoacetyl-CoA would be an unusually effective control factor for pyruvate carboxylase, particularly since 3-hydroxybutyryl-CoA is an activator of the enzyme. Alterations in the NADH/NAD ratio as reflected in the 3-hydroxybutyryl-CoA/acetoacetyl-CoA ratio could influence the rate of pyruvate carboxylation and gluconecogenesis. At present, the intramitochondrial levels of these acyl-CoA's are not well-established so the validity of this concept cannot be judged.

Phosphoenolpyruvate also appeared to be an attractive inhibitor of this enzyme but very recent studies have disclosed that the inhibitor is not phosphoenolpyruvate but an unidentified anionic contaminant which is present in crystalline samples of sodium phosphoenolpyruvate from one commercial source (Calbiochem).

The most attractive hypothesis for regulation of pyruvate carboxylation appears to be a multi-layered system of controls with the reaction rate based on available concentrations of pyruvate and/or $MgATP^{2\ominus}$ as modified by acetyl-CoA. The effects of the latter would in turn be modulated by the ratio of 3-hydroxybutyryl-CoA/acetyl-CoA. Such a system would provide both excellent sensitivity and flexibility.

This work was supported in part by NIH grant R 01 AM 12245 and AEC grant AT-1242.

Introduction

Extrapolation of results obtained from studies at the enzymic level suggests that control of pyruvate carobxylation and in part of gluconeogenesis may be vested in an interlocking series of controls. These include: limitation of pyruvate supply; activation by acetyl-CoA; inhibition by acetoacetyl-CoA which increases the apparent K_a for acetyl-CoA and destroys the allosteric activation of the enzyme by acetyl-CoA and influence of the redox state of the mitochondria exerted through the activator/inhibitor pair of ß-hydroxybutyryl-CoA/acetoacetyl-CoA.

As the initial step in conversion of pyruvate to carbohydrate in many gluconeogenic cells, the pyruvate carboxylase reaction might be expected to be one of the control sites for this metabolic sequence. Support for this hypothesis has been furnished by experiments with perfused liver by Williamson et al. (1969a), Exton et al. (1969) and Söling et al. (1968). Such studies have shown that under certain types of perturbation (e.g., the addition of fatty acids), the overall rate of synthesis of glucose from pyruvate or lactate increases and a "crossover" is observed between pyruvate and oxalacetate or malate. These observations are consistent with the view that the pyruvate carboxylase reaction is a rate-limiting reaction under these circumstances.

If the pyruvate carboxylase reaction is a control site, the factors which may influence the rate of this reaction are of interest. One particular approach to this problem has been based on the studies at the enzymic level. Such studies have the advantage that experimental conditions can be controlled rigidly, but possible control factors arising from this sort of investigation can serve only as suggestions for the testing of more complex systems. The present studies are an account of our attempts to identify substrates, activators and inhibitors which may influence the rate of reaction of pyruvate carboxylase from liver. Most of the studies have been carried out with the enzyme purified from chicken liver (Scrutton et al. 1969), but some of the studies have utilized analagous enzymes from rat and rabbit livers.

Pyruvate as a Possible Control Factor for Pyruvate Carboxalase

In any metabolic sequence, the relative availability of the initial substrate offers an opportunity for control of the overall process. If effective control is to be expected, the concentration of the substrate must vary through the range of concentrations where the rate of the initial enzymic reaction is most responsive, that is from values of about twice the K_m value downward. It is of interest to examine pyruvate and pyruvate carboxylase in this way. The K_m for pyruvate for the enzyme from chicken liver is approximately 0.3 mM when determined in the presence of less than 2 mM pyruvate and approximately saturating amounts of all other reaction components (Keech and Utter 1963). A more detailed examination of the response of this enzyme to pyruvate concentrations reveals that the kinetics are not Michaelis-Menten except in certain concentration ranges. Figure 1 shows that the double reciprocal plots of velocity versus pyruvate concentration at varying concentrations of $MgATP^{2-}$ are non-linear at higher concentrations of pyruvate. The K_m values obtained from linear portions of the curve (not shown here) at lower concentrations of pyruvate (2.0 mM-0.08 mM) are about 0.3 mM. It should be noted that non-linearity is not observed with all samples of the enzyme. When observed the departure from linearity depends upon the $MgATP^{2}$-concentration but in general the activation appears to begin at about 2 mM pyruvate and the leveling off at a minimum of 10 mM pyruvate. It seems likely that these concentrations are higher than might be expected to occur in vivo; therefore these effects may be of little importance physiologically. Taylor et al. (1969) have recently reported that the response of pyruvate carboxylase from sheep liver to pyruvate is also non-Michaelis-Menten.

Alterations in the concentration of other reaction components have relatively little effect on the apparent K_m for pyruvate. In Fig. 1, the K_m value obtained from the linear portions of the various curves (pyruvate between 2 and 0.1 mM vary only slightly with alterations in $MgATP^{2-}$ concentrations. The values are approximately 0.3 mM in all cases. The same is true if the K_m for pyruvate is determined at less than 1 mM pyruvate and different levels of acetyl-CoA as shown in Fig. 2. The data are presented as double reciprocal plots of initial velocities versus pyruvate concentration at levels of acetyl-CoA ranging from K_a values (2 μM) to 10 times that value. The data are restricted to the linear portions of the curve, that is, to concentrations of pyruvate of 1 mM and below. The apparent K_m values for pyruvate over this range of acetyl-CoA concentrations vary from approximately 0.4 to 0.45 mM. When the converse relationship is examined, it is found that the concentration of pyruvate has no significant effect on the apparent K_a for acetyl-CoA (not visualized in this report).

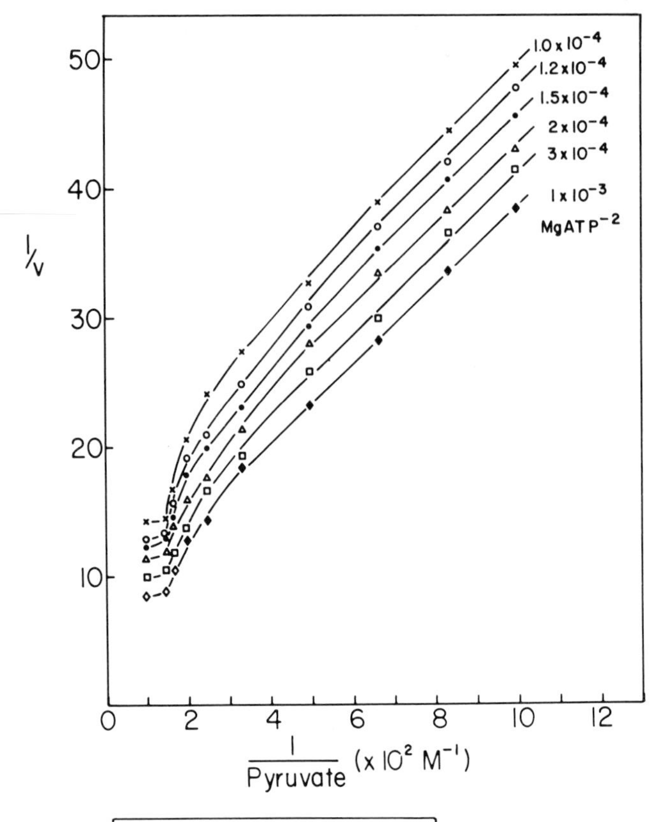

Fig. 1. Lineweaver-Burk Plot for Pyruvate (Pyruvate Carboxylase)

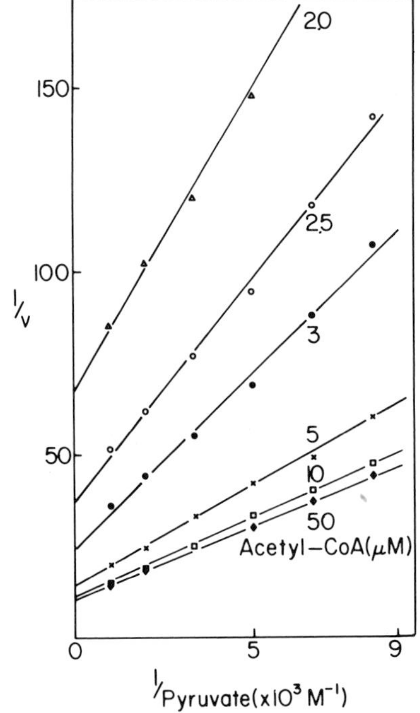

Fig. 2. Relationship of Acetyl-CoA and Pyruvate Concentrations on Pyruvate Carboxylase Activity

The lack of dependence on the concentrations of other reactants of the K_m for pyruvate is consistent with the reaction mechanism proposed earlier for this enzyme (Scrutton 1969).

$$\text{Biotin-Enz} + \text{ATP} + HCO_3^- \underset{Mg^{2+}}{\overset{\text{Acetyl-CoA}}{\rightleftharpoons}} CO_2^- \sim \text{Biotin-Enz} + \text{ADP} + P_i \quad (1)$$

$$CO_2^- \sim \text{Biotin-Enz} + \text{Pyruvate} \rightleftharpoons \text{Biotin-Enz} + \text{Oxalacetate} \quad (2)$$

According to this mechanism, the affinity of the enzyme for pyruvate should be independent of the concentration of other reactants, including acetyl-CoA. The anomalous results observed at high pyruvate concentrations (Fig. 1) suggest that under these conditions a somewhat more complex mechanism than the one presented above is obtained.

A K_m value for pyruvate in the range of 0.3 mM suggests that the concentration of this substrate may be one of the controlling factors for the pyruvate carboxylase reaction and the gluconeogenic pathway. Intramitochondrial concentration levels for pyruvate have not been well established but on the basis of tissue and serum levels of pyruvate, the 0.3 mM value would seem likely to be in the physiological range. It is interesting that Adam and Haynes (1969) have recently suggested that the rate of carboxylation of pyruvate by rat liver mitochondria is dependent on the rate of entry of pyruvate into the mitochondria and that the rates of entry can be influenced by prior treatment of the animal with various endocrine factors.

It should be emphasized that even if the rate of pyruvate carboxylation is limited by the concentration of pyruvate, the rate of the reaction can still be altered by changes in the concentration of other controlling factors. For example, in Fig. 2 the rate of the reaction at any particular concentration of pyruvate is obviously very much influenced by the acetyl-CoA concentration. In a similar fashion any control exercised through changes in the concentrations of activators, inhibitors or other substrates will be superimposed on rate limitations due to pyruvate concentrations.

Activation by Acetyl-CoA

Pyruvate carboxylases of animal origin show an absolute dependence on the presence of an acyl-CoA for catalytic activity and this property led to the suggestion that variations in the concentrations of acetyl-CoA might control the rate of pyruvate carboxylation and gluconeogenesis (Utter et al. 1964). If acetyl-CoA is to act effectively in this way, its in vivo concentrations must coincide with those effective in regulating the enzyme. Generally speaking, the concentration should be in the K_a range. Early studies with pyruvate carboxylase from chicken liver indicated K_a values of approximately 20 μM for acetyl-CoA for this enzyme (Keech and Utter 1963, Scrutton and Utter 1967), but more recent studies have shown that under optimal conditions the K_a is only about 2 μM. As discussed recently (Utter and Scrutton 1969), this very low value for the K_a for acetyl-CoA appeared to pose a problem because estimates of the in vivo concentrations of acetyl-CoA in rat liver ranged from 20 to 70 μM (Söling et al. 1968, Williamson et al. 1969b) with the intramitochondrial values perhaps running somewhat higher. These values are sufficiently high that pyruvate carboxylase would be unaffected by changes in acetyl-CoA concentration. That is, the enzyme would always be fully activated. This is particularly true because the effect of increasing acetyl-CoA concentration on initial velocity is highly cooperative with a Hill coefficient of about 3 (Scrutton and Utter 1967). The effective concentration range in such a highly cooperative relationship encompasses only about a 5-fold change in acetyl-CoA concentrations.

Recent studies with pyruvate carboxylases from rat and rabbit livers may furnish at least part of the answer to this apparent dilemma. When tested under identical conditions to those used for the enzyme from chicken liver, the responses to acetyl-CoA were quite different for the enzymes from the three different sources. The initial velocities (%Vm) are plotted against the log of acetyl-CoA concentrations in Fig. 3. It is apparent that the effective range of acetyl-CoA is much higher for the enzymes from rat and rabbit than for the avian enzyme and that the response curve is less steep for the mammalian enzymes, i.e., the interaction is less cooperative. Table 1 summarizes the K_a values and Hill coefficients for acetyl-CoA and the three species of pyruvate carboxylase. The K_a for the rat liver enzyme is about 20 μM and approximately 10-fold that of the avian enzyme and the Hill coefficient is 2.2 rather than 2.9 The rabbit liver enzyme resembles the rat liver enzyme rather closely. Other studies on the enzyme from turkey livers indicate that enzyme is very similar to the one obtained from chicken liver.

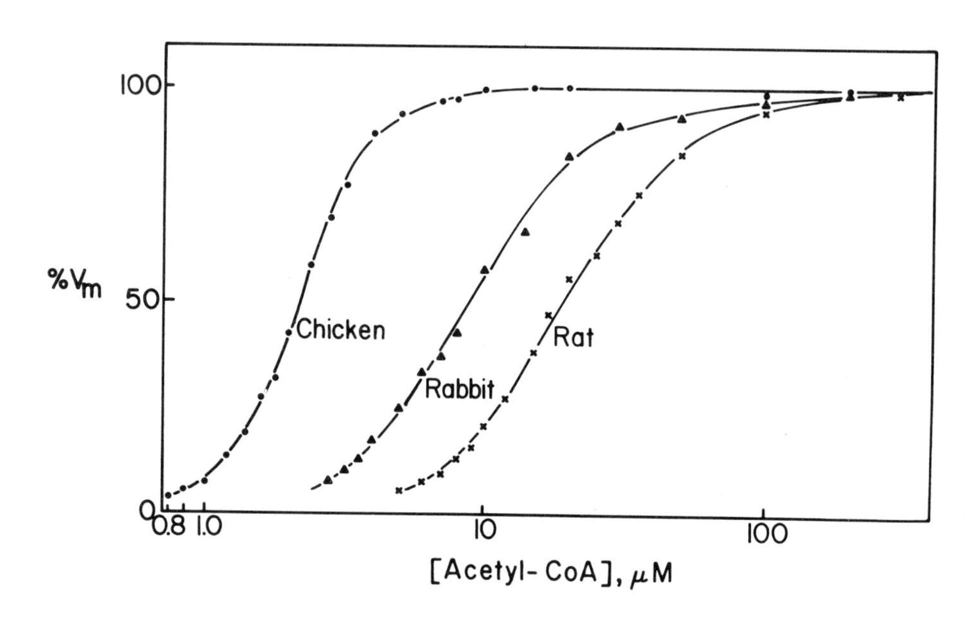

Fig. 3. Activation of Different Species of Pyruvate Carboxylase by Acetyl-CoA.

Species of enzyme	Acetyl-CoA K_a	Acetyl-CoA Hill Coefficient
	μM	
Chicken liver	2.2	2.95
Rat liver	19	2.2
Rabbit liver	9	2.0

Table 1. Comparison of activation of different pyruvate carboxylases by acetyl-CoA.

These results suggest that the interaction of acetyl-CoA varies considerably according to the species of enzyme involved. The considerably higher K_a observed for the rat liver enzyme (20 μM) comes much closer to the reported in vivo concentrations but still appears somewhat lower than those values. It should be noted, however, that the values reported in Table 1 are those obtained under optimal conditions and that a number of factors tend to increase the apparent K_a values for acetyl-CoA. For example with the chicken liver enzyme, decreasing the pH from 7.8 to 7.1 raises the K_a value 3 or 4 times (Scrutton and Utter 1967). Also a number of inhibitors, including those discussed in the next section, have the effect of increasing the apparent K_a for acetyl-CoA.

Inhibitors of Pyruvate Carboxylase

A number of compounds have been found to act as inhibitors of pyruvate carboxylase. These include: carboxylated forms of activators such as malonyl- and methylmalonyl-CoA; derivatives of acyl-CoA, such as acetylpantetheine and acetyl-desamino-CoA; oxidized CoA; sulfate ion; phosphoenolpyruvate (PEP); and acetoacetyl-CoA. The last two compounds on this list appeared to be the most interesting as possible physiologically important inhibitors and have therefore been investigated in greater detail. PEP was of interest because it is the product of the next reaction in the gluconeogenic sequence and would appear to offer a possibility for feedback inhibition: acetoacetyl-CoA because it was the only inhibitor of a list of some 20 tested which destroys the cooperative interaction of pyruvate carboxylase with acetyl-CoA. That is, in the presence of acetoacetyl-CoA the Hill coefficient for acetyl-CoA drops from 2.9 to 1.

Tests with PEP showed this inhibitor to be uncompetitive against pyruvate and a mixed inhibitor against acetyl-CoA and $MgATP^{2-}$. Against the latter substrate, the K_I for PEP was estimated approximately 0.45 mM. Unfortunately, it has recently become clear that the inhibition of pyruvate carboxylase is not due to PEP but to a contaminant which is present in certain commercial samples of this compound. All of the earlier studies have been carried out with the crystalline trisodium salt of PEP obtained from California Biochemicals Co. (A grade). Four different lots of this material have been found to contain the inhibitor although the level of contamination varies. On the other hand, samples of the trisodium salt of PEP obtained from Sigma Chemical Co. showed no inhibitory action on the pyruvate carboxylase reaction. Enzymatic analyses for PEP showed essentially identical PEP content in the various samples (>92%). Further examination of the California Biochemical Co. PEP discloses an anionic inhibitor which can be separated from PEP by DEAE-cellulose chromatography. The inhibitor has not yet been identified. The method of synthesis of PEP used by California Biochemicals Co. (private communication) is based on the method of Fischer and Baer (Ball 1952) and depends on the reaction of ß-chlorolactate with $POCl_3$ followed by removal of HCl by exposure to alcoholic KOH. It should be noted that Sigma Chemical Co. uses a different synthetic procedure for the preparation of PEP.

The elimination of PEP as a possible inhibitor of pyruvate carboxylase has focused our attention on acetoacetyl-CoA. As shown in Fig. 4, the presence of acetoacetyl-CoA with the chicken liver enzyme causes a substantial inhibition of the rate of the reaction at various levels of acetyl-CoA. At an acetyl-CoA level of 2×10^{-6}M (the K_a level), the presence of 0.545 mM acetoacetyl-CoA causes about 90 percent inhibition. The effect of acetoacetyl-CoA on the Hill coefficient ("n" value) and K_a for acetyl-CoA is shown in Fig. 5. The Hill coefficient falls rapidly in the presence of acetoacetyl-CoA until it reaches 1 at about 0.5 mM acetoacetyl-CoA. The K_a value for acetyl-CoA is markedly increased in the presence of acetoacetyl-CoA and is raised about 10-fold by the presence of a 0.5 mM concentration of the inhibitor. Acetoacetyl-CoA shows a cooperative effect in its inhibitory action with the enzyme with a Hill coefficient of about 2.8. The degree of cooperativity is independent of the concentration of acetyl-CoA.

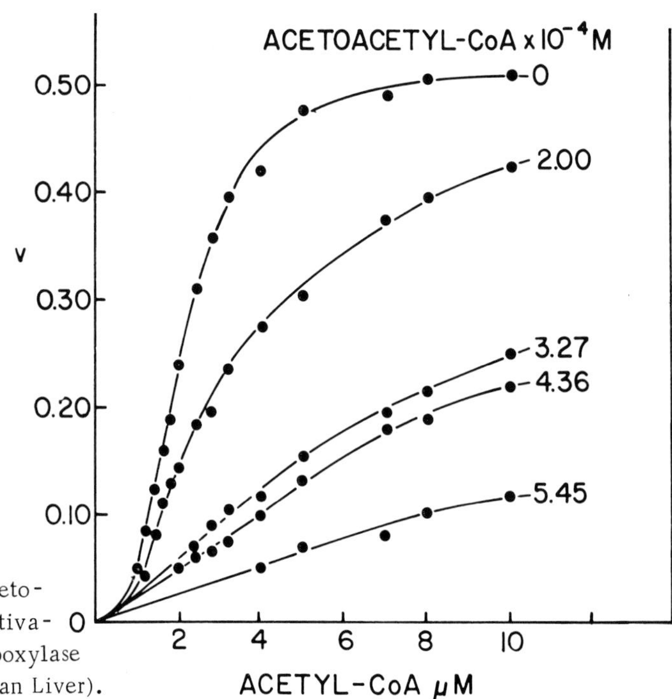

Fig. 4. Effect of Acetoactyl-CoA on the Activation of Pyruvate Carboxylase by Acetyl-CoA (Avian Liver).

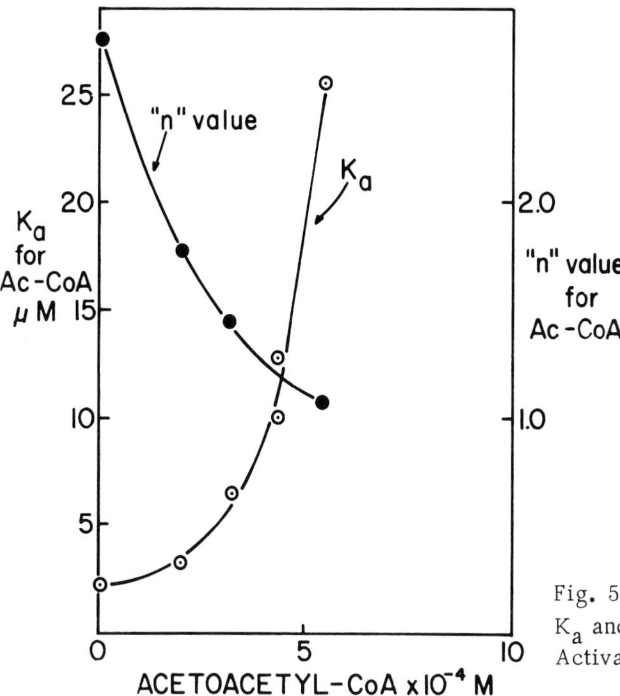

Fig. 5. Effect of Acetoacetyl-CoA on the K_a and Hill Coefficient of Acetyl-CoA Activation of Pyruvate Carboxylase.

The very dramatic effects of acetoacetyl-CoA on the apparent interaction of acetyl-CoA and the chicken liver enzyme appear to offer a way of studying the unsolved problem of the mechanism by which acetyl-CoA activates this enzyme. In addition, these observations raise the questions as to whether acetoacetyl-CoA may play a physiological role in the control of pyruvate carboxylation and whether the inhibitory effects of acetoacetyl-CoA apply to species of the enzyme other than that obtained from chicken liver. The latter question is answered by the data of Table 2 where it is shown that acetoacetyl-CoA inhibits the rat and rabbit liver enzymes as well as that from avian liver. The K_I values measured against $MgATP^{2-}$ at saturating concentrations of acetyl-CoA vary from 0.32 to 0.97 mM. The in vivo concentrations of acetoacetyl-CoA are not known and further evaluation of this physiological significance of acetoacetyl-CoA inhibition must await these data.

Species	Acetyl-CoA K_a	Acetoacetyl-CoA K_I
	μM	mM
Chicken liver	2.2	0.32
Rat liver	19	0.66
Rabbit liver	9	0.97

Table 2. Comparison of activating and inhibiting acyl-CoA for different species of pyruvate carboxylase.

One very interesting observation is that ß-hydroxybutyryl-CoA is an activator for pyruvate carboxylase. For the chicken liver enzyme, this compound has a K_a value of about 0.1 mM and a Hill coefficient of about 2.2. It seems likely that in mitochondria, the ß-hydroxybutyryl-CoA/acetoacetyl-CoA ratio may be influenced by the NADH/NAD ratio. This situation may offer an opportunity for the control of pyruvate carboxylation by the redox state of the mitochondria. It is interesting to note that Williamson et al. (1969a) have found that the NADH/NAD ratio increases when fatty acids are administered to perfused rat liver. Under these conditions gluconeogenesis and pyruvate carboxylation also appear to increase.

Acknowledgment

This work was supported by Grant AT (11-1)-1242 from the U. S. Atomic Energy Commission and grants 5R01AM12245, 5R01GM-13791 from the National Institutes of Health.

References

Adam, P. A. J., R. C. Haynes, Jr.: J. biol. Chem. 244 (1969) 6444

Biochemical Preparation, Vol. 2, Ed. by E. G. Ball and John Wiley, New York 1952, p. 25

Exton, J. H., J. G. Corbin, C. R. Park, J. biol. Chem. 244 (1969) 4095

Keech, D. B., M. F. Utter: J. biol. Chem. 238 (1963) 2609

Scrutton, M. C., D. B. Keech, M. F. Utter: J. biol. Chem. 240 (1965) 574

Scrutton, M. C., M. F. Utter: J. biol. Chem. 242 (1967) 1723

Scrutton, M. C., M. R. Olmsted, M. F. Utter: Methods in Enzymology, Vol. XIII, Ed. by J. M. Lowenstein, Academic, New York 1969, p. 235

Söling, H. D., B. Williams, D. Friedrichs, J. Kleineke: Europ. J. Biochem. 4 (1968) 364

Taylor, H., J. Nielsen, D. B. Keech: Biochem. biophys. Res. Commun. 37, (1969) 723

Utter, M. F., D. B. Keech, M. C. Scrutton: Advances in Enzyme Regulation, Vol. II, Ed. by G. Weber, Pergammon, Oxford 1964, p. 49

Utter, M. V., M. C. Scrutton: Current Topics in Cellular Regulation, Vol. I, Academic, New York 1969, p. 253

Williamson, J. R., E. T. Browning, R. Scholz: J. biol. Chem. 244 (1969a) 4607; 244 (1969b) 4617

Purification and Properties of Pyruvate Carboxylase from Rat Liver

C. D. Seufert, E. Herlemann, E. Albrecht and W. Seubert
Physiologisch-chemisches Institut der Universität Göttingen, FRG

Summary

Pyruvate carboxylase from rat liver was purified to almost homogenity. The 180-fold purified preparations have a specific activity of 22 units/mg protein. The enzyme from rat liver shows similar characteristics as already described for the enzyme from chicken liver. It differs from the latter enzyme in its response to acetyl-CoA, an allosteric activator of pyruvate carboxylase. Like isocitrate in the case of mammalian acetyl-CoA carboxylase, acetyl-CoA stimulates the transfer of enzyme bound CO_2 ("active CO_2") to the acceptor (eq. 2) catalyzed by pyruvate carboxylase:

$$(1)\quad ATP + CO_2 + \text{enzyme-biotin} \underset{}{\overset{Mg^{++},\ \text{acetyl-CoA}}{\rightleftharpoons}} \text{enzyme-biotin-}CO_2 + ADP + P_i$$

$$(2)\quad \text{enzyme-biotin-}CO_2 + \text{pyruvate} \overset{\text{acetyl-CoA}}{\rightleftharpoons} \text{enzyme-biotin} + \text{oxalacetate}$$

The activation curve shows a sigmoid shape in a concentration range between 0.5 to 2×10^{-4} M acetyl-CoA. The stimulatory effect of acetyl-CoA in the C^{14}-pyruvate-oxalacetate exchange assay (eq. 2) is associated with an increased maximum velocity and has no apparent effect on the Km-value for pyruvate. Incubation of purified pyruvate carboxylase under various conditions (ATP, Mg^{++}, $KHCO_3$, dilutions etc.) results in extreme differences of the catalytic properties when assayed with the conventional overall assay (eq. 1 and 2) and the C^{14}-pyruvate-oxalacetate exchange assay (eq. 2). The implication of these findings concerning different forms of the enzyme from rat liver will be discussed.

Introduction

The reaction limiting maximum gluconeogenesis from C_3-precursors in liver is located between pyruvate and phosphoenolpyruvate (Seubert et al. 1968, Exton and Park 1969, Williamson et al. 1969). Glucagon and epinephrine, acting through cyclic adenosine -3,5-monophosphate, accelerate gluconeogenesis at this level. As possible mechanisms responsible for acceleration of gluconeogenesis, changes of the intracellular concentrations of effectors of pyruvate carboxylase (Williamson et al. 1969), a shift of the redoxpotential of the NAD system to more negative (Williamson et al. 1969, Menahan and Wieland 1969) values and a change in the electrolyte concentrations have been considered (Williamson et al. 1969, Exton and Park 1968). Besides these secondary effects of glucagon on gluconeogenesis, the interconversion of the enzymes involved in phosphoenolpyruvate synthesis into different forms also had to be considered, as demonstrated for phosphorylase, glycogen synthetase and pyruvate dehydrogenase. In order to check a possible primary effect of glucagon at the level of pyruvate carboxylation, the enzyme involved in this process was purified from rat liver and its properties were studied.

Results

Pyruvate carboxylase was purified from freeze-dried, 14000 g-residue of liver of starved rats (Seufert et a.). The purification procedure includes fractionations with solid ammonsulfate and saturated ammonsulfate solutions, and a fractionated extraction of the residue obtained by the latter step. The purified preparations had specific activities of 11.5 - 22 U/mg protein. The purification effect was 150 - 200 times. Upon chromatography on DEAF sephadex the specific activity of the enzyme fell down to about 20% of the original activity when assayed in the overall assay (eq. 4 + 5). However, as is evident from Fig. 1, the impurities still present after fractionated extractions are eliminated at this step and the enzyme moves as a single band in the disc electrophoresis. The identity of the enzyme with a biotin-containing protein is supported by y co-electrophoresis with avidin, a protein binding specifically to biotin: as is evident from Fig. 1, in presence of avidin pyruvate carboxylase moves with the former protein.

Alberts et al. (1969) have demonstrated that acetyl-CoA carboxylase from *E. coli* is composed of three subunits:

$$(1)\quad ATP + HCO_3^- + biotin \xrightleftharpoons[]{Mn^{++},\ biotin\ carboxylase} CO_2\text{-}biotin + ADP + Pi$$

$$(2)\quad ATP + HCO_3^- + biotin\text{-}protein \xrightleftharpoons[]{Mn^{++},\ biotin\ carboxylase} CO_2\text{-}biotin\text{-}protein + ADP + Pi$$

$$(3)\quad CO_2\text{-}biotin\text{-}protein + CH_3\text{-}COSCoA \xrightleftharpoons[]{transferase} OOC\text{-}CH_2\text{-}COS\text{-}CoA + biotin\text{-}protein$$

The first subunit catalysis is a model reaction, the ATP dependent carboxylation of (+)-biotin (eq. 1). Another biotin containing subunit (biotin-carrier-protein) is carboxylated by the biotincarboxylase corresponding to the model reaction (eq. 2). A third subunit, called transferase, transfers the enzyme bound CO_2 to the acceptor acetyl-CoA. On the basis of these observations Alberts et al. (1969) have postulated that all biotin enzymes might be composed of similar subunits. According to their conclusions, the substrate specifity of biotin enzymes should be determined by the transferase subunit. The cross reaction of antibodies prepared to homogenous chicken liver acetyl-CoA carboxylase with the enzyme from rat liver, observed more recently by Majerus and Kilburn (1969) would be in accord with this postulate.

In order to differentiate between subunits of pyruvate carboxylase and to study a possible interconversion of a transferase subunit into forms with different catalytic properties, a new assay for pyruvate carboxylase was developed. This assay is based upon the second step of the overall carboxylation process (eq. 5), i.e. the transfer of CO_2 from enzyme bound carboxybiotin to pyruvate.

$$(4)\quad ATP + HCO_3^- \quad \text{biotin-enzyme} \xrightarrow[\text{acetyl-CoA}]{Mg^{++}} CO_2\text{-biotin-enzyme} + ADP + Pi$$

$$(5)\quad CO_2\text{-biotin-enzyme} + \text{pyruvate} \xrightarrow[\text{acetyl-CoA}]{} \text{oxalacetate} + \text{biotin-enzyme}$$

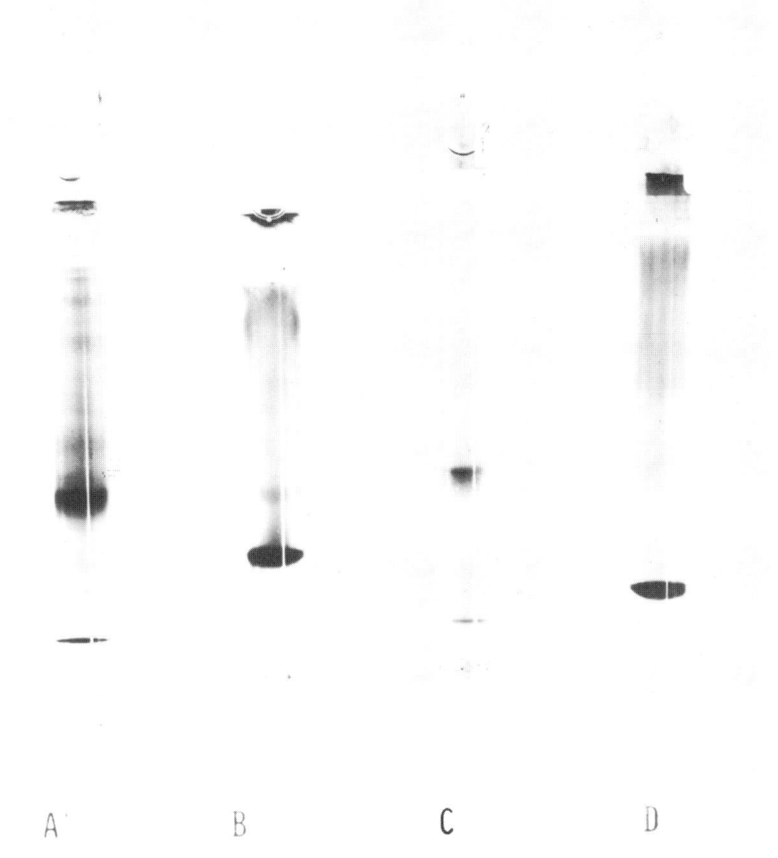

Fig. 1. Analysis of pyruvate carboxylase from rat liver at various purification stages.
A) 40 μg enzyme after fractionated extraction
B) 40 μg enzyme after fractionated extraction + 0,25 units avidin
C) 10 μg enzyme after chromatography on DEAE-sephadex
D) 10 μg enzyme after chromatography on DEAE-sephadex + 0,25 units avidin

This reaction can be followed by the ^{14}C-pyruvate-oxalacetate exchange. Radioactive oxalacetate formed was converted to citrate. The radioactivity of the citrate was determined after separation excess of radioactive pyruvate by high voltage electrophoresis according to Henning and Seubert (1964). Incorporation of ^{14}C-pyruvate into citrate was a measure of the enzyme activity in the exchange assay.

In Fig. 2, the kinetics of the exchange reaction with different amounts of the purified enzyme are illustrated. In absence of oxalacetate no incorporation of radioactivity into citrate is observed. As is evident, there is a straight line relationship between enzyme concentration and incorporation of radioactive pyruvate into oxalacetate (Fig. 2b). In contrast to the chicken liver enzyme (Scrutton et al. 1965), the enzyme from rat liver is also activated by acetyl-CoA in the second step of the overall process (eqn. 5). The activation curve shows a simoid shape in a concentration range between 50 and 200 μM (Fig. 3a). From the initial rate data for variation of acetyl-CoA concentration in the Hill plot a value of n=1.9 is calculated (Fig. 3b). This value is in the same order of magnitude as the value of n=2.4 calculated from the initial rates at various acetyl-CoA concentrations in the overall assay (Fig. 4a and b). Similar intersite interactions for both steps (eqn. 4 and 5) of the carboxylation process are deduced from these values.

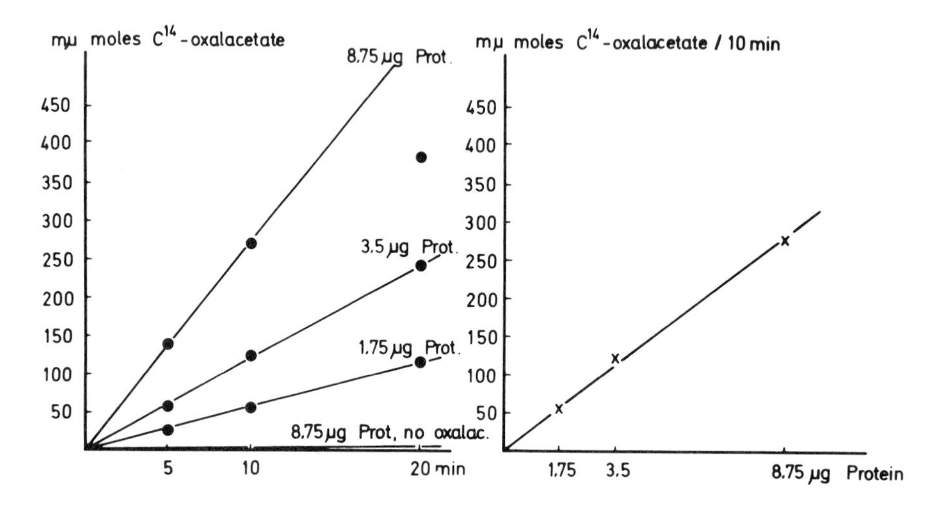

Fig. 2. Assay of pyruvate carboxylase activity by the 1-^{14}C-pyruvate-oxalacetate exchange (Seufert et al.).

A) kinetics of incorporation of ^{14}C-pyruvate with various amounts of enzyme

B) relation between enzyme concentration and incorporation into oxalacetate

The relationship between the activities of pyruvate carboxylase in the ^{14}C-pyruvate-oxalacetate exchange assay and the concentrations of ^{14}C-pyruvate gives with and without acetyl-CoA hyperbolic curves as shown in Fig. 5a. The stimulatory effect of acetyl-CoA is associated with an increased maximum velocity and has no apparent effect on the K_m-value for pyruvate (K_m for pyruvate in presence of acetyl-CoA = $3.5 \cdot 10^{-3}$ M). If the initial velocities are plotted against the pyruvate concentrations according to Hill, n-values of 1 are obtained (Fig. 5b).

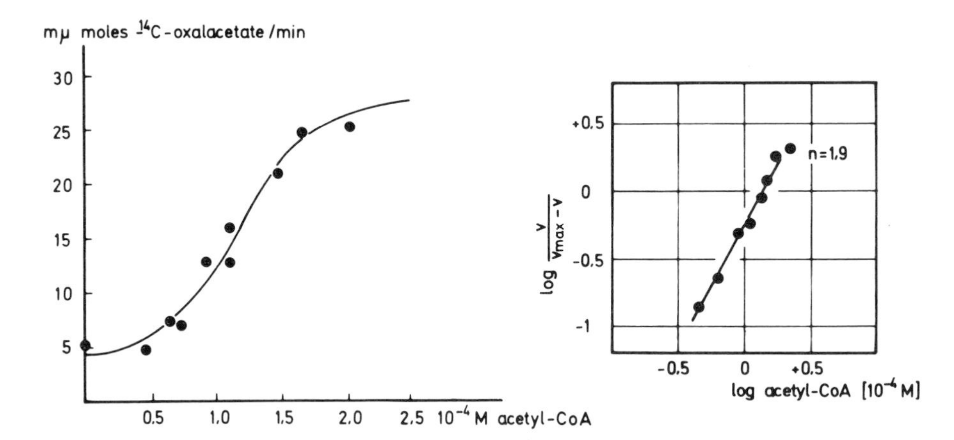

Fig. 3. Activation of pyruvate carboxylase by acetyl-CoA in the ^{14}C-pyruvate-oxalacetate exchange assay (Seufert et al.).

A) relation between acetyl-CoA concentration and ^{14}C-pyruvate-oxalacetate exchange

B) initial rate data for acetyl-CoA plotted as $\log_{10}$ (v/Vmax - v) with respect to $\lg_{10}$ acetyl-CoA

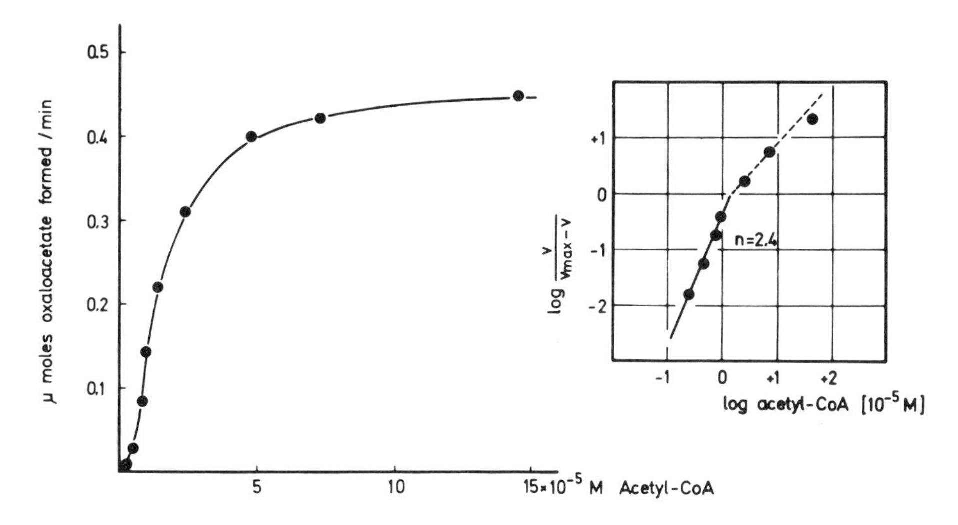

Fig. 4. Activation of pyruvate carboxylase by acetyl-CoA in the overall assay (Seufert et al. in press).

A) relation between acetyl-CoA concentrations and malate formation in the optical assay (Seufert et al.) 25 μg enzyme

B) initial rate data for acetyl-CoA plotted as $\log_{10}$ (v/Vmax - v) with respect to $\log_{10}$ acetyl-CoA.

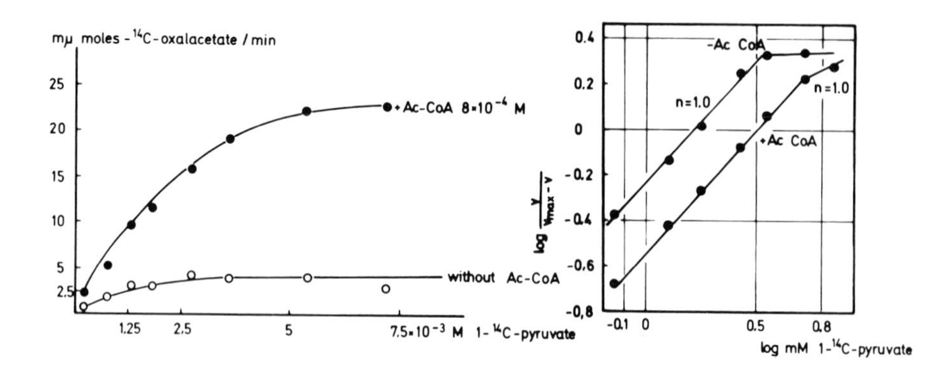

Fig. 5. Effect of pyruvate on pyruvate carboxylase activity in the exchange assay (Seufert et al.).
A) relation between incorporation in oxalacetate and pyruvate without acetyl-CoA and at saturating concentration of acetyl-CoA
B) initial rate data for pyruvate plotted as $\log_{10}$ acetyl-CoA

Activation of the transcarboxylase reaction has already been demonstrated for another biotin-containing enzyme, the acetyl-CoA carboxylase from chicken liver (Ryder et al. 1967, Stoll et al. 1968, Waite and Wakil 1963) and rat liver (Matsuhashi 1964). As demonstrated for pyruvate carboxylase from rat liver (Fig. 5), the allosteric effector isocitrate of acetyl-CoA carboxylase increases only the maximum velocity. According to Monod et al. (1965), mammalian acetyl-CoA carboxylase and pyruvate carboxylase from rat liver have therefore to be associated with the "V-system" of regulatory enzymes.

Oxalacetate is a strong inhibitor at higher concentrations (Fig. 6). The inhibition of the pyruvate-oxalacetate exchange by oxalacetate seems to be not competitive to acetyl-CoA (not shown in this report); with respect to pyruvate, a mixed type inhibition of oxalacetate was obtained.

Incubation of pyruvate carboxylase in presence of EDTA + Mg^{++} results in complete inactivation of the enzyme within a few minutes when assayed in the overall assay (eq. 4 and 5). The inactivation is dependent upon the presence of both Mg^{++} and EDTA. EDTA (Fig. 7) or magnesium (Fig. 8) are ineffective alone. When the enzyme activity is assayed after treatment with EDTA + Mg^{++} in the exchange assay (eq. 5), no activity loss can be observed (Fig. 8). Bicarbonate has a protective effect on the inactivation of pyruvate carboxylase. Also, ATP has some protective effect. Potassium stimulates the inactivation (Fig. 9).

The different catalytic properties of pyruvate carboxylase in the exchange assay and the overall assay after incubation with Mg^{++} + EDTA point to an inactivation of only the first step of the carboxylation process. In agreement with Alberts et al. (1969) participation of different subunits of pyruvate carboxylase or at least different catalytic sites of the enzyme in oxalacetate synthesis are deduced from these findings. Experiments are in progress to support a physiological significance of the above inactivation of rat liver pyruvate carboxylase in regulating gluconeogenesis from C_3-precursors.

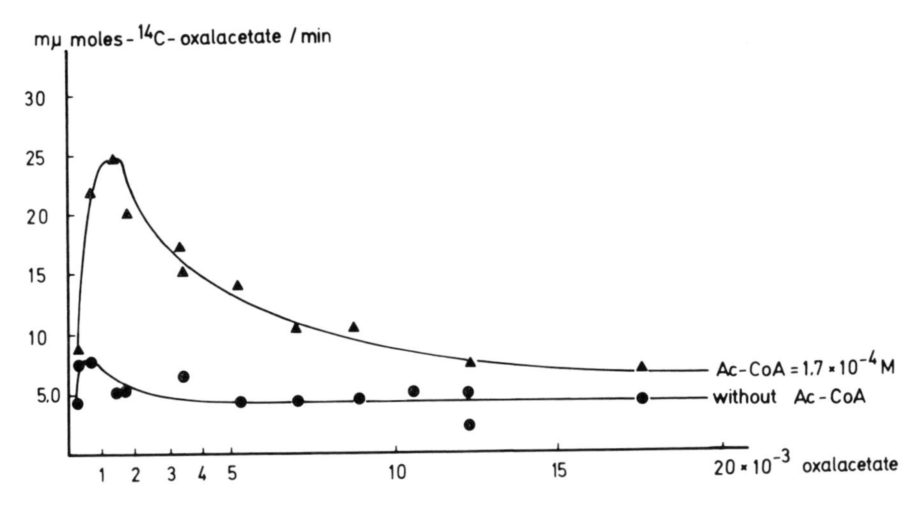

Fig. 6. Effect of oxalacetate on pyruvate carboxylase activity in the exchange assay (Seufert et al.).

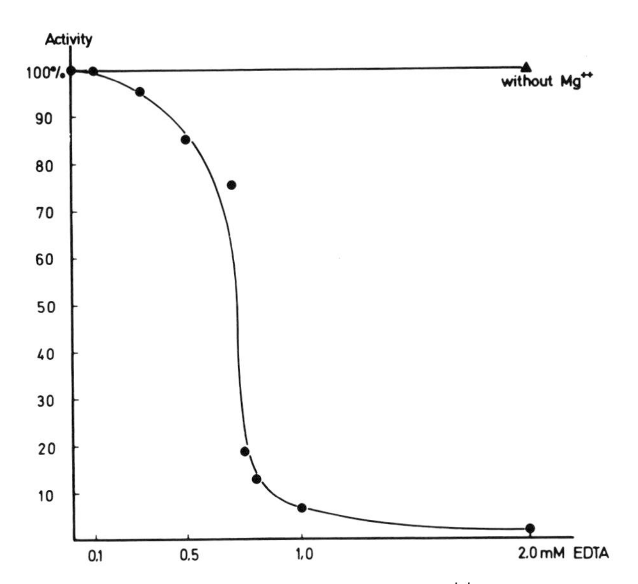

Fig. 7. Inactivation of pyruvate carboxylase by Mg^{++} + EDTA when assayed in the overall assay (Seufert et al.). 34 μg enzyme (spec. activity 16 U/mg) were incubated for 5 minutes with 10 mM $MgCl_2$ + EDTA as indicated in the figure. The activity was determined with the optical assay.

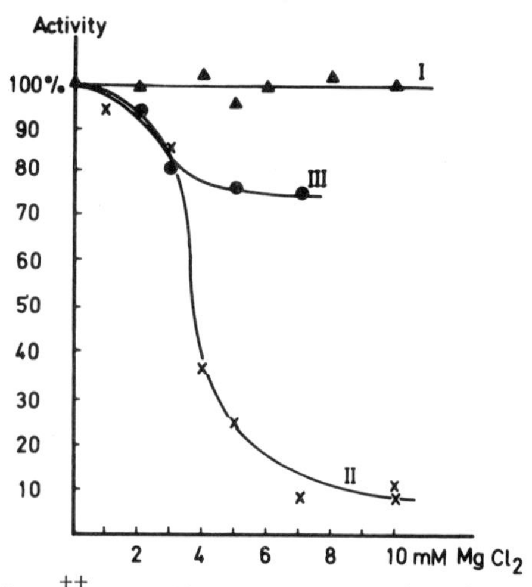

Fig. 8. Effect of Mg^{++} + EDTA on pyruvate carboxylase activity in the overall and exchange assays (Seufert et al.). Experimental conditions as described in Fig. 7. For the exchange assay only an aliquot of 1.2 µg enzyme was introduced into the assay mixture.

Curve I: 5mM EDTA + Mg^{++} as indicated. The activity was determined after preincubation by the exchange assay.

Curve II: Same conditions as Curve I, except that activity was determined in the overall assay.

Curve III: Same conditions as Curve II; EDTA omitted.

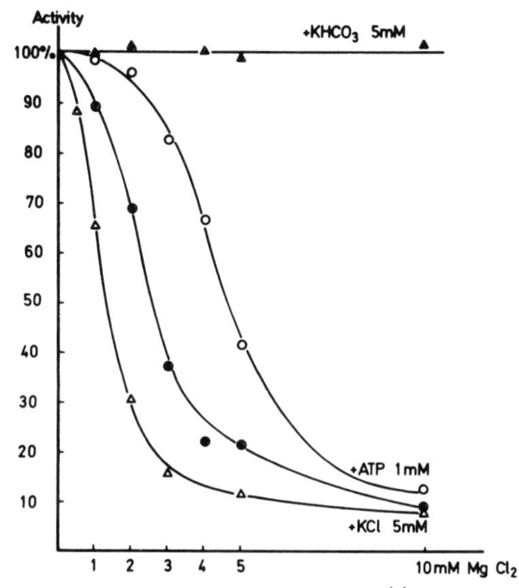

Fig. 9. Effect of various agents on the EDTA + Mg^{++} dependent inactivation of pyruvate carboxylase when assayed in the overall assay. Same experimental conditions as Fig. 7. Additions during the preincubation as indicated.

References

Alberts, A. W., A. M. Nervi, P. R. Vagelos: Proc. Natl. Acad. Sci. (Wash.) 63 (1969) 1319

Exton, J. H., R. C. Park: Advances in Enzyme Regulation. Ed. by G. Weber, Vol. 6. Pergamon, New York 1968, p. 391

Exton, J. H., R. C. Park: J. biol. Chem. 244 (1969) 1224

Henning, H. V., W. Seubert: Biochem. Z. 340 (1964) 160

Majerus, P. W., E. Kilburn: J. biol. Chem. 244 (1969) 6254

Matsuhashi. M., S. Matsuhashi, F. Lynen: Biochem. Z. 340 (1964) 263

Menahan, L. A., O. Wieland: Europ. J. Biochem. 9 (1969) 182

Monod, J., J. Wyman, J. P. Changeux: J. molec. Biol. 12 (1965) 88

Ryder, E., C. Gregolin, H. C. Chang, M. D. Lane: Proc. Natl. Acad. Sci. 57 (1967) 1455

Scrutton, M. C., D. B. Keech, M. F. Utter: J. biol. Chem. 240 (1965) 574

Seubert, W., H. V. Henning, W. Schoner, M. L'Age: Advances in Enzyme Regulation. Ed. by G. Weber, Vol. 6. Pergamon, New York 1968, p. 153

Seufert, D., E. Herlemann, E. Albrecht, W. Seubert, in preparation

Stoll, E., E. Ryder, J. B. Edwards, M. D. Lane: Proc. Natl. Acad. Sci. 60 (1968) 968

Announced Discussion

Comparative Studies with Pyruvate Carboxylase from Rat and Guinea Pig Liver

J. Kleineke and H. D. Söling
Abteilung für Klinische Biochemie, Medizinische Universitäts-Klinik Göttingen, FRG

Because of differences in the regulation of gluconeogenesis in rat and guinea pig livers, we were interested in kinetic parameters of pyruvate carboxylase from both sources.

Pyruvate carboxylase was partially purified as reported by Utter and Keech (1960) for the chicken liver enzyme. The preparation contained only traces of lactate dehydrogenase and NADH dehydrogenase activities. The purification factor was 12 to 15. The kinetic measurements were performed spectrophotometrically as described by Utter and Keech (1960). The apparent K_M's for ATP (Fig. 1) and pyruvate (Fig. 2) were similar for the rat and the guinea pig enzymes. There was a slight difference in the temperature dependency of the reaction velocity (Fig. 3).

The activation of pyruvate carboxylase by acetyl-S-CoA was studied at different p_H-values, as the K_a for acetyl-S-CoA had been found to be p_H dependent in studies with the chicken enzyme (Scrutton and Utter 1967).

For a given p_H-value, pyruvate carboxylase from rat and guinea pig liver exhibited exactly the same dependency on the concentration of acetyl-S-CoA (Fig. 4). Moreover, the dependency of the K_a for acetyl-S-CoA from the p_H-value was similar for both enzymes.

In further experiments, the effect of p_H changes on pyruvate carboxylase activity was studied at constant concentrations of either 2 mM $Mg \cdot ATP^{2-}$ or 0.5 mM $Mn \cdot ATP^{2-}$ (Fig. 5).

As can be seen with both cation-ATP complexes, the p_H optima for the guinea pig enzyme were found at lower p_H values compared with the rat enzyme: Assuming the intramitochondrial p_H to be more closely to seven than to eight, it may be that the guinea pig enzyme works under a more favorable condition in this respect. With the enzymes from both sources the p_H optima with $Mn \cdot ATP^{2-}$ were found to be lower than with $Mg \cdot ATP^{2-}$ Holten and Nordlie (1965), studying soluble and mitochondrial phosphoenolpyruvate carboxykinase from guinea pig livers, reported a far higher activity with Mn^{2+} compared with Mg^{2+} at p_H values below 7.4 when tested in direction of phosphoenolpyruvate formation. This would fit well with our findings with

+ Unusual Abbreviations: TRIS = tris-hydroxymethyl-amino methan

Enzyme Code Numbers: Lactate dehydrogenase (EC 1.1.1.27), Malate dehydrogenase (EC 1.1.1.37), NADH-dehydrogenase (EC 1.6.99.3), Phosphoenolpyruvate carboxykinase (EC 4.1.1.32), Phosphotransacetylase (EC 2.3.1.8), Pyruvate carboxylase (EC 6.4.1.1), Pyruvate kinase (EC 2.7.1.40).

guinea pig liver pyruvate carboxylase.

The findings of Holten and Nordlie (1965) and ours point to the special importance of Mn^{2+} ions or Mn-nucleoside triphosphate complexes for phosphoenolpyruvate synthesis.

Whether the different p_H optima of rat and guinea pig pyruvate carboxylase under the condition of constant concentrations of $Mn \cdot ATP^{2-}$ or $Mg \cdot ATP^{2-}$ can be related to the different type of regulation of gluconeogenesis remains to be clarified.

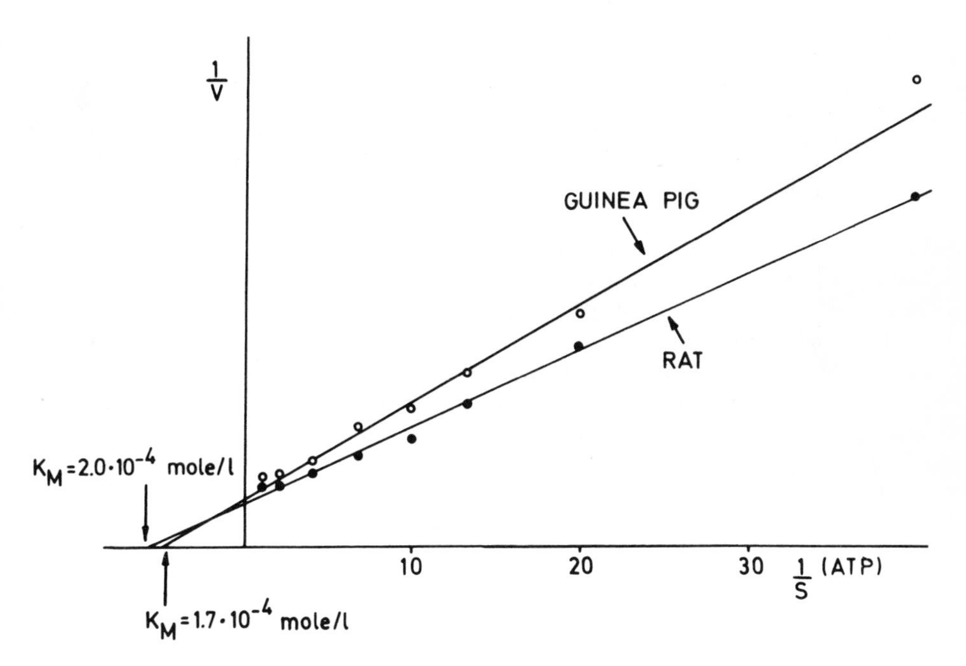

Fig. 1. Apparent K_M for ATP with pyruvate carboxylase from rat (●——●) and guinea pig (○——○) livers. The volume of the total reaction mixture was 1.5 ml. The varying concentrations of ATP were kept constant by an ATP regenerating system. The reaction mixture contained: TRIS-HCl (pH 7.4) 100 mM, $MgCl_2$ 7 mM, KCl 35 mM, $NaHCO_3$ 20 mM, phosphoenolpyruvate 15 mM, CoA-SH o.22 mM, acetylphosphate 2.4 mM, NADH 0.25 mM, malate dehydrogenase 1 μg/ml, pyruvate kinase 4.4 μg/ml, phosphotransacetylases 1.6 μg/ml and an aliquot of the pyruvate carboxylase preparation (0.10 mg of protein/ml). The ATP concentration was varied as indicated in the abscissa. The temperature was 30^0C.

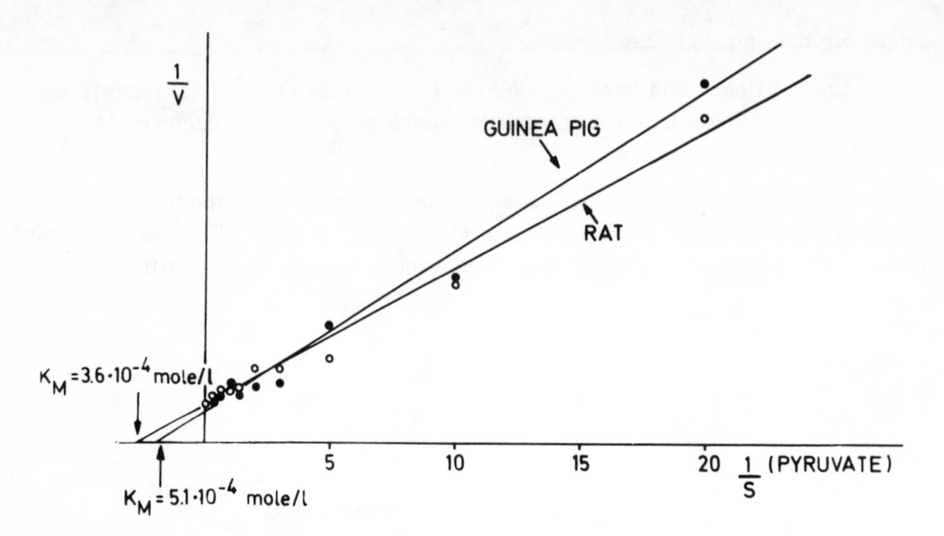

Fig. 2. Apparent K_M for pyruvate with pyruvate carboxylase from rat (o——o) and guinea pig (●——●) liver. The total reaction volume was 1.5 ml. The reaction mixture contained TRIS-HCl (pH 7.4) 100 mM, ATP 2.5 mM, $MgCl_2$ 7 mM, $NaHCO_3$ 20 mM, CoA-SH 0.22 mM, acetylphosphate 2.4 mM, NADH 0.25 mM, malate hydrogenase 1 μg/ml, phosphotransacetylase 1.6 μg/ml, and 0.15 mg of protein/ml of the pyruvate carboxylase preparation. The concentration of pyruvate was varied as indicated on the abscissa. The incubation temperature was 30^0C.

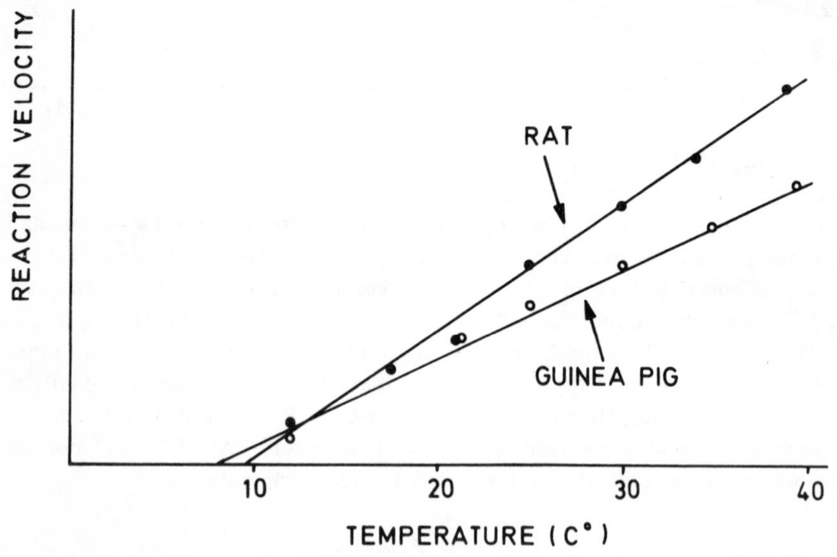

Fig. 3. Effects of temperature changes on the activity of pyruvate carboxylase from rat (●——●) and guinea pig (o——o) livers. The test conditions were the same as given for Fig. 2 with the exception that the concentation of pyruvate was constant (10 mM).

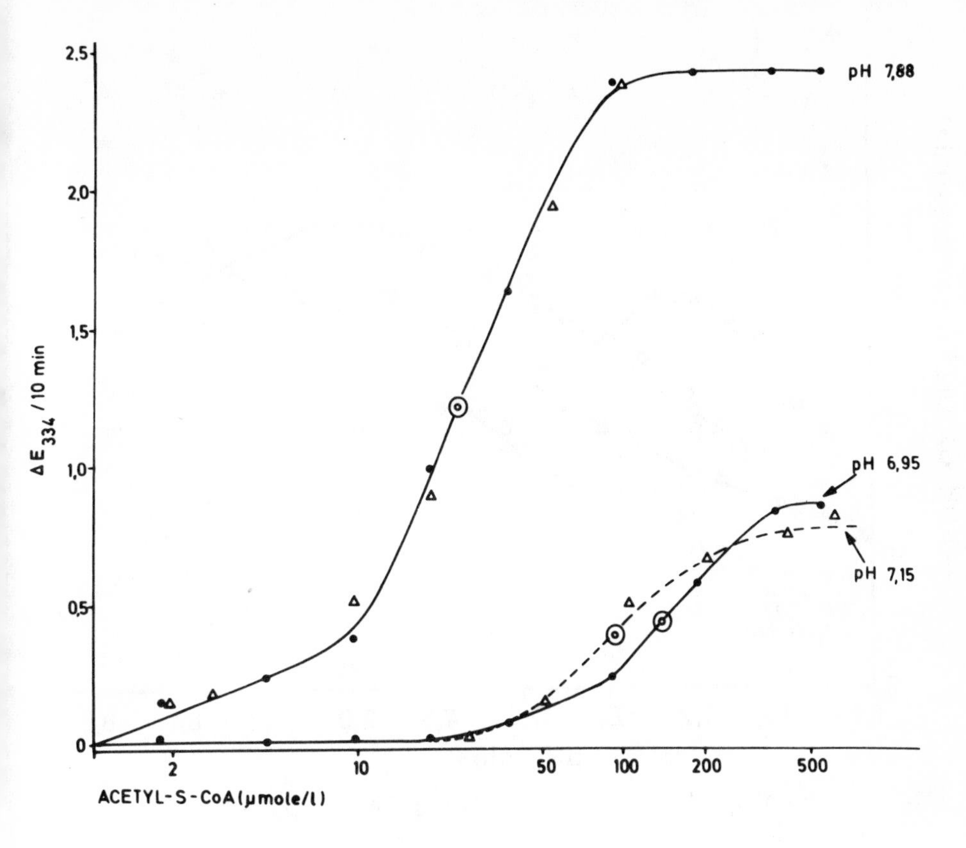

Fig. 4. Activation of pyruvate carboxylase activity from rat (△- - - -△) and guinea pig (●———●) liver by acetyl-S-CoA at different p_H values. The total reaction volume was 1.5 ml. Its composition was the same as given for Fig. 2. except that it contained a fixed concentration (10 mM) of pyruvate and that CoA-SH, acetylphosphate and phosphotransacetylase were omitted. Acetyl-S-CoA was added as indicated in the abscissa. The temperature was 30^0 C. The bigger circles indicate the point of half maximum velocity. The K_a at p_H 7.88 was $2.40 \cdot 10^{-5}$ M.

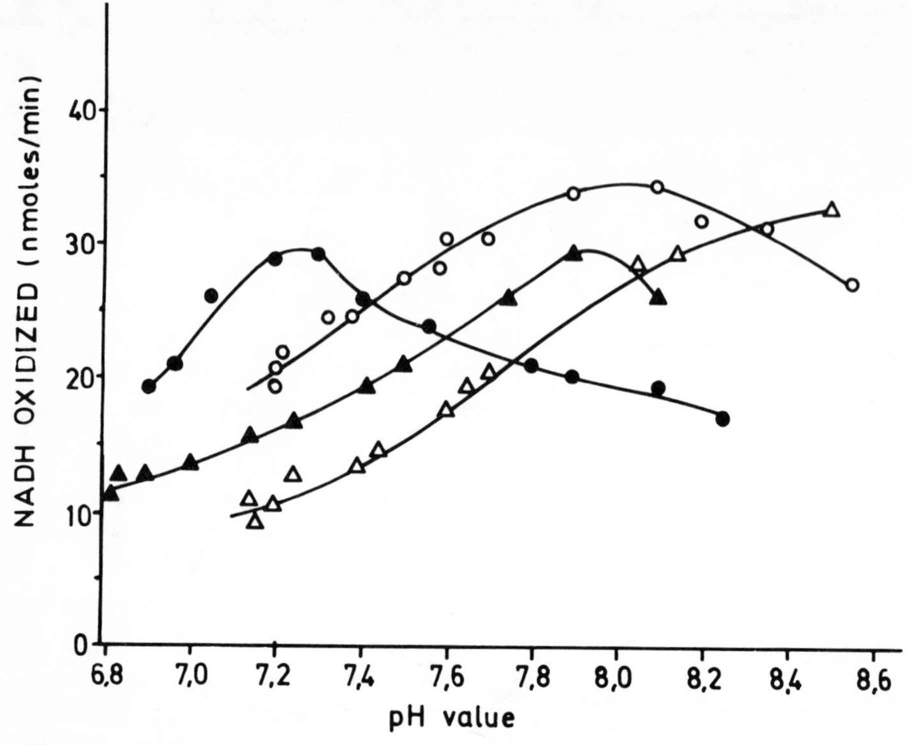

Fig. 5. Effects of p_H changes on the activity of pyruvate carboxylase from rat and guinea pig livers at constant concentrations of Mg · ATP^{2-} (2 mM) or Mn · ATP^{2-} (0.5 mM). The test volume was 1.5 ml. It contained (final concentrations): Triethanolamin-HCI 100 mM, $NaHCO_3$ 20 mM, NADH 0.25 mM, CoA-SH 0.22 mM, acetylphosphate 2.4 mM, malate dehydrogenase 1 μg/ml, phosphotransacetylase 1.6 μg/ml. For p_H values above 7.7, diethanolamine instead of triethanolamine was used. The temperature was 30 C (△——△) = rat enzyme with Mg·ATP^{2-}, (○——○) = rat enzyme with Mn · ATP^{2-}, (▲——▲) = guinea pig enzyme with Mg · ATP^{2-}, (●——●) = guinea pig enzyme with Mn · ATP^{2-}.

References

Holten, D. D., R. C. Nordlie: Biochemistry 4 (1965) 723

Scrutton, M. C., M. F. Utter: J. biol. Chem. 242 (1967) 1723

Söling, H. D., B. Willms, J. Kleineke: in: Regulation of Gluconeogenesis, 9th Conference of the German Society for Biological Chemistry, Reinhausen 1970, Thieme, Stuttgart 1970 p. 224

Utter, M. F., D. B., Keech: J. biol. Chem 235 (1960) PC 17

Utter, M. F., D. B. Keech: J. biol. Chem. 238 (1965) 2603

Discussion to Utter and Fung, and to Seufert, Herlemann and Seubert

J. R. Williamson: In order to work out the physiological significance of the control by ketone-CoA compounds it is necessary to know the range of values of these compounds in the intact liver. I made a fee measurements on the acetoacetyl-CoA content of rat liver. This is on the order of 5 to 10 nanomoles/g dry weight. If this is uniformly distributed throughout the cell the concentration would be about 2 to 4 μM, whereas if it is entirely located within the mitochondrial compartment, the concentration would be 25 to 50 μM. This is a little low for your K_i-value. On the other hand, the ß-hydroxybutyryl-CoA was unmeasurable. This is what one would anticipate because of the differences in the equilibrium constants between ß-hydroxybutyrate-dehydrogenase and the ß-hydroxybutyruyl-CoA-dehydrogenase (60 mV). Therefore, the ratio of ß-hydroxybutyrate/acetoacetate directed by the ratio of ß-hydroxybutyryl-CoA/acetoacetyl-CoA is about 10, so with a ß-hydroxybutyrate/acetoacetate ratio of unity there would be ten times more acetoacetyl-CoA than ß-hydroxybutyryl-CoA. Of course, I am very happy that Dr. Utter has come up with the postulate of the redox control of pyruvate carboxylase particularly in relation to the activation of pyruvate carboxylase in the presence of glucagon, where we have no change of acetyl-CoA but a change of the NAD - redox state towards a reduction, particularly in the mitochondria. But I have a question to Dr. Seufert: Do you consider the cyclic AMP might be interacting directly with pyruvate carboxylase?

Seufert: We have checked this, and we could not see any effects of cyclic AMP on the activation of pyruvate carboxylase.

Seubert: So far.

Utter: I would like to make a comment on Dr. Williamson's remark. We are concerned, of course, about the apparent discrepancy in levels of acetoacetyl-CoA needed to inhibit the isolated enzyme and the levels of this compound as determined by Dr. Williamson. We feel that other substances may have the effect of raising the apparent K_a for acetyl-CoA and lowering the apparent K_i for acetoacetyl-CoA for pyruvate carboxylase. PEP, or as we now know, the unidentified inhibitor in PEP has precisely these effects. Until we know what this material is or until we find something else with the same effects this possible solution to the apparent discrepancy in concentrations is just an interesting possibility.

Krebs: There should be a predictable relationship between the concentrations of acetoacetyl-CoA and acetyl-CoA because the activity of thiolase is rather high and might be sufficient to establish equilibrium. Dr. Williamson, are the concentrations which you found those expected for equilibrium?

J. R. Williamson: Calculations of the acetoacetyl-CoA content based on equilibrium of the thiolase show that it would be much smaller than the assay value, about 1×10^{-3} nmoles/g dry weight. My measured level is over a thousand times greater than the calculated level for equilibrium. I would suggest that this enzyme is not in equilibrium.

Seubert: I wonder if it is really possible to assay acetoacetyl-CoA. As far as I remember, acetoacetyl-CoA is rather unstable in acid solutions. I assume that you have done your assays on perchloric extracts.

J. R. Williamson: Of course, the assay was checked by recoveries. Actually we measured acetoacetyl-CoA on lyophilized tissue. It is unstable but we assay the extract within a short time after preparing the extract.

Wieland: I think there are really no valuable data at the moment for tissue concentrations of acetoacetyl-CoA or ß-hydroxybutyryl-CoA.

J. R. Williamson: I consider my values valuable, but I have not checked them against another independent assay method. I was using acetoacetyl-CoA dehydrogenase. An alternative method is to couple the thiolase with citrate synthase.

D H. Williamson: According to our studies thiolase is present in both compartments of rat liver (Williamson, D. H., M. W. Bates and H. A. Krebs (1968). Biochem. J. 108, 353): about 20-25% of the acetoacetyl-CoA thiolase is in the cytosol and the remainder in the particles. If thiolase is in equilibrium with acetyl-CoA and CoA, the calculated acetoacetyl-CoA concentrations would be in the picomolar range (Greville, G. D. and P. K. Tubbs, In Essays in Biochemistry, Vol. 4. Ed. by P. N. Campbell and G. D. Greville. Academic Press, New York 1968, p. 155). This would mean that they are virtually unmeasurable by present techniques. Alternatively, the thiolase may not be in equilibrium or some of the measured CoA or acetyl-CoA may be compartmented. One question to Dr. Utter: Did you exclude the possibility that the acetoacetyl-CoA was contaminated with acetoacetyl-glutathion?

Utter: Yes, the material was purified by chromatography in a manner which I am sure would remove any glutathione derivative.

I have a question for Dr. Seufert. I believe he said that potassium stimulated the inactivation of pyruvate carboxylase by magnesium ions. I wonder if he has any ideas concerning the mechanism of interaction of potassium and acetyl-CoA?

Seufert: Acetyl-CoA and potassium are necessary in the enzyme assay, but in the preincubation with EDTA and magnesium potassium inactivates. I have no idea about the mechanism.

Utter: I am sorry that I did not quite understand. Is there an interaction in the catalytic assay in which potassium increases the apparent K_a for acetyl-CoA?

Seubert: Dr. Weiss will come to this point. It increases.

Haeckel: I have a question to Dr. Kleineke: Do you think that the pyruvate-carboxylase is distributed in the same way in both species?

Kleineke: In both species we looked for mitochondrial and extra-mitochondrial pyruvate carboxylase. In both species we found our main activity in the mitochondrial fraction. In some case there was some activity of pyruvate carboxylase in the cytoplasm soluble fraction, but usually accompanied by an equal amount in glutamate dehydrogenase activity.

Seubert: I would like to take the occasion to really clarify the term "soluble" which we would like to see in quotation marks. It means that the enzyme is "soluble" under certain experimental conditions: Isotonic solution supplemented with monovalent cations. There is no doubt that the term "soluble" does not only include a "cytoplasmic" location. "Soluble" pyruvate cyrboxylase can also be located in another compartment (for reference see Adv. Enzyme Regul. 6, 159 (1968) and discussion to this paper). This is also the reason, why we prefer the term "extra-mitochondrial" and not "cytoplasmic", in order to include any possible compartment of the cell. Dr. Weiss will come back to this point again in his presentation.

J. R. Williamson: Could I take this opportunity to ask whether anyone has looked for pyruvate carboxylase in brain, studied its activity perhaps in relation to potassium and Mg^{++} concentrations?

Seubert: We have looked for pyruvate carboxylase in brain, muscle, and all kinds of tissues except adipose tissue, and as far as I remember, the activity in brain is very, very low.

Wieland: As far as I understood, Dr. Williamson asked especially for the distribution in brain, is this right?

J. R. Williamson: Salganicoff and Koeppe (J. Biol. Chem. 243 (1968) 3416) did show relatively high activities in rat brain. Of course, it is very low when compared with liver. I asked only if anyone has isolated pyruvate carboxylase of brain and studied it kinetically.

Seubert: I just wanted to ask Dr. Henning: I think that several years ago you did the distribution studies? As far as I remember we found very low activity in brain. The main activity we found in the liver, the kidney, and also a low activity in muscle. But this activity is only a few percent of that in liver.

Wieland: This would agree with Dr. Böttger's findings in my laboratory (Böttger, Wieland, Brdiczka and Pette, Eur. J. Biochem. 8, 113, 1969). He has measured the distribution of pyruvate carboxylase in different rat tissues and he found a small, but distinct activity in brain.

Ballard: We find that the pyruvate carboxylase in brain is entirely particulate and is antigenically identical to the liver enzyme.

Wieland: May I ask Dr. Utter: At the present time there is a big run on interconvertible enzyme systems. Do you believe that pyruvate carboxylase could perhaps represent such an interconvertible enzyme system?

Utter: It would be dangerous to say that this might not be so but some years ago, Dr. Landau spent some months in my laboratory and he spent some time trying to demonstrate such an interconversion for the chicken liver enzyme. He was unsuccessful but did find a stabilizing effect of ATP on the enzyme which has been useful in the purification process.

Seubert: We can confirm this "activation" by ATP. We also realized later, that this effect was due to a stabilization. Concerning the inactivation by Mg^{++} we were quite happy to have an opposing effect as compared with pyruvate-dehydrogenase which is activated by Mg^{++}. Unfortunately it turned out, that EDTA is necessary for inactivation of pyruvate carboxylase. At the moment we therefore cannot deduce from these results a physiological significance, unless we can substitute the EDTA by an effector which is present in the cell. I have a question to Dr. Utter: What explanation have you for the effect of pyruvate on the activity of pyruvate carboxylase?

Utter: I assume you are referring to the substrate activation by pyruvate. One possible explanation for such an observation is the presence of a second site for pyruvate which comes into play only at high concentrations of pyruvate but we have no real evidence for this.

Seufert: Some weeks ago Dr. Keech wrote a paper on sheep pyruvate carboxylase. In this paper he reported two different K_m values at different pyruvate concentrations.

Utter: We had a curve with three possible components so we could have calculated three different K_m values for pyruvate. The curves are difficult to interpret but when I talk about a second site for pyruvate as I was a moment ago, I really am saying a second K_m in the same way as is Dr. Keech.

Seufert: Dr. Keech had an alternate explanation for his data; that in his enzyme preparation two enzymes might exist. This is not your opinion?

Utter: We have no evidence on this point.

On the Intracellular Location of Pyruvate Carboxylase

G. Weiss, B. Ohly, H. Brod and W. Seubert
Physiologisch-chemisches Institut der Universität, Göttingen, FRG

Summary

The concept of the exclusive location of pyruvate carboxylase in mitochondria is based upon the analysis of a subcellular fraction representing a mixture of fragments, and on the exclusion of a cytoplasmic location of this enzyme. Preferential extraction of pyruvate carboxylase — as compared to glutamate dehydrogenase — with isotonic solution supplemented by monovalent cations has tempted us to postulate a location of pyruvate carboxylase in different compartments. Additional experimental support for this assumption is the following:

1) An enrichment of the activities of pyruvate carboxylase — as compared to glutamate dehydrogenase — with the peroxisomes upon isopycnic fractionation. A peroxisomal location, however, could be excluded: treatment of the fraction with digitonin only solubilized D-aminoacid oxidase, but not pyruvate carboxylase.
2) Inhibition of pyruvate carboxylase by potassium (conc. range from 100 to 150 mM) by lowering the affinity of acetyl-CoA to the enzyme.
3) Acceleration of CO_2-fixation in kidney cortex slices by lowering intracellular K^+ with ouabain, an inhibitor of the Na^+, K^+- pump (no effect was observed in liver). The effect of strophantine was only observed with pyruvate and lactate. In presence of succinate no stimulation of the CO_2-fixation by strophantine could be demonstrated, indicating an exclusively mitochondrial carboxylation process. This assumption could be further supported by a study of CO_2-fixation in liver: no CO_2-fixation could be demonstrated in presence of succinate because of the cytoplasmic location of malic enzyme in this tissue.
4) Different inhibitory effects of phenylpyruvate, an inhibitor of pyruvate carboxylase, on the CO_2-fixation in kidney cortex slices in presence of various precursors of glucose.

A search for a marker enzyme of the postulated new compartment is in progress.

Introduction

The concept of the exclusive location of mammalian pyruvate carboxylase in mitochondria is based upon the analysis of a subcellular fraction from chicken liver and rat liver isolated between 700 and 10,000g (Keech and Utter 1963, Freedman and Kohn 1964, Shrago and Lardy 1966, Böttger et al. 1969, Marco et al. 1969), and on the exclusion of a cytoplasmic location of this enzyme (Böttger et al. 1969, Marco et al. 1969). As shown by De Duve and Baudhuin (1966) and Leighton et al. (1968), however, the fraction usually designated by biochemists as a mitochondrial fraction represents a mixture of all kinds of subcellular units. No conclusions on the intracellular location of pyruvate carboxylase can therefore be drawn from the analysis of such a "mitochondrial" fraction. Also the exclusion of a cytoplasmic location of pyruvate carboxylase does not justify the postulate of an exclusive mitochondrial location of this enzyme.

According to Leighton et al. (1968), glutamate dehydrogenase is located within the mitochondria. Preferential extraction of pyruvate carboxylase — compared to glutamate dehydrogenase — with isotonic solutions supplemented by monovalent cations has therefore tempted us to postulate a location of pyruvate carboxylase in addition to mitochondria in another compartment (Seubert et al. 1968, Dugal and Seubert 1969). Additional experimental support for this assumption is presented in the following information.

Results and Discussion

Preferential extraction of pyruvate carboxylase with isotonic solutions supplemented by monovalent cations could be demonstrated from rat liver (Seubert et al. 1968), chicken liver (Dugal and Seubert 1969) and kidney cortex of rat. In order to get some indications on the location of this "soluble" fraction of pyruvate carboxylase, the behavior of acid phosphatase, a marker enzyme of the lysosomes, of D-amino acid oxidase, a marker enzyme of the peroxisomes and of glutamate dehydrogenase was studied under identical conditions. As is evident from Fig. 1, extraction of liver with isotonic solution supplemented by monovalent cations, results also in preferential extractions of acid phosphatase and amino acid oxidase as compared to glutamate dehydrogenase, the marker enzyme of mitochondria. Pyruvate carboxylase behaves like acid phosphatase. A possible location of pyruvate carboxylase in the lysosomes was therefore investigated.

Tritosomes were isolated according to Leighton et al. (1968) after treatment of the rats for three days with triton. Accumulation of this detergent in the lysosomes lowers the specific density of this subcellular unit. They can be easily separated from peroxisomes and mitochondria after this treatment in a suitable sucrose gradient by flotation. Analysis of the isolated tritosomes did not show activity of pyruvate carboxylase. A location of this enzyme in the lysosomes can therefore be excluded.

De Duve has first focused attention to the fact that peroxisomes and the ability to synthesize glucose from C_3-precursors are restricted in mammalian systems to liver and kidney (De Duve and Baudhuin 1966). In our search for the location of the "soluble" pyruvate carboxylase we therefore investigated a possible location of this enzyme in the peroxisomes.

Enrichment of the activities of pyruvate carboxylase — compared to glutamatic dehydrogenase — could be demonstrated upon separation of the subcellular particles of rat liver according to De Duve et al. (1955) into the nuclear fraction (N), the heavy mitochondria (M) and the light mitochondria (L). As is evident from Table 1, the ratio glutamate dehydrogenase activity/pyruvate carboxylase activity in the individual subcellular fractions is shifted in favor of the latter enzyme in the M-fraction. In contrast to this behavior of pyruvate carboxylase, 3-hydroxy-acyl-CoA dehydrogenase

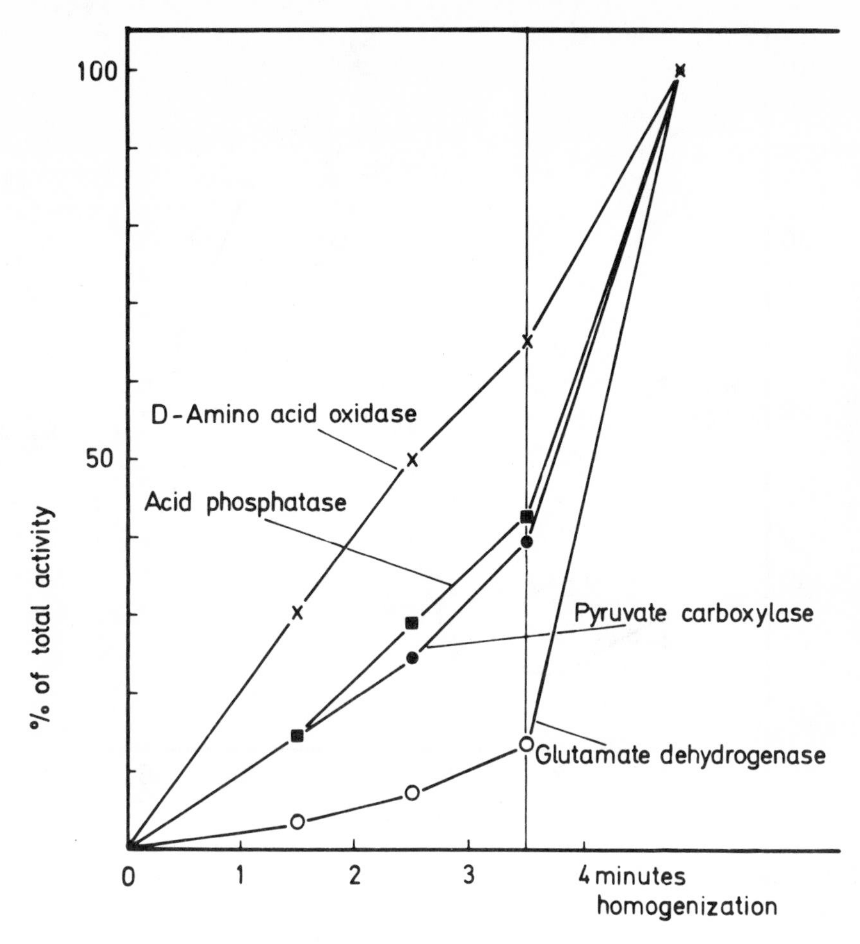

Fig. 1. Solubility of pyruvate carboxylase and marker enzymes of mitochondria, lysosomes and peroxisomes in isotonic buffered solutions supplemented with K^+. Composition of the medium: 0.25 M sucrose, 0.02 M Trap-buffer, pH 7.2, 5mM KCL. One hundred percent enzyme activity refers to an extraction with the same medium supplemented by 0.1% Na-desoxycholate. Assays: D-aminoacid oxidase according to Baudhuin et al. (1964), acid phosphatase according to Gianetto and De Duve (1955). For additional assays, see Table 1.

shows a distribution like that of mitochondrial glutamate dehydrogenase. Also separation of the particulate fraction from kidney cortex resulted in enrichment of pyruvate carboxylase in the M-fraction. Upon isopycnic subfractionation of the N+M-fraction of kidney cortex in a discontinuous sucrose gradient according to De Duve et al. (1955), the pyruvate carboxylase peak moved with D-amino acid oxidase (Fig. 2), while 3-hydroxy-acyl-CoA dehydrogenase stayed — as expected — with glutamate dehydrogenase. Assays of the activity ratio glutamate dehydrogenase/pyruvate carboxylase showed a further shift in favor of the latter enzyme in fraction No. 5 enriched with D-amino acid oxidase (Table 2). This shift

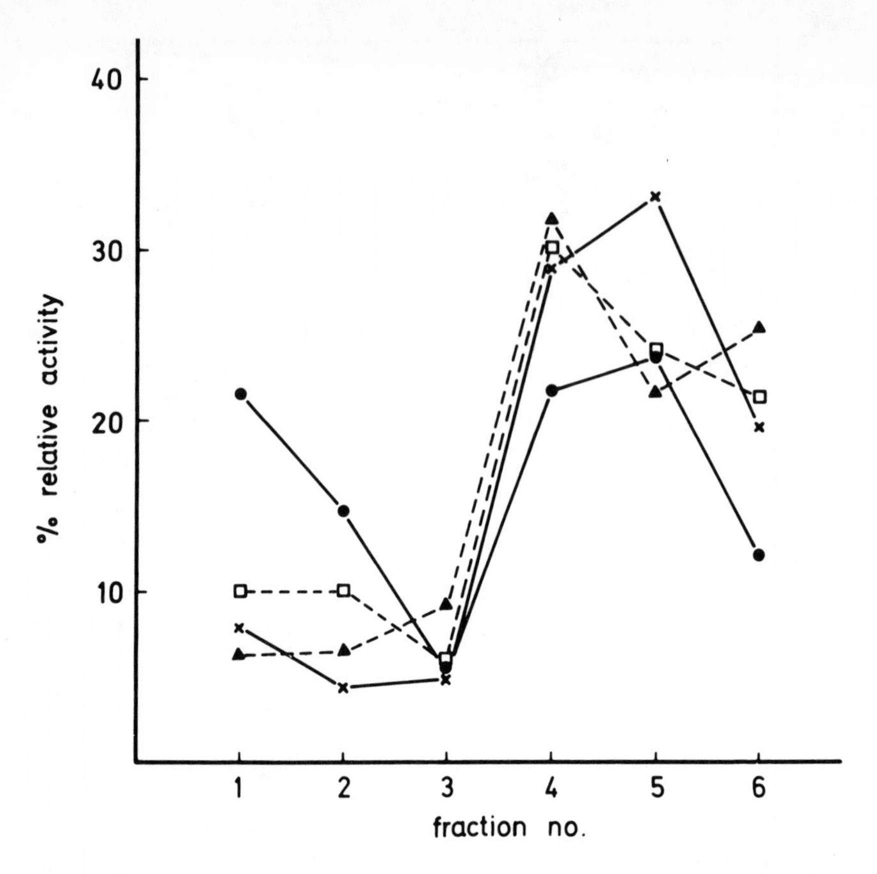

Fig. 2. Isopycnic fractionation of N + M-fraction according to De Duve et al. (1955). The N + M-fraction from 5.5 g rat kidney in 7 ml 0.5 M sucrose was set on a discontinuous sucrose gradient consisting of (from top to bottom):

7 ml 1.3 M sucrose
5 ml 1.5 M sucrose
5 ml 1.65 M sucrose
5 ml 1.8 M sucrose
5 ml 2.0 M sucrose

All sucrose solutions were buffered with 0.02 M glycylglycerine pH 7.5. Separation into various subfractions was achieved by a 3-hour centrifugation at 27,000 r.p.m. in a SW27 rotor of the Spinco model L2-65 B preparative ultracentrifuge. Fractionation into 6 fractions (0.5 M - 2.0 M) according the above conditions was carried out according to Leighton et al. (1968). For enzyme assays, see Table 1 and Fig. 1.

x = pyruvate carboxylase
● = D-aminoacid oxidase
□ = glutamate dehydrogenase
▲ = 3-hydroxyacyl-CoA dehydrogenase

enzyme	enzyme activity (units/ g fresh wt. of liver)			
	homogenate	nuclear fraction (10,850 g x min.)	heavy mitoch. fraction (24,450 g x min.)	light mitoch. fraction (348,000 g x min.)
glutamate dehydrogenase	126.8	13.0	39.9	38.2
pyruvate carboxylase	6.0 (21.1)	0.6 (20.3)	2.8 (14.2)	1.3 (29.4)
3-hydroxyacyl-CoA dehydro-genase	43.7 (2.90)	5.6 (2.34)	17.9 (2.23)	16.8 (2.28)
citrate synthase	3.70 (34.2)	0.67 (19.3)	1.07 (37.3)	1.04 (36.5)

Table 1. Distribution of pyruvate carboxylase, glutamate dehydrogenase, citrate synthase and 3-hydroxyacyl-CoA dehydrogenase in subcellular fractions of rat liver. Tissue fractionation was done according to (De Duve et al. 1955). Enzymes were assayed after homogenization of the various fractions (1 min) with a sucrose-desoxycholate medium (0.25 M sucrose, 0.02 M Trap pH 7.2, 5 mM KCL, 0.1% Na-desoxycholate). Assays: Pyruvate carboxylase according to Henning and Seubert (1964), citrate synthase according to Ochoa et al. (1955), 3-hydroxyacyl-CoA dehydrogenase according to Lynen and Ochoa (1953), α-glutamate dehydrogenase with α-ketoglutarate supplemented with 2 mM ADP according to Hogeboom and Schneider (1953). Ratio glutamate dehydrogenase/enzymes; enzymes in parentheses.

Enzyme	Units Total Activity						
	fraction						
	N + M	1	2	3	4	5	6
glutamate dehydrogenase	138.9	11.0	11.3	6.9	33.2	26.3	23.7
pyruvate carboxylase	11.97 (11.6)	0.98 (11.2)	0.54 (20.9)	0.60 (11.5)	3.47 (9.6)	3.96 (6.6)	2.34 (10.1)
D-amino acid oxidase	6.37 (21.8)	1.22 (9.0)	0.82 (13.8)	0.31 (22.3)	1.22 (27.2)	1.33 (19.8)	0.71 (33.4)
3-hydroxyacyl-CoA dehydrogenase	133.4 (1.04)	7.98 (1.38)	8.30 (1.36)	11.63 (0.59)	39.86 (0.83)	27.07 (0.97)	31.04 (0.76)

Table 2. Isopycnic fractionation of N + M-fraction of rat kidney cortex. In parentheses are ratios glutamate dehydrogenase/enzyme. For experimental conditions, see Fig. 2.

could repeatedly be demonstrated not only for the N + M-fraction of kidney cortex, but also from the liver. For further purification of this subcellular unit, the N + M-fraction from kidney cortex has been prefered because of the greater stability of pyruvate carboxylase in hypertonic sucrose solutions (0.5-1.8 M) employed for separation of the individual fractions.

Treatment of the N + M-fraction of kidney cortex with digitonine (4.2 mg digitonine 20 mg protein) solubilized to 40% of D-amino-acid oxidase. Pyruvate carboxylase and glutamate dehydrogenase were solubilized by this treatment between 4 and 7%. This finding has to be interpreted either by an exclusion of a peroxisomal location of pyruvate carboxylase or by different solubilities of peroxisomal enzymes as compared to D-amino acid oxidase. The latter possibility has been verified for another peroxisomal enzyme, urate oxidase. Experiments are in progress to clarify this possibility in respect to pyruvate carboxylase.*

The concept of a carboxylation of pyruvate in <u>different</u> compartments was unexpectedly supported by parallel studies on the pathway of carbon in gluconeogenesis from pyruvate and succinate. Fixation of radioactive carbon dioxide in kidney cortex shows almost identical incorporation rates in presence of lactate and succinate (Table 3). Netto synthesis of glucose is inhibited by phenylpyruvate, an inhibitor of pyruvate carboxylase, only with C_3-precursors as carbon source for glucode synthesis. This means that the intermediates of the citric acid cycle are converted to phosphoenolpyruvate by at least two pathways in kidney-cortex: 1) By a direct conversion via fumarate, malate, oxalacetate. 2) By an oxidative decarboxylation of malate formed to pyruvate, and its further conversion via the abbreviated dicarboxylic acid cycle to phosphoenol pyruvate. The different effect of phenylpyruvate on the netto synthesis of glucose from sucinate (Table 3) and the fixation of radioactive CO_2 in the presence of succinate (Table 5) may be explained by sufficient activities of succinodehydrogenase, fumarase and malate dehydrogenase — compared with phosphenolpyruvate carboxykinase — to supply the latter enzyme with its substrate oxalacetate, even when pyruvate carboxylase is inhibited.

Stimulation of CO_2-fixation into glucose by succinate is only evident in kidney cortex, but not in liver slices (Table 3). These different results are due to different locations of malic enzyme in kidney cortex and liver. According to Brdiczka et al. (1969) the malic enzyme is located in the kidney mainly in the particulate fraction. In liver the enzyme has been found exclusively in the cytoplasm (Hsu and Lardy 1967). The low incorporation of radioactive CO_2 in liver with succinate has therefore to be correlated with the cytoplasmic location of malic enzyme. Most likely insufficient activities of malic enzyme are responsible for the low incorporation. Isotope studies of Müllhofer (1970) support this assumption: the radioactivity of 2-C^{14}-pyruvate is <u>not</u> converted to the 1-position of pyruvate, as should be expected after its further reaction via the citric acid cycle.

---

* Note added in press: Peroxisomal location of pyruvate carboxylase could be excluded by differential fractionation and zonal centrifugation.

substrate	kidney cortex slices			liver slices
	glucoseformation (μ moles/hour/g)		$^{14}CO_2$-fixation (Cpm/min./g)	$^{14}CO_2$-fixation (Cpm/min./g)
	no inhibitor	phenylpyruvate 10^{-3}M		
pyruvate	36.1 ± 1.3 (6)	15.0 ± 0.6 (6)	32,610 ± 570 (8)	17,870 ± 570 (5)
lactate	-	-	21,370 ± 690 (7)	20,121 ± 500 (5)
succinate	37.7 ± 2.1 (5)	42.4 ± 1.4 (5)	20,460 ± 290 (8)	11,450 ± 1,060 (5)
no substrate	3.9 ± 0.6 (4)	4.0 ± 0.8 (4)	1,970 ± 100 (8)	10,540 ± 470 (5)

Table 3. Gluconeogenesis and CO_2-fixation in rat kidney cortex and liver slices. The formation of glucose was measured according to Henning et al. (1966). $^{14}CO_2$-fixation was measured according to L'age et al. (1968) in the presence of 10 μM KCL. Number of experiments in parentheses. Values are expressed as means ± s.e.m.

substrate	fixation of $^{14}CO_2$ (Cpm/min./g wet wt.)		
	control	ouabain 10^{-4} M	P
pyruvate	32,610 ± 570 (8)	37,520 ± 680 (8)	< 0.001
lactate	21,390 ± 690 (7)	24,540 ± 450 (8)	< 0.01
succinate	20,460 ± 290 (8)	19,900 ± 410 (8)	-
no substrate	1,970 ± 100 (8)	2,960 ± 90 (8)	< 0.001

Table 4. Effect of ouabain on CO_2-fixation in rat kidney cortex slices. For experimental conditions see Table 3. Number of experiments in parentheses. Values are expressed as means ± s.e.m.

Fractionation of particulate malic enzyme from kidney cortex in the ultracentrifuge showed in contrast to pyruvate carboxylase a preferential location in the L-fraction of subcellular fragments (not shown in this report). Carboxylation of pyruvate as an intermediate in gluconeogenesis from succinate and C_3-precursors was therefore expected to occur in different compartments. Experimental support for this assumption was obtained by a study of the effect of ouabain on the CO_2-fixation in kidney cortex in the presence of C_3-precursors and succinate. As summarized in Table 4, fixation of radioactive bicarbonate can be stimulated by ouabain in the presence of pyruvate and lactate but not in presence of succinate. The effect of ouabain on CO_2-fixation could only be demonstrated in kidney cortex but not in liver slices, indicating an inhibition of active Na^+, K^+-transport in kidney.

substrate	fixation of $^{14}CO_2$ (Cpm/min./g wet wt.)		
	no inhibitor	phenylpyruvate (10^{-3} M)	% inhibition
no substrate	1,800 ± 50 (8)	590 ± 20 (8)	70
lactate	25,840 ± 350 (8)	2,710 ± 60 (8)	90
pyruvate	26,410 ± 390 (8)	4,940 ± 150 (8)	80
succinate	21,000 ± 240 (8)	12,130 ± 80 (8)	40

Table 5. Effect of phenylpyruvate on CO_2-fixation in rat kidney cortex slices. For experimental conditions see Table 3. Number of experiments in parathenses. Values are expressed as means ± s.e.m.

In order to correlate the loss of intracellular potassium with a control of pyruvate carboxylase rather than a control of the transport of the various glucogenic precursors employed, the effect of potassium on the activities of pyruvate carboxylase was investigated. In Fig. 3a, the activation of pyruvate carboxylase by acetyl-CoA at different concentrations of K^+ is illustrated: with decreasing concentrations of the cation in the range from 140 (physiological concentrations) to 80 mM the affinity of acetyl-CoA to the enzyme increases, resulting in an activation of the enzyme at suboptimal concentrations of acetyl-CoA (Fig. 3b). When the lowering of intracellular potassium by inhibition of the Na^+, K^+-pump is related to the observed control of CO_2-fixation with pyruvate and lactate, the different effects of ouabain on CO_2-precursors, respectively, only can be interpretated as carboxylation processes occurring in different compartments.

Additional support for a pyruvate carboxylation in different subcellular compartments could be obtained by a study of the effect of phenylpyruvate on the CO_2-fixation in kidney cortex in the presence of pyruvate, lactate and succinate (Table 5). While carboxylation from endogenous substrate and from C_3-precursors was inhibited to the same extent, inhibition of CO_2-fixation in presence of succinate was significantly reduced under identical conditions. Different access of the inhibitor to different compartments participating in carboxylation of pyruvate may be the cause of the diverging results.

The physiological significance of pyruvate carboxylation in different compartments may be related to different mechanisms of gluconeogenesis from lactate and other C_3-precursors up to the stage of oxalacetate. This possibility is supported by isotope studies of Müllhofer and Kuntzen . According to their results, carbon atom 2 of lactate is converted to the 4-position of glucose not to the same extent as pyruvate (relative activities 0.05 and 0.32, respectively). This points to a different degree of cycling of carbon from lactate and pyruvate via the citric acid cycle. Incomplete equilibration of pyruvate and oxalacetate formed from lactate with the mitochondrial pools of the former substrates could be the cause of the different labeling patterns. Additional support for this assumption includes:

1) The insufficient equilibration of radioactive lactate with aspartate (Müllhofer and Kuntzen, Müllhofer et al. 1969a, Williamson et al.), as should be expected on the basis of the postulated pathway of carbon in gluconeogenesis from lactate (for a summary see Seubert et al. 1968).
2) The slow equlibration of tritium from L-lactate-2-T with NAD (Müllhofer et al. 1969b).
3) The low activity of liver lactic dehydrogenase in direction of pyruvate formation (14 units/g wet wt. at pH 7,4, 2 mM lactate and 2 mM NAD).

Final proof for the above hypothesis necessitates further purification and characterization of the subcellular unit containing "soluble" pyruvate carboxylase.

Acknowledgment

This work was supported by the Deutsche Forschungsgemeinschaft.

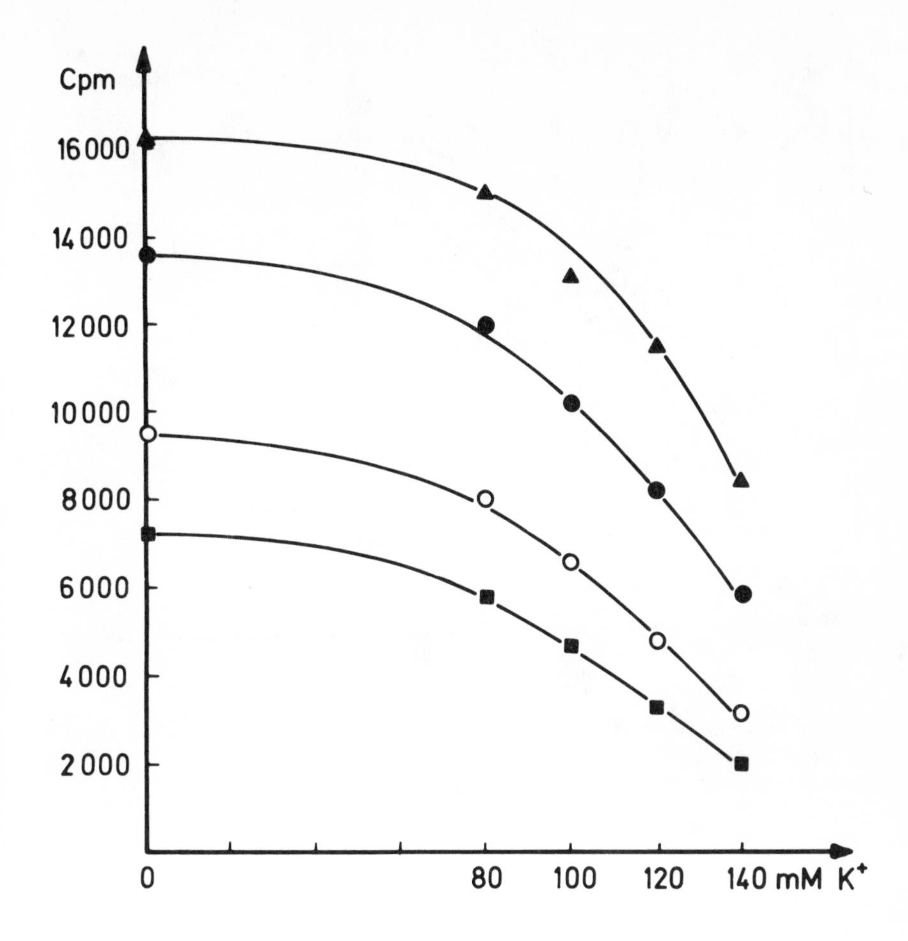

Fig. 3 a and b. Effect of K^+ on acetyl-CoA activation of pyruvate carboxylase in crude extracts of rat liver. a) Relation between enzyme activity and acetyl-CoA concentration at various concentrations of K^+.

▲ without K^+
● 80 μM K^+
○ 100 μM K^+
□ 120 μM K^+
■ 140 μM K (physiological concentration)

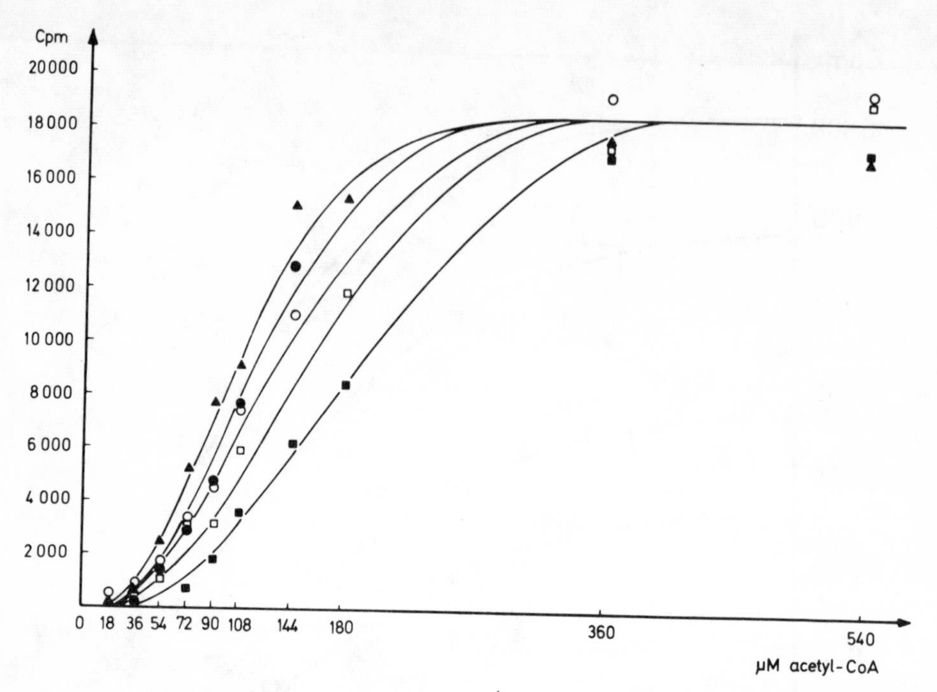

b) Relation between enzyme activity and K^+-concentration at suboptimal concentrations of acetyl-CoA

■ 90 µM acetyl-CoA
○ 108 µM acetyl-CoA
● 144 µM acetyl-CoA
▲ 180 µM acetyl-CoA

References

Baudhuin, R., H. Beaufay, Y. Rahman-Li, O. Z. Sellinger, R. Wattiauy, P. Jaques, C. De Duve: Biochem. J. 92 (1964) 179

Böttger, J., O. Wieland, D. Brdicka, D. Pette: Europ. J. Biochem. 8 (1969) 113

Brdiczka, D.,D. Pette: Hoppe-Seylers Z. physiol. Chem. 350 (1969) 19

De Duve, C., B. C. Pressman, R. Gianetto, R. Wattiaux, F. Appelmans: Biochem. J. 60 (1955) 604

De Duve, C., P. Baudhuin: Physiol. Rev. 46 (1966) 323

Dugal, B., W. Seubert in 20 Colloquium d. Gesellschaft f. Biol. Chem. 1969, Berlin-Heidelberg-New York 1969, p. 389

Freedman, A. D., L. Kohn: Science 145 (1964) 58

Gianetto, R., C. De Duve: Biochem. J. 59 (1955) 433

Henning, H. V., W. Seubert: Biochem. Z. 340 (1964) 160

Henning, H. V., B. Stumpf, B. Ohly, W. Seubert: Biochem. Z. 344 (1966) 274
Hogeboom, G. H., W. C. Schneider: Biochem. J. 204 (1953) 233
Hsu, R. J., H. A. Lardy: J. biol. Chem. 242 (1967) 520
Keech, D. B., M. F. Utter: J. biol. Chem. 238 (1963) 2609
L'Age, M., H. V. Henning, B. Ohly, W. Seubert: Biochem. biophys. Res. Commun. 31 (1968) 241
Leighton, F., B. Poole, H. Beaufay, P. Baudhuin, J. W. Coffey, St. Fowler, C. De Duve: J. Cell Biol. 37 (1968) 482
Lynen, F., S. Ochoa: Biochem. et Biophys. Acta 12 (1953) 299
Marco, R., J. Sebastian, A. Sols: Biochem. biphys. Res. Commun. 34 (1969) 725
Müllhofer, G., O. Kuntzen, S. Hesse, Th. Bücher: FEBS-Letters 4 (1969a) 33
Müllhofer, G., O. Kuntzen, S. Hesse, Th. Bücher: FEBS-Letters 4 (1969b) 47
Müllhofer, G., O. Kuntzen: Hoppe-Seylers Z. physiol, Chem. 351, (1970), 277
Müllhofer, G., O. Kuntzen, this symposium
Ochoa, S. in S. P. Colowick, N. O. Koplan: Methods in Enzymology, Vol. I, Academic, New York 1955, p. 685
Seubert, W., H. V. Henning, W. Schoner, M. L'Age: Advances in Enzyme Regulation. Ed. by G. Weber, Vol. 6. Pergamon, New York 1968, p. 153
Shrago, E., H. A. Lardy: J. biol. Chem. 241 (1966) 663
Williamson, J. R., this symposium

Announced Discussion

F. J. Ballard
C.S.I.R.O. Division of Nutritional Biochemistry, Adelaide, Australia

This problem of intracellular distribution of pyruvate carboxylase has interested us for some years. In our publications we have reported activities of this enzyme both in particulate and cytosol fractions of rat liver and adipose tissue and have inferred that these activities either are or could be distinct proteins and subject to different controls. We have had some support for this concept from dietary studies with rat liver and from differential changes in the two activities that were found in adipose tissue and mammary gland.

Immunochemical studies with malate dehydrogenase, aspartate aminotransferase and P-enolpyruvate carboxykinase have been used to confirm the existence of distinct activities of these enzymes in the mitochondria and cytosol. It seemed to us that this technique would be useful to resolve the controversy regarding the sub-cellular distribution of pyruvate carboxylase. We therefore isolated mitochondria from rat liver, carefully washed them to remove all the cytosol fraction and purified pyruvate carboxylase from this fraction by techniques described by Utter for the chicken liver enzyme. Rat liver mitochondria pyruvate carboxylase had a specific activity of about 14 units/mg protein. Antibodies prepared against this enzyme in rabbits were titrated against liver extracts. An example of antibody-antigen titrations with active antibody and control serum against mitochondrial and a 100,000g supernatant of rat liver is shown in Fig. 1a. It is apparent that the titration curves with the two extracts are identical and that the control γ-globulin did not cause inactivation. In order to obtain a statistical evaluation of the similarity between antibody-antigen reactions with both mitochondrial and 100,000g supernatant enzymes, we measured the amount of enzyme inactivated by a constant portion of antibody. In this experiment we tested mitochondrial and supernatant activities from all tissues that had previously been reported to contain pyruvate carboxylase in high activity. It can be seen from Table 1 that approximately 60 milliunits of enzyme were inactivated by 20 µl of antibody and that no statistical differences were detected between the amounts of enzyme inactivated from either subcellular compartments from the various tissues.

One possible error in this experiment could occur if the supernatant fraction contained two activities, one leached from the mitochondria and thus identical to the mitochondrial enzyme, together with a second activity that exhibited less inactivation with the antibody and was derived from the cell cytosol. The error would occur if sufficient antibody had been added to titrate only the enzyme of the first type. This possible contingency is partially eliminated by the titration curve in Fig. 1a, but a more convincing type of experiment is shown in Fig. 1b. In this experiment, increasing amounts of enzyme are added to a constant quantity of antibody. If the soluble fraction contained two distinct antibodies, one of which exhibited less immunochemical activity, the equivalence points in this case would occur with a lower amount of enzyme activity. It can be seen from Fig. 1b that such an interpretation is impossible.

We conclude from these experiments that the soluble activities detected in several rat tissues are either the result of mitochondrial damage or are so chemically similar to the mitochondrial enzyme to be antigenically identical. The recent report of Böttger et al. (1969) supports the first alternative and we believe that we were incorrect in interpreting the soluble activitiy of pyruvate carboxylase as a separate enzyme derived from the cell cytosol. I would also like to mention that recent immunochemical studies by Wallace, Watson and Keech on the adaptive pyruvate carboxylase of sheep liver also indicate antigenic identity of "soluble" and mitochondrial activities.

Reference

Böttger, I., O. Wieland, D. Brdiczka, D. Pette: Europ. J. Biochem. 8 (1969) 113

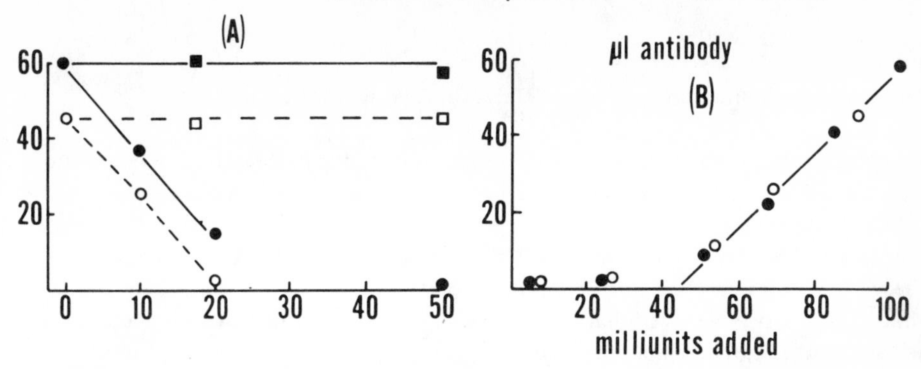

Fig. 1a. Addition of either control antiserum (square symbols) or antiserum prepared from rabbits immunized with mitochondrial pyruvate carboxylase (round symbols) to 100,000g supernatants (open symbols) or mitochondrial extracts (closed symbols) prepared from rat liver.

Fig. 1b. Addition of increasing amounts of pyruvate carboxylase derived from 100,000g supernatants (O) or mitochondria (●) from rat liver against a constant amount of antibody. The activity of pyruvate carboxylase remaining in the supernatant was measured after removal of antibody-antigen precipitates by centrifugation.

Inactivation of Pyruvate Carboxylase by a Constant Amount of Antibody

Tissue	Fraction	milliunits inactivated per 20 μl antibody
Liver	particulate	64.9 ± 5.1 (8)
	soluble	54.3 ± 3.0 (7)
Kidney	particulate	62.1 ± 4.1 (8)
	soluble	59.4 ± 6.9 (8)
Mammary gland	particulate	69.7 ± 8.4 (6)
	soluble	59.6 ± 8.1 (7)
Adipose tissue	particulate	55.3 ± 5.1 (7)
	soluble	68.1 ± 7.8 (7)

Table 1. Soluble and particulate fractions were prepared by centrifugation of 0.25 M sucrose homogenates at 100,000g for 30 minutes. The particles were freeze dried and pyruvate carboxylase extracted. Antibody was added to portions of each fraction and after standing for two hours, the solutions were centrifuged and the activity remaining in the supernatant assayed. Values are the means ± s.e.m. with the number of animals in parentheses. Analysis of variance indicates no significant differences.

Announced Discussion

P. Walter
Medizinisch-chemisches Institut der Universität Bern, Switzerland

In our laboratory, Mr. Anabitarte made some interesting observations in connection with the activities of glutamate dehydrogenase (=GDH) and of pyruvate carboxylase (=PC) in supernatant fractions of rat liver homogenates. First, we repeated the experiments of Seubert's group by using their methods and media for the preparation of the various cytosols (Henning et al. 1966, Seubert et al. 1968). In agreement with their findings, we observed that only in the medium containing sodium acetate the PC activity was greater than the GDH activity (Table 1; all media contained 1 mM versene and 1 mM glutathion). We have carried out several experiments with the TEA-containing medium and we found varying amounts of enzyme activities, but the GDH/PC ratio was always about three. We prefer to measure the GDH in the presence of ADP because we get more reproducible values in this manner. Interestingly, in Experiment 4, with the sodium acetate-containing medium, the GDH had lost its sensitivity to ADP.

In Table 2, an experiment is shown demonstrating the instability of GDH in the medium containing sodium acetate. Up to the step "Combined Supernatants", the method of Seubert's group using the sucrose medium was followed. Also according to Seubert, part of the "Combined Supernatant" was centrifuged and a GDH activity of 1.09 and 2.14, when assayed without and with ADP respectively, was obtained. However, when a concentrated solution was added to another part of the "Combined Supernatant" and when this mixture was rehomogenized and centrifuged more than half of the GDH activity had disappeared, whereas the PC activity was slightly increased. As in Experiment 4 of Table 1, this GDH had also lost its sensitivity to ADP. In the control experiment, where a part of the "Combined Supernatant" was rehomogenized and centrifuged without added sodium acetate, practically no changes occurred in the enzyme activities as compared to Supernatant I. It appears, therefore, that it is difficult to use experiments with the sodium acetate containing medium for the quantitative evaluation of mitochondrial damage which occurred during the preparation of the supernatant. Whether some loss of GDH activity also occurs in the other media cannot be decided at the present time.

Homogenizing medium	Exp. Nu.	Enzyme Activities					
		Cytosol			Total Homogenate		
		PC	GDH		PC	GDH	
			-ADP	+ADP		-ADP	+ADP
			μmoles/min x g liver				
0.3 M Sucrose	1	0.5	1.3	3.3	5.7	20.2	87.4
0.24 M Sucrose and 20 mM TEA pH 7.2	2	1.05	1.59	3.27	6.83	44.9	206.0
	3	0.7	0.8	2.1	5.8	47.5	181.8
0.28 M Sucrose and 50 mM Na-acetate	4	0.8	0.5	0.6	7.1	34.2	101.4

Table 1. Activities of Glutamate Dehydrogenase and Pyruvate Carboxylase in Rat Liver.

References

Henning, H. V., B. Stumpf, B. Ohly, W. Seubert: Biochem. Z. 344 (1966) 274

Seubert, W., H. V. Henning, W. Schoner, M. L'Age: Advances in Enzyme Regulation. Vol. 6 (1968) 153

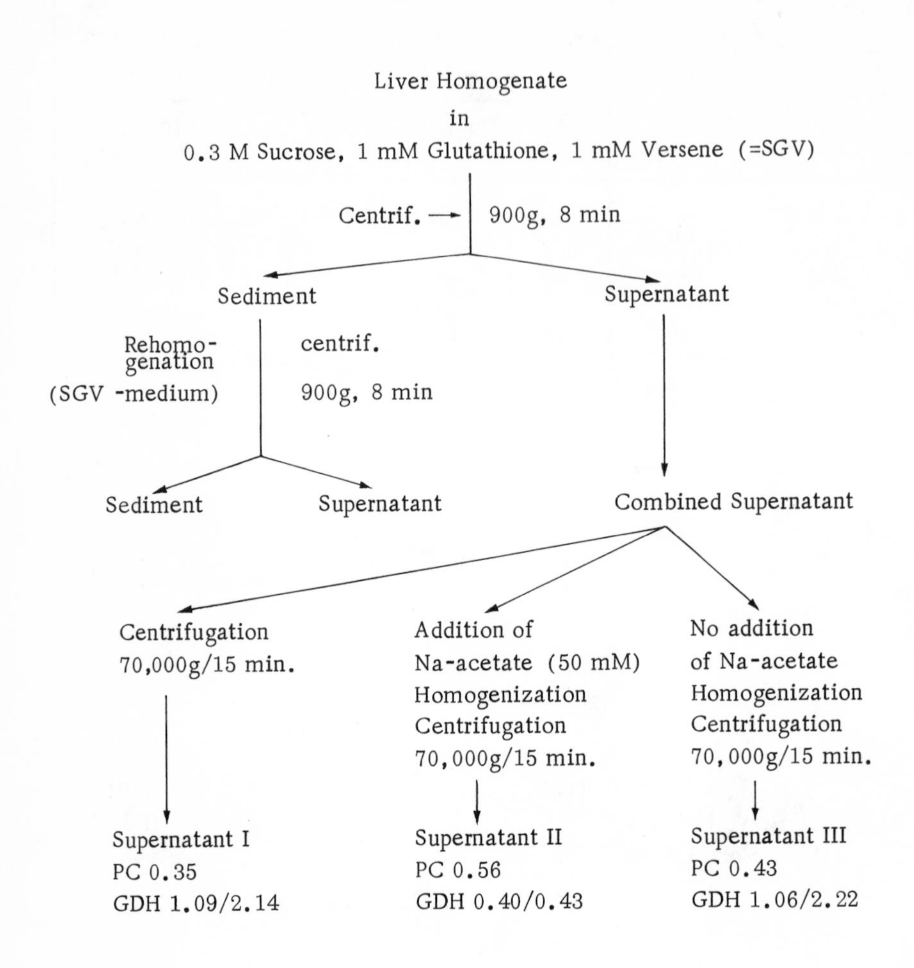

Table 2. Activities of Glutamate Dehydrogenase and of Pyruvate Carboxylase in Cytosol of Rat Liver.

Discussion to Weiss, Ohly, Dugal and Seubert

Seubert: First, I would like to answer these data. The paper you refer to is about three years old. Concerning the assay of glutamate dehydrogenase, we originally employed the reaction from glutamate to α-ketoglutarate without ADP as a measure of the enzyme activity. In this direction the activity of glutamate dehydrogenase is much lower. Presently, we are assaying with α-ketoglutarate and in presence of ADP. Of course we now find higher values, even higher than Dr. Walter has shown. But the percentage of glutamate dehydrogenase solubilized by the different extraction procedures is the same, independent of the assay method used.

Concerning the sodium acetate effect: We no longer use this medium. We use the sucrose-Triethanolamine medium adjusted by potassium hydroxide to pH 7.2. (Adv. Enzyme Regul. 6,153 , 1968).

Third: These solubility studies were just an indication for a possible location of pyruvate carboxylase in different compartments. They have tempted us to continue our studies on the location of pyruvate carboxylase in different subcellular compartments. The solubility studies were only starting points. We have repeated these extraction studies several times and not just by one person. I was quite anxious about this controversy.

Recently, Mr. Dugal studied the solubility of the enzyme from chicken liver in our laboratory. He found a "soluble" pyruvate carboxylase as compared to glutamate dehydrogenase (20. Coll. Ges. f. Biol. Chemie, Springer, Berlin, 389, 1969). The final proof for the location of pyruvate carboxylase, however, can only bring the isolation and characterization of this subcellular unit. This is the aim which we are pursuing now.

Exton: With regard to the intracellular location of pyruvate carboxylase we have made a few investigations together with Drs. D. H. Morgan and T. H. Claus. We used techniques to separate cell organelles based essentially on those of De Duve. We isolated lysosomes and peroxisomes and measured the activity of pyruvate-carobxylase in relation to other marker enzymes. We have not found any evidence for a specific localization of the enzyme in these organelles, but these are very preliminary results.

Seubert: Did you already have pure peroxisomes?

Exton: Right, this is the crucial point.

Seubert: Yes, this is the point. We spent about half a year to learn this.

Söling: I would like to comment on Dr. Ballard's data: I am very hesitant at the present time to make a decision on whether there are different pyruvate carboxylases. One could argue that 1) if the antigenic site of pyruvate carboxylase is not at the transferase subunit of the enzyme but at the CO_2 fixing subunit, and 2) if the data and speculations of Alberts (Alberts et al., Proc. Nat. Acad. Sc. 63, 1319, 1969) that the CO_2 fixing subunit is identical for different biotin enzymes would be consistent with all other findings, that one could quite well have the same antigenic behavior for different enzymes. This could be valid also for different pyruvate carboxylases. I mention this only to build a golden bridge for those believing in different pyruvate carboxylases.

Schäfer: I should like to ask if there is anyone in the audience who could comment on the biosynthesis of the enzyme. Is it not possible that during biosynthesis of the enzyme which occurs outside of the particles there is a certain amount of enzyme already activated, whereas the major portion of active holo-enzyme is essentially built inside of the particles?

Wieland: That is a good question. This could perhaps explain the specific changes in the so-called extra-mitochondrial part of the enzyme under various experimental conditions.

Söling: Generally it is accepted that the matrix enzymes are synthesized outside the mitochondria. If pyruvate carboxylase were a true matrix enzyme (as seems to be the case according to the experimental results available), one could assume that it is synthesized outside and carried by some way into the matrix space.

Ballard: If it was synthesized outside and passed into the mitochondria, it is conceivable that a residual activity remains outside the mitochondria and this activity would be antigenically the same as the mitochondrial enzyme. The situation with PEP-carboxykinase could not fit this model since the mitochondrial and cytosol enzymes are unrelated.

Seubert: I would like to add a comment to Dr. Söling's suggestion that the immunochemical assay might not be sufficient to differantiate between different forms of the enzyme. Majerus and Kilburn (J. Biol. Chem. 244, (1969) 6254) were able to show that antibodies against purified chicken liver acetyl-CoA carboxylase cross-react with the enzyme from rat liver and, as far as I remember, the precipitation was the same for both enzymes. Since the immuno-chemical assay apparently is not sufficient to differentiate between biotin enzymes from different species, it seems possible that it is also not sufficient to differentiate between different forms of a biotin enzyme of the same species.

Ballard: We have not measured the inactivation of other biotin-containing carboxylases by the pyruvate carboxylase antibody, but we will certainly do so, etc.

Seubert: I would suggest to you to examine a possible cross-reaction with acetyl-CoA carboxylase or crotonyl-CoA carboxylase, or with pyruvate carboxylase from another species.

Utter: I would like to ask Dr. Weiss a question concerning the concentration of pyruvate carboxylase in heavy and light mitochondria. The latter classification has been used to describe other functions of mitochondria such as respiratory control. Has any criterion such as this been compared with pyruvate carboxylase in a distribution study?

Weiss: Not yet so far. First we want to purify our compartment if it exists.

Utter: May I ask a further question: How many kinds of mitochondria do you think there might be?

Seubert: I must agree with you; there might be different populations of mitochondria.

Wieland: So we come together.

Seubert: But if there are different populations of mitochondria, you also must postulate different physiological functions.

Exton: In relation to my earlier comment about the localization of pyruvate carboxylase, let me say that we did not find any specific localization in the lysosomes or peroxisomes neither did we find that the carboxylase peak perfectly overlapped the succinate-cytochrome-C-reductase peak.

Seubert: May I ask you again: Did you study the distribution of cytochrome-C-reductase and pyruvate-carboxylase?

Exton: Yes, and they did not overlap exactly.

Seubert: The studies of Dr. Weiss (Fig. 2) on the distribution of 3-hydroxyacyl-CoA dehydrogenase, glutamate dehydrogenase and pyruvate carboxylase after isopycnic fractionation are in agreement with your results.

Guder: Some years ago Brdiczka from Pette's group (D. Brdiczka et al. Herbsttagung der Gesellschaft für Physiologische Chemie, Marburg 1966, abstract II, 5) has shown that there are different populations of mitochondria under different thyroid states. These populations had different enzyme patterns and possibly represented mitochondria of different intracellular localization or different age. Possibly the results of Dr. Seubert could be explained on a similar basis.

Hanson: John Ballard and I have looked at the distribution of pyruvate carboxylase in rat adipose tissue and found that about 30% of the activity is in the cytosol. Recently Gul and Dils (Biochem. J. 111 (1969) 263) have given support for this distribution using marker enzymes and their work appears convincing. However, I should indicate that Dr. Ballard will present data later this morning showing that pyruvate carboxylase in adipose tissue cytosol is immuno-cnemically identical with the mitochondrial enzyme. This kind of evidence although not conclusive, makes us doubt our earlier distribution studies.

Seubert: I would like to stress once again the physiological significance of a location of pyruvate carboxylase in different compartments. In this connection I would like to return to Dr. Müllhofer's finding that C-2 of pyruvate and lactate is converted to the C-4-position of glucose to different degrees. This means that the pyruvate formed from lactate does not equilibrate with the pyruvate pool of the mitochondria. If pyruvate formed from lactate had to go through the mitochondrial pyruvate one would expect an identical labeling pattern in glucose with lactate and pyruvate as precursors.

Regulation of Pyruvate Metabolism in Rat Liver Mitochondria

P. Walter and J. W. Stucki
Medizinisch-chemisches Institut der Universität Bern, Switzerland

Summary

The activities of pyruvate carboxylase and of pyruvate dehydrogenase in intact rat liver mitochondria were measured under various experimental conditions. In agreement with previous findings, the pyruvate dehydrogenase activity was found to be strongly inhibited by octanoate and palmitoyl-carnitine. Pyruvate carboxylase activity was not altered in the presence of octanoate, when high concentrations of pyruvate were used. However, with low concentrations of pyruvate, octanoate and palmitoyl-carnitine stimulated the carboxylation of pyruvate by 53% and 61% respectively.

As reported earlier, elevated ADP levels in the medium caused a strong inhibition of pyruvate carboxylase activity. A partial reversal of this inhibition was obtained upon the addition of octanoate, but not with octanoyl-L-carnitine. Solubilized pyruvate carboxylase could also be inhibited by ADP but here, octanoate caused no reversal. From experiments with atractylate, dinitrophenol, oligomycin and from measurements of oxygen uptake and adenine nucleotide concentrations, it is concluded that the pyruvate carboxylase activity is regulated mainly by the level of intramitochondrial ADP. In the case of octanoate, the reversal of the ADP inhibition of pyruvate carboxylase is related to the production of intramitochondrial AMP by fatty acid activation leading to a lowering of the ADP plus ATP pool. The increased utilization of ADP for oxidative phosphorylation in the presence of octanoate may also play a role in reducing the concentration of intramitochondrial ADP.

Oleate was also found to partially reverse the inhibition of ADP on pyruvate carboxylation, but this fatty acid probably acts by a different mechanism than octanoate.

It can be concluded that some of the main factors regulating pyruvate metabolism in mitochondria are substrate availability, inhibition of pyruvate dehydrogenase by fatty acids and acyl-carnitines, and intramitochondrial levels of acetyl-CoA and adenine nucleotides.

Introduction

Liver mitochondria contain two important enzymes which catalyze metabolic conversions of pyruvate. The pyruvate carboxylase is the enzyme for the transformation of pyruvate to oxalacetate and the pyruvate dehydrogenase is the enzyme for the reaction of pyruvate to acetyl-CoA. Of the two products formed by these enzymes, only oxalacetate can be used by the liver cell as substrate for gluconeogenesis. Mitochondrial pyruvate metabolism might therefore be a site where metabolic regulation could be expected and there is a lot of evidence indicating this to be the case. Especially the physiological role of fatty acids is very much discussed as fatty acids have been shown to inhibit the oxidation of pyruvate to acetyl-CoA and also to stimulate gluconeogenesis under various experimental conditions. The stimulating effects of glucagon and c-AMP on gluconeogenesis also may be connected with the regulation of pyruvate metabolism. These results have been reviewed by Scrutton and Utter (1968).

In the following presentation some factors which are important for the regulation of the pyruvate metabolism regarding in vitro experiments with rat liver mitochondria will be summarized and their possible physiological significance will be discussed.

Materials and Methods

Except where noted, the following incubation conditions were used:

a) Incubation mixture of controls: 4 mM ATP, 10 mM $MgSO_4$, 6.6 mM potassium phosphate buffer pH 7.4, 10 mM $KHCO_3$, 10 mM pyruvate, 6.6 mM TEA buffer pH 7.4, sucrose solution, mitochondria of 0.5 g of liver. Final volume 3.0 ml. 250 μg of creatine kinase was added to all incubations containing creatine. The incubations were carried out in shaking flasks at 37°C. The reactions were started by adding the mitochondria after an equilibration time of 2.5 minutes and stopped after 8-10 minutes by the addition of perchloric acid. The metabolites were determined in the protein-free neutralized extracts according to Walter et al. 1966, Mehlman et al. 1967 and Walter et al. 1970.

Results and Discussion

In Table 1, the results are summarized of an experiment which was carried out as described under Materials and Methods. For the calculation of the PC activity, it is important to realize that all of the oxalacetate formed by the PC reaction accumulates as one intermediate or another of the tricarboxylic acid cycle, independent of how fast the cycle turns. Therefore, by adding up the amounts of all the tricarboxylic acid intermediates, an accurate value of the PC activity can be obtained. The amounts of isocitrate, aconitate and oxalacetate in these experiments were found to be less than 0.1 μ moles and were therefore neglected. Succinate was not measured in this experiment and, as will be seen in later experiments, would amount to about 0.3 μ moles. Another method for obtaining values for the PC activity in mitochondria consists of measuring the amount of $^{14}CO_2$ incorporated into the acids of the tricarboxylic acid cycle. However, as can be seen from the results of Table 1, this method yielded too low values. The reason for this was: 1) that some of the ^{14}C-bicarbonate was cleaved off again when products were decarboxylated in the cycle and, 2) because no correction was made for the dilution of the added radioactive bicarbonate by metabolically produced non-radioactive bicarbonate. The addition of octanoate increased the incorporation of $^{14}CO_2$ but did not change the amount of pyruvate carboxylated. This can be explained on the basis of earlier work (Walter et al. 1966) where octanoate was shown to enhance malate formation by direct reduction of oxalacetate and furthermore because less non-radioactive bicarbonate was produced due to the inhibition of pyruvate decarboxylation by octanoate. This strong inhibition of

Metabolites	Control	+ Octanoate 2 mM
	μ moles/3 ml	
Citrate found	3.3	2.2
α-Ketoglutarate found	0.1	0.1
Fumarate found	0.9	1.0
Malate found	5.7	6.5
Total $^{14}CO_2$ incorporated	7.0	8.7
Pyruvate carboxylated	10.0	9.8
Pyruvate used	18.1	10.6
Pyruvate oxidized	8.1	1.0

Table 1. Measurement of Pyruvate Carboxylase Activity in Intact Rat Liver Mitochondria. The incubations were carried out as described in "Materials and Methods", except that radioactive bicarbonate was used. The "total $^{14}CO_2$ incorporated" was calculated on the basis of the specific radioactivity of the $KH^{14}CO_3$ added. The "Pyruvate carboxylated" represents the sum of the measured acids of the citric acid cycle. "Pyruvate oxidized" was obtained by substracting "Pyruvate carboxylated" from the "Pyruvate used."

the conversion of pyruvate to acetyl-CoA by octanoate was also apparent in the experiment of Table 1. The amounts of pyruvate oxidized were calculated by subtracting the pyruvate carboxylated from the total pyruvate used.

High and probably saturating amounts of pyruvate were used in the experiment of Table 1. The experiment summarized in Table 2 was performed with more physiological amounts of pyruvate as about 0.6 μmoles were added every minute yielding a final concentration of maximal 0.2 mM. Upon addition of octanoate or palmitoyl-carnitine, pyruvate oxidation was again inhibited, however, in contrast to the previou experiment the fatty acids caused here a stronger accumulation of malate and a definite increase in pyruvate carboxylation. It is quite likely that the PC is not saturated by the pyruvate concentrations used in this experiment and thus the increased pyruvate carboxylation could therefore be the consequence of the inhibited pyruvate oxidation. On the other hand, fluctuations of the intramitochondrial levels of acetyl-CoA, which is an allosteric activator of PC, may also be of importance. The acetyl-CoA levels were not measured here but it is quite possible that the use of lower concentrations of pyruvate resulted in lower and, for the PC, limiting levels of acetyl-CoA. The addition of fatty acids could lead to higher acetyl-CoA levels and could also consequently in this manner cause the observed stimulation of pyruvate carboxylation. In perfusion experiments with isolated rat livers, a lower ratio of lactate used to glucose formed has been observed in the presence of fatty acids than in its absence (Teufel et al. 1967, Williamson et al. 1968). This "sparing effect" may very well

Metabolites	Additions to System		
	Control	Octa-noate 6 μmoles	Palmitoyl-carnitine 0.9 μmoles
		μmoles/ 3 ml	
Citrate found	1.35	0.70	1.11
Succinate found	0.20	0.30	0.12
Fumarate found	0.41	0.71	0.78
Malate found	1.12	3.14	3.09
Pyruvate carboxylated	3.17	4.85	5.10
Pyruvate used	5.89	5.44	5.81
Pyruvate oxidized	2.72	0.59	0.71

Table 2. Effect of Octanoate and Palmitoyl-carnitine on Pyruvate Metabolism in Mitochondria with Limiting Amounts of Pyruvate. The reaction mixture was the same as described in "Materials and Methods" with the following changes: Pyruvate was added during the incubation in portions of 0.6 μmoles at one minute intervals. The first addition was made at the same time as the mitochondria were added and the last one minute before the end of the incubation. Palmitoyl-L-carnitine was also added in ten portions of 0.09 μmoles together with the pyruvate. All reaction mixtures contained 20 mg bovine serum albumin and the pyruvate used was labeled in the 2-position with ^{14}C. Labeled pyruvate was used in order to measure the succinate and fumarate (Walter and Stucki, 1970).

be based on the effect we observed here of fatty acids on the mitochondrial pyruvate metabolism.

Harris et al. (1969) have shown that under certain conditions lactate inhibits the accumulation of pyruvate in the mitochondria. We have added up to 100 times more lactate than pyruvate to incubations as described in Tables 1 and 2, but we could not observe any changes in pyruvate metabolism in our system.

It was observed by Walter et al. (1966) that elevated ADP levels in the incubation medium caused an inhibition of PC activity in rat liver mitochondria. It was found that at a concentration of 0.17 mM of ADP in the medium the inhibition amounted to about 50%. As these ADP levels are in the same order of magnitude as the intracellular levels of this nucleotide, the effect was further investigated. In the experiment of Table 3, the medium ADP was held high by the addition of creatine plus creatine kinase, and an inhibition of pyruvate carboxylation of 76% was observed. Upon addition of octanoate, this inhibition was partially reversed whereas the change in the ADP levels of the medium was small and not reproducible. This effect of octanoate

Metabolites	Additions to System			
	Control	Octanoate 0.8 mM	Creatine 13 mM	Octanoate Creatine
	μmoles/3 ml			
Citrate found	3.51	2.86	1.05	1.31
α -Ketoglut. found	0.24	0.12	0.07	0.02
Succinate found	0.27	0.36	0.21	0.31
Fumarate found	0.62	0.77	0.10	0.33
Malate found	3.95	4.65	0.81	1.99
Pyruvate carboxyl.	8.59	8.76	2.24	3.96
Pyruvate used	17.15	12.82	9.0	7.45
Pyruvate oxidized	8.56	4.16	6.76	3.49
ADP found	0.28	0.31	1.16	1.08
AMP found	0.30	0.45	0.68	0.66

Table 3. Effect of Creatine and Octanoate on Pyruvate Metabolism in Rat Liver Mitochondria. This experiment is taken from Walter and Stucki, (1970). The incubations were carried out as described under "Materials and Methods", except that pyruvate-2-^{14}C was added.

was also observed with concentrations of 0.3 mM of the fatty acid (Walter and Stucki, 1970). With 0.8 mM octanoate, the reversal was apparent after three to six minutes of incubation (Walter and Stucki, 1970). The addition of the uncoupler dinitrophenol caused a strong inhibition of pyruvate carboxylation in the system without creatine (Table 4) as well as with creatine. The opposite effects of dinitrophenol and octanoate show that the fatty acid exerts its action on the inhibited PC activity by a mechanism other than uncoupling. Oligomycin, which blocks the phosphorylation of ADP inside the mitochondria, caused the strongest inhibition (Tables 4 and 5). Atractylate, on the other hand, which is known to inhibit ADP and ATP transport through the inner membrane of the mitochondrion, reversed the inhibition caused by the addition of ADP (Table 5).

When mitochondria were broken up and when the solubilized crude PC was assayed in the presence of creatine and creatine kinase, elevated levels of ADP were also found to inhibit the PC (Fig. 1). However, neither octanoate nor higher levels of acetyl-CoA could reverse this inhibition. All of these results and especially those with atractylate and with the solubilized enzyme point to the possibility that the levels of intramitochondrial adenine nucleotides are regulating the PC activity. It should be noted here that the intramitochondrial adenine nucleotides do not always have the same AMP: ADP:ATP ratio as those in the medium because the transport through the inner membrane is not passive but mediated by a translocase and furthermore because only ADP and ATP is transported in and out of the mitochondria whereas AMP is not (Pfaff et al.

1969, Duée and Vignais 1969, Winkler et al. 1968). It might be added that the PC has been found to be localized in the matrix space of the mitochondria (Böttger et al. 1969, Marco et al. 1969).

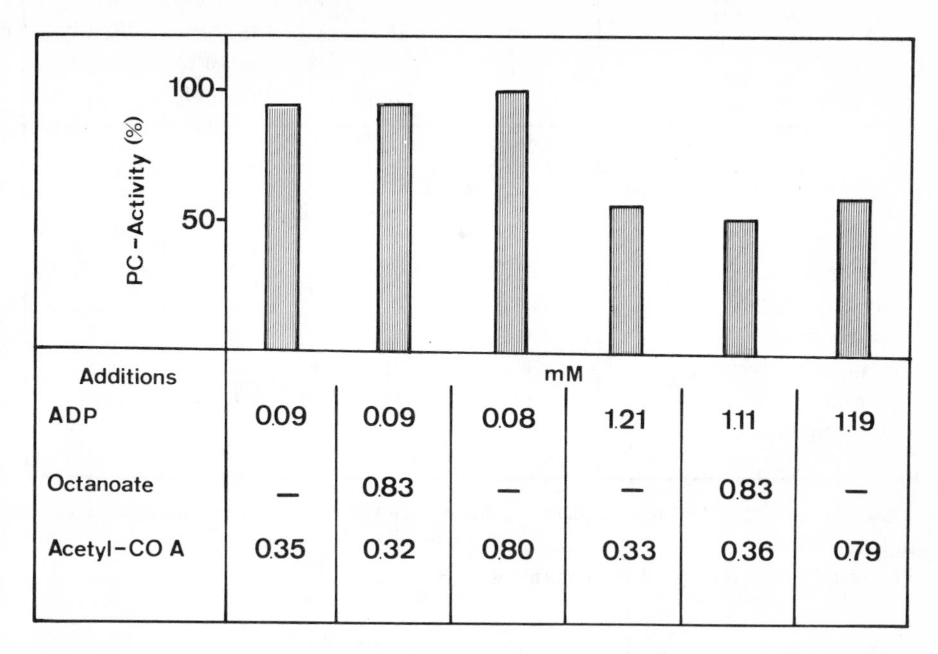

Additions	mM					
ADP	0.09	0.09	0.08	1.21	1.11	1.19
Octanoate	—	0.83	—	—	0.83	—
Acetyl-CO A	0.35	0.32	0.80	0.33	0.36	0.79

Fig. 1. Influence of ADP, Octanoate and Acetyl-CoA on the Solubilized Pyruvate-carboxylase. Detailed experimental results and incubation conditions discussed in Walter and Stucki (1970). The ADP concentration was held constant throughout the 30 minute incubation by the addition of ATP, creatine and creatine kinase. The same is true for the acetyl-CoA concentration, where CoA, acetyl phosphate and phosphotransacetylase were used.

The intramitochondrial adenine nucleotides were measured under various conditions after five minutes of incubation (Table 6). In this experiment the mitochondrial pellets were separated from the media by the Millipore technique according to Lehninger (Winkler et al. 1968). The relative rates of PC were taken from the experiments of the previous tables. The results show that there is a good correlation between the intramitochondrial ADP levels and the relative rates of PC activity. In some cases the relative changes of these two parameters are not proportional and it may be that other factors such as the ATP to ADP ratio and changes in the intramitochondrial water content may also play a role. It is also possible that in some cases the ATP concentration is too small to saturate the PC. It has been shown previously that the size of the intramitochondrial adenine nucleotide pool is dependent on the incubation conditions and the fluctuations observed in the experiment of Table 6 are in good agreement with the literature (Meisner and Klingenberg 1968).

Metabolites	Additions to System			
	Control	Octanoate 1.6mM	Dinitrophenol 0.7mM	Oligomycin 0.7 μg/ml
	μmoles/3 ml			
Citrate found	5.2	3.0	0.6	0.3
α-Ketoglut. found	0.3	0.3	0.1	0.1
Succinate found	0.3	0.3	0.1	0.1
Fumarate found	0.7	1.0	<0.1	<0.1
Malate found	4.3	5.8	0.1	0.3
Pyruvate carboxyl.	10.8	10.4	0.9	0.8
Pyruvate used	18.4	11.3	6.0	3.3
Pyruvate oxidized	8.4	1.1	5.1	2.5

Table 4. Effect of Octanoate, Dinitrophenol and Oligomycin on Pyruvate Metabolism in Liver Mitochondria. This experiment is taken from Walter and Stucki, (1970). The same conditions were used as in Table 3.

In order to explain the ability of octanoate to reverse the inhibition by added creatine, one could conclude that the fatty acid has an atractylate like effect and inhibits ADP and ATP transport through the inner membrane of the mitochondria. However, this conclusion is likely to be wrong because octanoate increased the coupled respiration (Table 7), whereas atractylate inhibited this process (not shown in this report).

Additions (All with creatine)	Pyruvate carboxylase activity
	μmoles/3 ml
None	2.9
0.013 mM Atractylate	7.7
0.7 μg/ml Oligomycin	0.7

Table 5. Effect of Atractylate and Oligomycin on Pyruvate Carboxylase Activity in Intact Mitochondria. Detailed results on the amounts of the citric acid cycle intermediates discussed in Walter and Stucki (1970). Thirteen mM creatine was present in all incubations.

Additions to System	Intramitochondrial adeninenucleotides				Relative rate of pyruvate carboxylase
	AMP	ADP	ATP	Total	
mM	n moles/mg protein				% of control
None (control)	1.06	5.90	2.93	9.89	100
0.8 Octanoate	6.40	4.83	3.10	14.33	100
13 Creatine 0.013 Atractylate	0.80	4.93	3.07	8.80	88
13 Creatine 0.8 Octanoate	4.63	6.80	1.87	13.30	46
13 Creatine 0.8 Octanoyl-L-carnitine	1.43	9.63	2.13	13.19	28
13 Creatine	1.37	11.13	2.30	14.80	26
13 Creatine 1.3 μg/ml Oligomycin	1.83	14.83	1.77	18.43	8

Table 6. Correlation Between Intramitochondrial Adeninenucleotide Levels and Pyruvate Carboxylase Activity in Intact Rat Liver Mitochondria. The values for the relative rates of PC are taken from the previous tables and Walter and Stucki 1970). The Millipore technique was used for the separation of the pellets from the media. Experimental details discussed in Walter and Stucki (1970).

On the other hand, the increased rate of oxidative phosphorylation in the presence of the fatty acid may lead to a lower steady state concentration of intramitochondrial ADP and thus could be responsible for the release of the inhibition of the PC.

Another factor which is important for the explanation of the octanoate effect is the fact that octanoate and ATP are converted to the acyl-CoA derivative and AMP, respectively, and thereby cause an increase of AMP and a decrease in ADP and ATP levels in the matrix. This effect is more important than the lowering of ADP by increased oxidative phosphorylation. This follows from a comparison of the experiments with octanoate and with octanoyl-L-carnitine. With the latter substrate, no AMP was produced and coupled respiration increased more than with octanoate but caused only 15% lowering of the intramitochondrial ADP level which was not enough to relieve the inhibition of the PC, whereas octanoate led to a 39% decrease in ADP and a 77% stimulation of PC.

In Table 8, some preliminary results on the effect of oleate are summarized. Creatine was present in all of the incubations, and, as judged by the amounts of the main products malate and citrate, 0.15 mM oleate caused a more than 30% increase in PC activity. At higher concentrations of the fatty acid, the PC activity decreased which was probably due to uncoupling. Intramitochondrial AMP was not increased as a result of oleate addition and also other analytical data indicate that oleate probably acts by a different mechanism than octanoate. Wojtczak and Zaluska (1967) have reported that low concentrations of oleate inhibit the incorporation of ^{14}C-ATP into mitochondria. It is quite possible that their effect and our observations are based on some action of oleate on the inner membrane of the mitochondria.

The experiment with oleate was performed in the absence of albumin. However, upon addition of an oleate-albumin complex (7 moles to 1 mole), we also observed a more than 20% reversal of PC inhibition.

Actyl-carnitines were also tested. So far, these compounds did not release the ADP inhibition of PC to the same extent as was found with oleate or octanoate. Oleyl-L-carnitine was found to be the most potent inhibitor of pyruvate oxidation.

According to Rose et al. (1964), liver tissue contains between 0.40 and 0.87μ equivalents of fatty acids per gram wet weight. Almost all of these fatty acids have chain lengths between 16 and 18 carbons. As our effects on the activation of PC were obtained with 0.15 mM oleate, an effect of fatty acids on the inner membrane of the mitochondria could also be possible in the liver cell. Moreover, recent work by Pfaff et al. (1969) have shown that the rate of oxidative phosphorylation and of adenine nucleotide transport are in the same order of magnitude. Therefore, small changes of the translocase activity could lead to important changes of the intramitochondrial adenine nucleotide pattern and therby influence PC activity. According to Font and Gautheron (1969), coenzyme A was also found to inhibit the adenine nucleotide translocase in pig heart mitochondria and there may be other as yet unknown substances in the cell having similar effects.

Additions to system (All contain creatine)	Oxygen uptake	
	No Oligomycin	With Oligomycin
mM	μg atoms O/min	
None	0.48	0.14
0.02 Octanoate	0.71	0.18
0.80 Octanoate	0.83	0.22
None	0.58	0.16
0.80 Octanoyl-L-carnitine	1.02	0.22

Table 7. Stimulation of Oxygen Uptake by Octanoate and Octanoyl-L-carnitine. The oxygen measurements were performed with a Clark Electrode (Yellow Springs Instr., Ohio, U.S.A.). The incubation mixtures were the same as described in "Materials and Methods", except that mitochondria of only 0.3 g of liver were used. Experimental details in Walter and Stucki (1970).

Metabolites	Oleate addition (mM)			
	None	0.05	0.15	0.30
	µMoles/3 ml			
Citrate found	1.52	1.46	1.76	1.36
Malate found	0.72	0.93	1.25	1.31
Citrate + Malate	2.24	2.39	3.01	2.67
Pyruvate used	12.79	12.73	12.45	10.93
Acetoacetate found	2.24	2.10	2.04	1.92

Table 8. Influence of Oleate on Pyruvate Carboxylase Activity of Rat Liver Mitochondria. The conditions used were the same as described under "Materials and Methods." Oleic acid was added as a 25 mM solution in butane-1.3-diol at the beginning of the experiment. All incubations contained 13 mM creatine.

The experiments with octanoate are probably of little direct physiological significance because short chain fatty acids do not normally appear as substrates in the liver cell. These experiments did show, however, how changes in the intramitochondrial adenine nucleotide pattern can strongly influence the rate of pyruvate carboxylation, and furthermore that extra-mitochondrial adenine nucleotide patterns must not necessarily reflect those in the matrix space.

Acknowledgment

This work was supported by the Schweizerischer Nationalfonds zur Förderung der wissenschaftlichen Forschung and by F. Hoffmann-La Roche Basel.

References

Böttger, I., O. Wieland, D. Brdiczka, D. Pette: J. Europ. Biochem. 8 (1969) 113

Duée, E. D., P. V. Vignais: J. biol. Chem. 244 (1969) 3920

Font, B., D. Gautheron: Bull. Soc. Chim. Biol. (Paris) 51 (1969) 1613

Harris, E. J., J. R. Manager: Biochem. J. 113 (1969) 617

Marco, R., J. Sebastian, A. Sols: Biochem. biophys. Res. Commun. 34 (1969) 725

Mehlman, M. A., P. Walter, H. A. Lardy: J. biol. Chem. 242 (1967) 4594

Meisner, H., M. Klingenberg: J. biol. Chem. 243 (1968) 3631

Pfaff, E., H. W. Heldt, M. Klingenberg: Europ. J. Biochem. 10 (1969) 484

Rose, H., M. Vaughan, D. Steinberg: Amer. J. Physiol. 206 (1964) 345

Scrutton, M. C., M. F. Utter in Annual Review of Biochemistry. Ed. by P. D. Boyer. Vol. 37, Annual Reviews, Palo Alto, Calif. 1968, p. 249

Teufel, H., L. A. Menahan, J. C. Shipp, S. Böning, O. Wieland: Europ. J. Biochem. 2 (1967) 182

Walter, P., V. Paetkau, H. A. Lardy: J. biol. Chem. 241 (1966) 2523

Walter, P., J. W. Stucki: Europ. J. Biochem. 12, (1970) 508

Williamson, J. R., E. T. Browning, M. S. Olson: Advances in Enzyme Regulation. Ed. by G. Weber. Vol. 6, Pergamon, New York 1968, p. 67

Winkler, H. H., F. L. Bygrave, A. L. Lehninger: J. biol. Chem. 243 (1968) 20

Wojtczak, L., H. Zaluska: Biochem. Biophys. Res. Commun. 28 (1967) 76

Discussion to Walter and Stucki

Krebs: How does the activity of pyruvate carboxylase, as assayed by your method, compare with the enzymic assay of the enzyme?

Walter: In our incubation system with whole mitochondria we find a pyruvate carboxylase activity of about 4 moles per 60 mg mitochondrial protein at 37^0 which is equivalent to about 4 moles per g wet weight of liver. In sonicated mitochondria or in sonicated homogenates we measure between 7 and 8 moles of activity per g wet weight of liver at 30^0. We therefore have a difference in the factor of at least two. We do not know what is limiting the pyruvate carboxylase activity inside the mitochondria. It may be that in our incubation systems the intramitochondrial ADP concentration is too high or that the ATP/ADP ratio is too low for maximal pyruvate carboxylase activity.

Ballard: Are you suggesting that pyruvate carboxylation is regulated by ADP levels rather than the ratio of ATP to ADP?

Walter: I think that both are regulating. From our experiments, ADP seems to correlate best with the activity. However, if you look at these results more closely, you will see that the changes in ADP content are not always proportional to the changes in pyruvate carboxylase activity. In these cases the ADP/ATP ratio may be of importance.

Utter: Were all these experiments done with 4 mM ATP added outside?

Walter: Yes.

Utter: What is the influence of lower amounts of ATP which might change your intramitochondrial ATP/ADP ratios? Have you had any experience in variations of ATP concentrations added?

Walter: We did experiments where we added less ATP outside. In these experiments, the activity of pyruvate carboxylase decreased somewhat, but we have never measured the intra-mitochondrial ATP content as a function of ATP added outside.

Hanson: Don't you think it somewhat confusing to use the term pyruvate carboxylase when in your whole mitochondria incubation system you are measuring pyruvate carboxylation? You are certainly not measuring meximum velocities of this enzyme when it is not solubilized. This may explain the much lower rate that you find relative to what we get in a preparation of freeze-dried mitochondria in which the pyruvate carboxylase has been solubilized.

Walter: Well I think we are measuring pyruvate carboxylase activity in the intact mitochondria. This may be not all of it, but we measure in any case some pyruvate carboxylase activity.

Hanson: I would say what you are measuring is carboxylation of pyruvate itself limited by whatever factors may be interfering with the mitochondrial system. Your values are far lower than what we find on a situation in which the enzyme is soluble.

Exton: I calculated from your data the ATP/ADP ratios and the correlation of pyruvate carboxylase activity appeared to be best with the ATP/ADP ratios. It was much better than ATP and rather better than ADP.

Walter: I feel that we really have not enough data to be able to fully evaluate the relative importance of the ADP level and of the ATP/ADP ratio for the regulation of the pyruvate carboxylase activity.

Seubert: Did you study the fixation of CO_2?

Walter: Only in the experiment shown in Table 1.

Ballard: Did you measure PEP formation as synthesis of this intermediate need not be measured as pyruvate carboxylation? PEP synthesis might explain your apparently lower rates of pyruvate carboxylation at high ATP concentrations.

Walter: Well, in rat liver the formation of PEP is known to be extremely low. Therefore, we did not measure it.

J. R. Williamson: I noticed that your total nucleotides varied from experiment to experiment. Extramitochondrial fluid is carried over into the mitochondrial fraction. This can cause very large corrections of the nucleotides when they are present at high concentrations in the incubation fluid.

Walter: All the results on the amounts of adenine nucleotides in Table 6 were obtained with the same mitochondrial preparation. Each value represents the average of three incubations and the fluctuations between these three values were very small. For full details, see Walter and Stucki (1970). The observed differences in the total adenine nucleotide pools between the different incubation mixtures have been reproduced in other experiments and are therefore not likely to be separation artefacts. The direction and size of these changes are in good agreement with the results obtained by the group of Klingenberg.

J. R. Williamson: You mentioned the interesting effect of oleate having an atractyloside-like action. Could you comment on this? After addition of fatty acids there is a stimulation of respiration which is quite contrary to an atractyloside-like action. So, how can you inhibit the translocase step and still get an increase in respiration?

Walter: When we added oleate to our incubations with mitochondria we did not obtain an increase in respiration. The observed increase in respiration which you observed in your perfusion experiments upon addition of oleate (Williamson, J. R., Scholz, R., Browning, E. T.: J. Biol. Chem. 244 (1969) 4617) is probably the result of the conversion of oleate to oleylcarnitine which, as was shown for other acylcarnitines should be a better substrate for mitochondrial oxidation than the free acid. Your calculations of your perfusion results indicated that this increased respiration was coupled to phosphorylation. Increased amounts of ADP must therefore be able to cross the mitochondrial membrane and this makes it unlikely that oleate has a strong effect on the translocase in this system. However, a weak effect can by no means be excluded. There is at least a theoretical possibility that oleate could prevent this increased rate of oxidative phosphorylation from being even greater by weakly inhibiting the translocase and by thereby making the ADP the limiting factor for the stimulated coupled respiration. This weak inhibition of the translocase could also lead to a lower matrix ADP level and could consequently cause a stimulation of the pyruvate carboxylase activity.

This is very hypothetical and a simpler answer to all this would certainly be that in vivo oleate has no effect on the translocase.

Activities of Pyruvate Carboxylase, Phosphoenolpyruvate Carboxykinase and Pyruvate Kinase of New Zealand Obese[+] (NZO) Mice Livers During Different Phases of Diabetes after Starvation, Cortisol and Insulin Treatment.

B. Störmer and W. Staib
Institut für Physiologische Chemie der Universität Düsseldorf

Summary

We studied the behavior of activities of pyruvate carboxylase, phospoenolpyruvate carboxykinase and pyruvate kinase in the livers of New Zealand, obese (NZO) mice under various conditions. With increasing age, blood glucose and the enzyme activities of phosphoenolpyruvate carboxykinase and pyruvate carboxylase increased, while the pyruvate kinase decreased. We find a similar action of the enzyme activities of phosphoenolpyruvate carboxykinase and pyruvate carboxylase after starvation and corresponding decreasing enzyme activities of pyruvate kinase. Cortisol shows a significant stimulating effect on phosphoenolpyruvate carboxykinase after six hours and so does insulin ten hours after injection. Insulin effects a decrease of phosphoenolpyruvate carboxykinase and an increase of pyruvate kinase on fed mice as well as on starved mice after 20 hours. Pyruvate carboxylase is not influenced by these hormones.

Introduction

Insulin has been reported to be a suppressor of gluconeogenic enzymes, which are elevated in diabetes, starvation and after glucocortiocoid administration (Shargo et al. 1963, Foster et al. 1966, Westman 1968).

NZO-mice (Brunk 1967) available to us are characterized by developing diabetes, obesity, hyperglycemia and hyperinsulinemia with increasing age caused by pluri-genetic disposition. With these conditions provided and others (e.g. starvation, glucocorticoid and insulin application), we studied the enzyme activities of pyruvate carboxylase, phosphoenolpyruvate carboxykinase and pyruvate kinase.

Materials and Methods

Female New Zealand, obese mice 3-9 months old were used. All animals, unless indicated otherwise, were fed ad libitum on a fat rich diet (Intermast GmbH Bockum-Hövel, FRG). The fasting period was either 12 or 22 hours. When cortisol was used, 5 mg per 100 g of body weight was injected intraperitonially and the mice were killed six hours later. Insulin at a dose of 10 U per 100 g of body weight was applied to mice subcutaneously after two hours of fasting; they were killed ten hours later. When they were given a second injection of insulin after a fasting period of 12 hours, they were killed ten hours later.

+ We are grateful to Dr. Strasser, Farbwerke Hoechst, for giving us the NZO-mice.

Blood Glucose and Glycogen

The blood samples were obtained by cutting the tip of the tail. Blood glucose was measuref by the o-toluidine-method (Hoffman 1967). When glucose tolerance tests were performed, 250 mg glucose per 100 g of body weight were given intraperitoneally. Blood specimen were taken at various intervals between 0 and 180 minutes after glucose application. Glycogen was assayed as described by Stetten et al. (1958).

Enzyme Assays

Pyruvate carboxylase (EC 6.4.1.1.), phosphoenolpyruvate carboxykinase (EC 4.1.1.32) and pyruvate kinase (EC 2.7.1.40) were assayed as described by Störmer (1970). Protein was determined by a method of Lowry et al. (1951).

Results and Discussion

Growing older, NZO-mice develop a diabetes detectable by blood glucose level and glucose tolerance test.

At the age of three months, blood glucose is around 150 mg%; at six and nine months respectively, blood glucose reaches 200 mg%. The glucose tolerance of mice six and nine months old are nearly identical and are significantly higher than that of three month old mice ($p < 0.0005$). Three-months old mice show blood glucose concentration of 300 mg% 20 minutes after glucose application and reach the starting point after 180 minutes. The blood glucose of six and nine months old mice, however, rises to about 600 mg% and 400 mg% blood glucose remains after 180 minutes. Diabetic metabolism is shown by the activities of the typical key enzymes of gluconeogenesis (Krebs et al. 1965). Between the third and sixth month, the activities of pyruvate carboxylase and phosphoenolpyruvate carboxykinase increase significantly ($p < 0.0005$), while that of pyruvate kinase decreases significantly ($0.0025 < p < 0.0005$).

Other experiments carried out on starved, three-months old, NZO mice showed highly increased activities of pyruvate carboxylase and phosphoenolpyruvate carboxykinase ($p < 0.0005$), while that of pyruvate kinase decreased ($0.0025 < p < 0.0005$). In contrast with our former studies of rats (Störmer et al., 1970), the additional application of cortisol potentiated the effect of starvation on the activity of phosphoenolpyruvate carboxykinase of NZO mice livers ($p < 0.0005$). Pyruvate carboxylase and pyruvate kinase did not show any evident effect after six hours of hydroxycortisone application. This is in contrast to corresponding controls of starved mice, both three and six months old.

As far as phosphoenolpyruvate carboxykinase is concerned, we found similar results with six months old NZO mice which were starved and cortisol treated ($p < 0.0005$), but the differences in the activities of pyruvate carboxylase and pyruvate kinase were no longer significant.

Ten hours after injecting insulin into starved mice, we found elevated enzyme activities of phosphoenolpyruvate carboxykinase, which obviously exceeded the already maintained level by starvation activated phosphoenolpyruvate carboxykinase ($p < 0.0005$). Then, after 20 hours and a further application of 4 U. insulin, phosphoenolpyruvate-carboxykinase decreased evidently and reached a nearly normal level. This becomes clearer when one considers that after 20 hours phosphoenolpyruvate carboxykinase without insulin effect would have got a further activation by the gaining stress of starvation. Corresponding to these results the enzyme activities of pyruvate kinase increased 20 hours after insulin application ($p < 0.005$). Pyruvate carboxylase, on the other hand, is not influenced by insulin application.

After giving injection of 4 U. insulin to fed animals, we saw no effect on enzyme activities until after ten hours; however, 20 hours after insulin injection and after a second application of 4 U. insulin, phosphoenolpyruvate carboxykinase decreased markedly ($0.01 < p < 0.005$) and pyruvate kinase showed an evident progress of its enzyme activities ($0.025 < p < 0.0125$). Pyruvate carboxylase did not change.

If one administers insulin to starved mice for ten hours and cortisol for six hours in combination, enzyme activities of phosphoenolpyruvate carboxykinase reach a level which is situated higher than that of untreated, starved animals ($p < 0.0005$). As said before, insulin alone, as well as cortisol alone potentiated the activities of phosphoenolpyruvate carboxykinase after ten hours (insulin) and six hours (cortisol). Pyruvate carboxylase and pyruvate kinase were the same as before.

At the same time, we measured the enzyme activities and examined the glycogen contents of mice livers under the above conditions. The results showed a clear rise of glycogen over the normal level after 20 hours ($0.005 < p < 0.0025$) and after two injections of 4 U. insulin given to fed mice. The already extremely low data of fasted animals decreased additionally and significantly until ten hours after insulin application ($0.01 < p < 0.005$) and rose again after 20 hours ($p < 0.0005$). Hydrocortisone led to an obvious addition of glycogen in the mice liver of starved animals ($p < 0.0005$).

References

Brunk, R. in: Heffter-Heubner Handbuck der experimentellen Pharmakologie, Handbook of Experimental Pharmacology, Vol.: Insulin, 55 (1967)

Foster, D. O., P. O. Ray, H. A. Lardy: Biochemistry 5 (1966) 555

Hoffmann, W. S.: J. biol. Chem. 120 (1967) 51

Krebs, H. A., R. N. Speake, R. Hems: Biochem. J. 94 (1965) 712

Lowry, O. H., N. J. Rosenbrough, A. L. Farr, R. J. Randal: J. biol. Chem. 193 (1951) 265

Shrago, E., H. A. Lardy, R. C. Nordlie, D. O. Foster: J. biol. Chem. 238 (1963) 3188

Stetten, M. R., H. M. Katzen, O. Stetten: J. biol. Chem. 232 (1958) 475

Störmer, B., W. Janssen, H. Reinauer, W. Staib, S. Hollmann: Hoppe-Seyler s Z. Physiol. Chem. 351 (1970) 296

Westman, S.: Diabetologia 4 (1968) 141

The Development of Gluconeogenic Function in Rat Liver +

F. J. Ballard and Helen Philippidis
Fels Research Institute, Temple University School of Medicine, Philadelphia, Pennsylvania, U.S.A. and C.S.I.R.O., Division of Nutritional Biochemistry, Adelaide, Australia

Summary

Earlier work has shown that the incorporation of pyruvate into glucose and glycogen occurs at negligible rates in slices from fetal rat liver and that this overall pathway appears at birth. Measurement of gluconeogenic enzymes during the transition from the fetus to the newborn suggested that the absence of cytosol P-enolpyruvate carboxykinase limits gluconeogenesis in the fetus. Although a mitochondrial activity of P-enolpyruvate carboxykinase is active in fetal rat liver, it seemed unlikely that this enzyme was participating in gluconeogenesis. This was particularly obvious in fetal guinea pig liver which, like the fetal rat liver, does not have a functional gluconeogenic pathway, but contains a very active mitochondrial P-enolpyruvate carboxykinase. Purification of P-enolpyruvate carboxykinase from the cytosol fraction of rat liver permitted the conclusive demonstration that the cytosol and mitochondrial enzymes were distinct, but did not explain why fetal liver slices could not carry out gluconeogenesis.

Experiments in vivo confirmed that gluconeogenesis was absent in fetal rat liver and suggested a block between pyruvate and P-enolpyruvate. Glucagon injected into fetal rats increased the activity of hepatic cytosol P-enolpyruvate carboxykinase 10-15 fold, an increase also found in the incorporation of pyruvate-^{14}C into glycogen in slices. However, in vivo experiments under these conditions indicate no glyconeogenesis. This discrepancy between results in vitro and in vivo has been shown to be caused by the ether anesthesia in vivo and resultant hypoxia. It appears that ether causes changes in both cytosol redox and adenine nucleotide balance and that these changes may be sufficient to completely prevent gluconeogenesis. This proposal has been further investigated in newborn animals where anoxia also prevents gluconeogenesis. Analyses suggest the diversion of energy during hypoxia from gluconeogenesis to maintain redox state and nucleotide balance.

+ Unusual Abbreviations: PEP - phosphoenolpyruvate

Enzyme Code Numbers: PEP-carboxykinase (EC 4.1.1.32), pyruvate carboxylase (EC 6.4.1.1), glucose-6-phosphatase (EC 3.1.3.9), fructose diphosphatase (EC 3.1.2.11), aspartate aminotransferase (EC 2.6.1.1), and ß-hydroxybutyrate dehydrogenase (EC 1.1.1.30).

Introduction

Gluconeogenesis in mammalian liver functions as an integral part of the regulatory system for the maintenance of blood glucose. The pathway has special significance in animals such as ruminants which do not ordinarily obtain carboxhdrate from the diet since the entire requirement of hexose must be synthesized *de novo* (Ballard et al. 1969). Another extreme is found in fetuses of animals such as the rat, which are presented with a constant supply of glucose in the uterine circulation, and as such rat would not need to synthesize glucose or to regulate their blood glucose concentration. Out interest in gluconeogenesis in fetal and newborn animals has centered on these ideas. We believe that the change from a condition in which there is no requirement for glucose synthesis to one in which the pathway is essential due to the low carbohydrate nature of the diet affords a unique system to study the control or regulation of gluconeogenesis. Our earlier experiments consisted of surveys of the activities of gluconeogenic enzymes during development and measurements of gluconeogenesis from radioactive precursors in liver slices.

Experiments in which rat liver slices have been incubated with pyruvate-^{14}C indicate substantial conversion of this precursor to both glucose and glycogen in post-natal animals. Some incorporation of label was detected in fetal liver, but the rate was 5% or less of that found in the newborn (Ballard and Oliver 1963, Ballard and Oliver 1965). Assays of the key gluconeogenic enzymes in fetal and newborn liver have indicated that fructose diphosphatase (Ballard and Oliver 1962), glucose-6-phosphatase (Ballard 1964) and pyruvate carboxylase (Ballard and Hanson 1967b) all increase in activity at birth, but the increases are only about two-fold, and as such are not likely to account for the 20 to 30-fold increase in the overall pathway. Over this time period the cytosol activity of PEP carboxykinase increases at least 50 fold (Ballard and Hanson 1967b) leading us to speculate that it is this increase in activity which is responsible for the increase in gluconeogenesis at birth. Therefore, PEP carboxykinase would be rate-limiting for gluconeogenesis in fetal rat liver.

There is some activity of mitochondrial PEP carboxykinase in fetal liver and this does not change appreciably at birth (Ballard and Hanson 1967b). We are uncertain whether this enzyme participates in gluconeogenesis in fetal liver, but in chickens and pigeons. Where this is the only form of PEP carboxykinase in liver, the enzyme must certainly be involved in the overall pathway (Gevers 1967, Nordlie and Lardy 1963).

Our goal in this report was to examine the possible controls that might be regulating gluconeogenesis in livers of fetal and newborn rats. We have chosen both to follow the development of gluconeogenesis in these animals and to prevent gluconeogenesis in one-day-old animals with hypoxia. During these manipulations, we have, wherever possible, used techniques that are applicable to the intact animal and have looked primarily at changes in the relative proportions of adenine nucleotides, changes in redox and alterations in the activity of cytosol PEP-carboxykinase.

Development of Gluconeogenesis in Rat Liver In Vivo

Although experiments with liver slices and the direct measurements of hepatic enzymes give evidence on the neonatal development of gluconeogenesis, it is not likely that complete regulation of a pathway by the activity of a single enzyme occurs in the intact animal. Such a process would be accompanied by large increases in the precursors of that enzyme which in turn would regulate their own synthesis and add secondary controls to the system. We wished to test these ideas in vivo and have utilized two techniques to realize this aim: 1) the fetus is injected with trace amounts of radioactive precursors and the fate of these is followed with time. It is expected that the rate of labeling of a product such as glucose would give an indication of the rate of the gluconeogenic pathway. 2) the injection of substrate amounts of precursors followed by measurements of pathway intermediates.

The techniques used for the isotope injection experiment have been described by Philippidis and Ballard (1969). The results in Table 1 were obtained after injection of pyruvate-3-^{14}C into fetal rats which were killed 25 minutes later. In these experiments approximately 4% of the injected radioactivity was recovered in the fetal liver and of this no counts were found in glucose. The hepatic lactate pool contained substantial radioactivity with lesser amounts in aspartate, ß-hydroxybutyrate, alanine and pyruvate (Philippidis and Ballard 1969). Comparable experiments with newborn animals showed extensive labeling of glucose and some counts in aspartate, ß-hydroxybutyrate, alanine and pyruvate. This experiment clearly shows the development of gluconeogenesis in vivo. In addition to the change in glucose labeling, we note a ten-fold increase of lactate labeling. This suggests a more reduced state of the cytosol NAD pool in fetal liver.

If the block in gluconeogenesis from pyruvate occurs prior to oxaloacetate, as has been shown in certain perfusion studies with liver from adult rats (Williamson et al. 1969, Exton and Park 1966, Veneziale et al. 1968), administration of a gluconeogenic precursor that is not metabolized through pyruvate should produce gluconeogenesis in fetal rats. Although aspartate is a relatively poor gluconeogenic precursor, this amino acid is transaminated in liver from fetal and newborn rats (Hanson and Ballard 1968). The aspartate-^{14}C injection studies in Table 2 show similar results to those found with radioactive pyruvate. Glucose is not labeled in the newborn animal, suggesting that the block in gluconeogenesis in fetal liver is not prior to oxaloacetate. Additional evidence is also provided for a highly reduced redox by the recovery of 17.5% of hepatic radioactivity in the malate pool of the fetal animal, while only 1.4% of the radioactivity is found in malate in the newborn. We interpret these data as conversion of aspartate-derived oxaloacetate to malate via malate dehydrogenase:

$$\text{malate} + NAD^{+} \rightleftharpoons \text{oxaloacetate} + NADH + H^{+}$$

Additional experiments with lactate-^{14}C as gluconeogenic precursor also show no conversion to glucose in fetal liver but rapid conversion in the newborn (Philippidis and Ballard 1969).

% Liver Radioactivity	Fetus	Newborn
Glucose	<0.5	47 $\pm$ 6
Lactate	31 $\pm$ 11	3.2 $\pm$ 0.7
Pyruvate	1.4 $\pm$ 0.9	1.2 $\pm$ 0.2

Table 1. Distribution of Label from Pyruvate-^{14}C in Livers of Intact Rats.
A dose of 0.5 μCi of pyruvate-3-^{14}C was injected into fetal or newborn animals. After 25 minutes the radioactivity in liver intermediates was determined according to Philippidis and Ballard (1969). Values are the percent of the total radioactivity in liver and represent the mean $\pm$ s.e.m. for determinations in three experiments.

Measurements of cytosol and mitochondrial redox in freeze-clamped livers of fetal and newborn rats show results that are consistent with the isotope findings. The NAD^+ to NADH ratio as calculated from the lactate-pyruvate couplet gives a value of 79 in fetal liver and 667 in liver from newborn animals. The mitochondrial NAD pool is also relatively reduced as compared to the newborn animal (Table 3). However, there is a very large increase in the total concentrations of ketone bodies between the fetus and newborn. We note the following: 1) experimental manipulations do not alter the level of total ketones in fetal liver, and 2) liver slices from fetal rats have a very poor ability to synthesize ketone bodies. For these reasons, and notwithstanding a substantial activity of ß-hydroxybutyrate dehydrogenase in fetal liver mitochondria (Ballard, unpublished), we have not used the ß-hydroxybutyrate to acetoacetate couplet in the fetus as an index of mitochondrial redox. Measurements of adenine nucleotides are of lesser value than the redox calculations since they do not allow any determinations of intracellular distribution. Therefore, any change in mitochondrial adenine nucleotides would be masked by the greater pool of cytosol nucleotides. The fetal-newborn results, however, do indicate a considerable alteration in the ATP/ADP ratio towards a condition of greater phosphorylation in the newborn.

% Liver Radioactivity	Fetus	Newborn
Glucose	<0.5	39 $\pm$ 4
Malate	17.5 $\pm$ 4.3	1.4 $\pm$ 0.8
Aspartate	37 $\pm$ 12	33 $\pm$ 7

Table 2. Distribution of Label from Aspartate-^{14}C in Livers of Intact Rats.
A dose of 0.5 μCi of L-aspartate-U-^{14}C was injected into fetal or newborn rats. Other details are given in Table 1.

	Fetus	Newborn
Lactate	5.7 ± 0.8	1.00 ± 0.20
Pyruvate	0.050 ± 0.008	0.074 ± 0.017
Lactate/pyruvate	114	13.5
Cytosol NAD^+/NADH	79	667
ß-hydroxybutyrate	0.053 ± 0.014	0.45 ± 0.05
Acetoacetate	0.050 ± 0.005	0.52 ± 0.06
ß-hydroxybutyrate/acetoacetate	1.06	0.87
Mitochondrial NAD^+/NADH		23
ATP	1.25 ± 0.07	2.52 ± 0.13
ADP	0.41 ± 0.04	0.36 ± 0.04
AMP	0.20 ± 0.03	0.25 ± 0.01
ATP/ADP	3.05	7.0

Table 3. Metabolites in Liver of Fetal and Newborn Rats. Values are expressed as μmoles per g. tissue with the mean + s.e.m. for six animals. NAD^+-to-NADH ratios were calculated as described by Williamson et al. (1967).

The highly reduced environment in fetal liver and the relatively low phosphorylation of adenine nucleotides might be caused by either hypoxia or perhaps by lipid oxidation In the fetus, it is unlikely that lipid mobilization to the liver is occurring at a substantial rate since this tissue synthesises lipids in large amounts (Ballard and Hanson 1967. Hypoxia, on the other hand, might be expected in fetal tissues if oxygen transfer acros the placenta was inadequate. Although we have no information on the rat, the oxyger

Precursor Injected	Fetus	Newborn
None	0.033	0.035
Aspartate, 100 μmoles	0.022	0.049
Lactate, 200 μmoles	0.023	0.046
Pyruvate, 200 μmoles	0.031	0.699

Table 4. PEP Concentrations in Liver After Injection of Precursor. The concentration of PEP was measured in livers 15 minutes after the injection of precursor (Philippidis and Ballard 1969). Values are the means of three determinations.

	Fetus -glucagon	Fetus +glucagon	Newborn
PEP carboxykinase	0.070 ± 0.013	1.01 ± 0.10	2.6 ± 0.1
Pyruvate-^{14}C into glycogen, slices	1.08 ± 0.18	12.7 ± 1.0	24 ± 3
Pyruvate-^{14}C into glycogen, in vivo	3.0 ± 0.6	13.4 ± 1.2	173 ± 6
Glucose-^{14}C into glycogen, in vivo	12.9 ± 2.0	12.9 ± 1.3	-

Table 5. Effects of Glucagon on Gluconeogenesis in Fetal Rats. PEP carboxykinase activity is expressed in units/g, slice incorporation in μmoles/g/2 hr. and incorporations in vivo in dpm/mg/hr. Animals which were injected with glucagon were given 250 μg five hours prior to experimentation. Values are the means ± s.e.m. for at least six animals.

concentration in fetal blood of other species is always lower than in the maternal blood. In humans this difference is very marked with concentrations of 14.4 ml/100 ml in the uterine artery and 8.4 ml/100 ml in the umbilical vein (Stenger et al. 1965). Corresponding values of 13.5 ml/100 ml and 9 ml/100 ml have been reported for the goat (Kaiser and Cummings 1958). Since placentas differ structurally from one species to another (Starck 1959), it is difficult to apply

	Minimal ether	60 min. ether
Lactate/pyruvate	5.7/0.050	14.6/0.040
Cytosol NAD^+/NADH	79	25
ATP/ADP	3.05	2.15
Pyruvate-^{14}C to glycogen	3.0 ± 0.6	0.37 ± 0.05
Pyruvate-^{14}C to glycogen, glucagon injected	13.4 ± 1.2	0.26 ± 0.06

Table 6. Effect of Ether on Redox, Adenine Nucleotides and Gluconeogenesis in Fetal Rats. Metabolite concentrations are expressed as μmoles/g and incorporation results as dpm per mg liver per hour after an injection of 0.5 μCi of Pyruvate-3^{14}C. Values are the means of at least six determinations ± s.e.m.

	Air	N_2
Lactate	1.04 ± 0.06	4.38 ± 0.46
Pyruvate	0.067 ± 0.006	0.063 ± 0.009
Cytosol NAD^+/NADH	614	139
ß-Hydroxybutyrate	0.341 ± 0.048	0.534 ± 0.043
Acetoacetate	0.502 ± 0.047	0.063 ± 0.016
Mitochondrial NAD^+/NADH	29.7	2.4
Adenylate charge	0.820	0.598

Table 7. Effect of Nitrogen Atmosphere on Levels of Liver Metabolites After Aspartate Injection. Animals were either exposed to air or to a nitrogen atmosphere. After five minutes, the one day old rats were injected with 20 μmoles L-aspartate. At 20 minutes, the livers were freeze-clamped and the levels of liver metabolites determined. Values are the means ± s.e.m. for six animals. Adenylate charge was calculated as described by Atkinson (1965) and NAD^+-to-NADH ratios according to Williamson et al. (1967).

information concerning the efficiency of oxygen transfer through the placenta between species. However, we presume that if a placental gradient of similar magnitude occurs in the rat, the fetal oxygen tension would also be low and would result in a relative hypoxia. This could produce the reduced redox and low degree of adenine nucleotide phosphorylation that are evident from the freeze-clamping experiments.

Phosphoenolpyruvate levels are approximately equal in liver from both fetal and newborn rats (Table 4). If large amounts of aspartate, pyruvate or lactate are injected into fetal livers, the concentrations of PEP are unaffected. Similar experiments on newborn animals result in increases in PEP, especially when pyruvate is the precursor. We interpret this experiment as support for the limitation of gluconeogenesis prior to PEP in fetal liver. However, it is not conclusive since the low PEP levels in the fetal liver might be a result of rapid removal of this compound.

Glucagon Experiments

Injection of glucagon into fetal rats produces a large increase in PEP carboxykinase activity within five hours (Yeung and Oliver 1968). If this enzyme is limiting gluconeogenesis in fetal rat liver, an increase in its activity should be accompanied by increases in the overall pathway. As can be seen in Table 5, both pyruvate-^{14}C incorporation into glycogen in liver slices and pyruvate-^{14}C incorporation into glycogen in intact fetuses are substantially increased under these conditions. We have also measured the incorporation of glucose-^{14}C into glycogen in vivo as a control to account for the glycogenolytic action of glucagon. These results show no differences between incorporation into glycogen with or without glucagon and in other experiments we found only slightly less glycogen in glucagon-treated fetuses than in control animals. Presumably, the low activity of phosphorylase accounts in part for this rather negligible glucagon effect on glycogenolysis (Ballard 1964).

Ether Anesthesia and Gluconeogenesis

Ether anesthesia of pregnant rats causes decreased maternal blood pressure as well as hypoventillation and vasoconstriction. These effects would all tend to potentiate fetal hypoxia. It is apparent from Table 6 that the cytosol compartment of fetal liver becomes more reduced if the mother is kept under ether anesthesia for a prolonged period. The ATP to ADP ratio is also lowered. Measurements of gluconeogenesis carried out in intact fetuses under these conditions show a depression of the glycogen labeling to 15% of that found in untreated fetuses. Furthermore, if the fetal rats are treated with glucagon in order to stimulate gluconeogenesis, the ether anesthesia results in complete suppression of the glucagon effect. We consider these ether experiments as indicating a potentiation of the normal fetal conditions so that even the minimal gluconeogenesis seen in fetal liver has been abolished.

Gluconeogenesis is stimulated or hindered in the following ways in the various experimental conditions:

A) Fetus treated with ether.

The liver under this condition has essentially no cytosol PEP-carboxykinase; even when this enzyme is induced with glucagon, there is still no glyconeogenesis. Obviously cytosol PEP-carboxykinase is not limiting the pathway. The highly reduced redox in the cytosol would prevent lactate conversion to pyruvate and prevent malate conversion to oxaloacetate. The cytosol PEP-carboxykinase would not, therefore, be provided with substrate. Presumably the mitochondrial PEP-carboxykinase is not functional due to a lowering in the ATP to ADP and thus the GTP to GDP ratios. Although the ATP to ADP ratio as measured in the total cell is only reduced 30%, it is quite likely that if one effect of prolonged ether anesthesia is hypoxia, this will result in a depression of ATP levels principally in mitochondria.

B) Normal Fetus.

The low rate of glyconeogenesis found in the liver of normal fetuses is limited by low ATP/ADP ratios and low PEP-carboxykinase activity. The high cytosol redox appears to direct pyruvate to lactate (Table 1) and oxaloacetate to malate (Table 2).

C) Fetus with PEP-carboxykinase induced.

Since glyconeogenesis and PEP-carboxykinase activity increased under this treatment while there are no changes in cytosol redox or adenine nucleotides, it appears that the increase in cytosol PEP-carboxykinase is responsible for the increase in the overall pathway. Should glucagon be producing a lipolytic effect in fetal liver, synthesis of acetyl CoA or mitochondrial reducing equivalents might also contribute to the increase in gluconeogenesis.

D) Newborn

In the newborn the increase in the ATP to ADP ratio would favor pyruvate carboxylation and the phosphorylation reactions in the cytosol. The relatively oxidized mitochondrial redox should hinder malate generation from oxalacetate but since glyconeogenesis is markedly increased in this condition the low redox is not limiting glyconeogenesis. The fall in the cytosol redox would clearly favor oxaloacetate generation and lactate utilization but would hinder gluconeogenesis at the glyceraldehyde-phosphate dehydrogenase step. In the newborn rat, PEP-carboxykinase activity is much greater than in the normal fetus but only two times greater than in the glucagon-treated fetus. The 15-fold increase in glycogen labeling from pyruvate-^{14}C between the glucagon-treated fetus and the newborn would suggest that the major cause of the stimulus in the overall pathway is due either to a more oxidized cytosol or to greater phosphorylation of the adenine nucleotides.

These experiments have permitted us to use four separate conditions, each of which has a different rate of hepatic gluconeogenesis: ether-treated fetuses, normal fetuses, glucagon-treated fetuses and newborn animals. There are marked differences in redox state and adenine nucleotides between the conditions and it appears that different controls in gluconeogenesis are operative. The following experiments approach the same goals but in an opposite manner. We have attempted to reduce gluconeogenesis with hypoxic conditions and simultaneously measure changes in redox and adenine nucleotides.

Hypoxia Studies in Newborn Rats

It is well known that newborn mammals can withstand total anoxia for a considerable time. With one day old rats maintained at 35^0C in a humidicrib, the animals can survive for 30 minutes, but older animals will live for only two or three minutes. Since the principle use of oxygen in mammalian cells is to support the generation of ATP via oxidative phosphorylation, and this process occurs in mitochondria, we have assumed that anoxia will result in a fall in the rate phosphorylation of mitochondrial adenine nucleotides. This in turn would probably limit gluconeogenesis since the pathway requires the conversion of six molecules of nucleotide triphosphate to nucleotide diphosphate for each molecule of glucose synthesized.

In the studies reported below we have measured both the levels of hepatic intermediates and the rate of incorporation of radioactive precursors into the total glucose and glycogen pools of the newborn rat. For the determination of gluconeogenesis, three isotopes have been used: L-aspartate-U-^{14}C, L-lactate-U-^{14}C or L-serine-U-^{14}C. Twenty micromoles of carrier amino acid were injected with the aspartate-^{14}C or serine-^{14}C and 100 μmoles of L-lactate with the lactate-^{14}C. At various times after injection, the animals were frozen in liquid nitrogen, pulverized and the frozen powder extracted twice with perchloric acid. These solutions were neutralized and portions of extract were passed through columns of Dowex-1-formate over Dowex-50. This technique provides quantitative yields of glucose plus glycogen from the whole animal. Figure 1 shows the incorporation of either lactate or aspartate into glucose plus glycogen. The data are expressed as percent of the injected radioactivity found in glucose plus glycogen and while these values are an underestimate of true gluconeogenesis due to recycling through the Cori cycle, we believe that they offer a satisfactory measure of gluconeogenesis. It can be seen from Fig. 1 that the incorporation in animals exposed to a normal atmosphere proceeds linearly with time from both aspartate and lactate. With animals that were kept in the absence of air for five minutes before injection of precursor and were maintained under a nitrogen atmosphere throughout the experiment, there was no measurable gluconeogenesis. Similar results were found with serine-^{14}C as precursor but are not presented in this report.

We have made two assumptions for the evaluation of these data: 1) as mentioned above, the recycling is not substantial, and 2) most of the gluconeogenesis observed occurs in the liver. Since we wished to follow the changes in intermediates in gluconeogenic tissue that occur under the various hypoxic states, we have freeze-clamped the livers of animals that had been injected with gluconeogenic precursors and measured the concentrations of lactate, pyruvate, ß-hydroxybutyrate, acetoacetate and adenine nucleotides in liver extracts. Hepatic intermediates measured in freeze-clamped livers of animals 15 minutes after precursor injection show marked differences between air-exposed and nitrogen-exposed animals (Table 7). As predicted by the primary action of an oxygen-poor environment, the degree of adenine nucleotide phosphorylation is depressed in animals exposed to nitrogen. In addition to this change in adenine nucleotides, anoxia causes marked reductions in mitochondrial and cytosol redox. These results were expected since reoxidation of NADH is ordinarily dependent on the electron transport chain and this is limited by oxygen availability.

In order to obtain information on the relative importance of adenine nucleotides and redox as controls for the overall gluconeogenic activity of liver, we have constructed various hypoxic states that are intermediate between fully oxygenated with normally active gluconeogenesis and fully anoxic with completely suppressed gluconeogenesis. The animals are maintained in these controlled atmospheres for five minutes to attain some degree of a steady state and then the precursor is injected. The results of experiments of this type are shown in Fig. 2. With lactate as precursor, gluconeogenesis in the whole animal falls gradually as the proportion of nitrogen in the atmosphere is increased, until, at 12% air, gluconeogenesis is proceeding at only 15% of the rate observed in oxygenated animals. With aspartate or serine as precursor, the qualitative changes are similar, although the gluconeogenic rates found under an atmosphere of 12% air are somewhat greater, at 18% and 30%, respectively.

When hepatic intermediates were measured under these conditions of varying hypoxia, a gradual change was not usually evident. With either aspartate or serine as precursors (Fig. 3a and 4a), the degree of phsophorylation of adenine nucleotides was unchanged down to an air content of 12%, so that only after animals were kept under

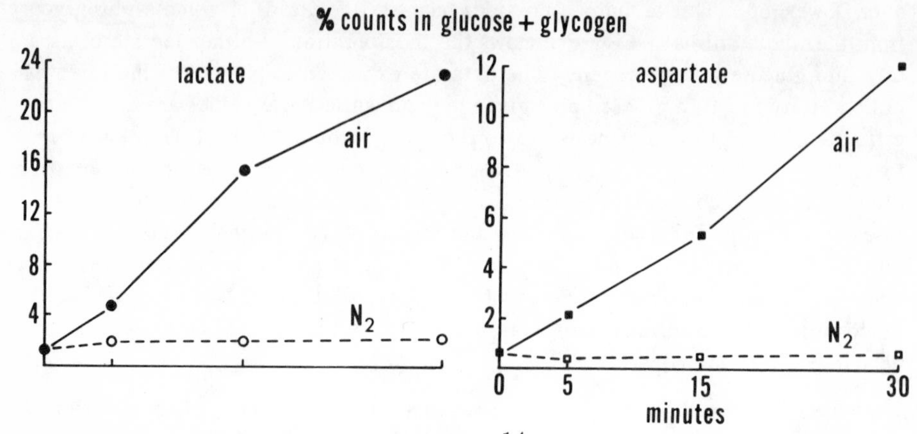

Fig. 1. Either 100 μmoles of L-lactate-U-^{14}C (0.5μ Ci) or 20 μmoles of L-aspartate-U-^{14}C (0.5μ Ci) were injected into one-day-old rats. After varying times under a nitrogen atmosphere or in air the animals were frozen in liquid nitrogen. Glucose plus glycogen was extracted as described in the text. Incorporation is expressed as the precent of injected dose in glucose + glycogen.

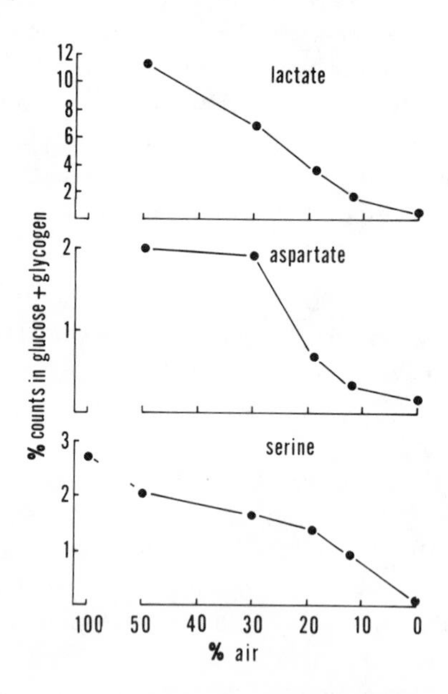

Fig. 2. Lactate (100 μmoles, 0.5 μCi), aspartate (20 μmoles, 0.5 μCi) or serine (20 μmoles, 0.5 μCi) were injected into one-day-old rats. After 15 minutes exposure to atmospheres which contained different proportions of air and nitrogen, the animals were frozen in liquid nitrogen. Glucose plus glycogen was extracted as described in the text.

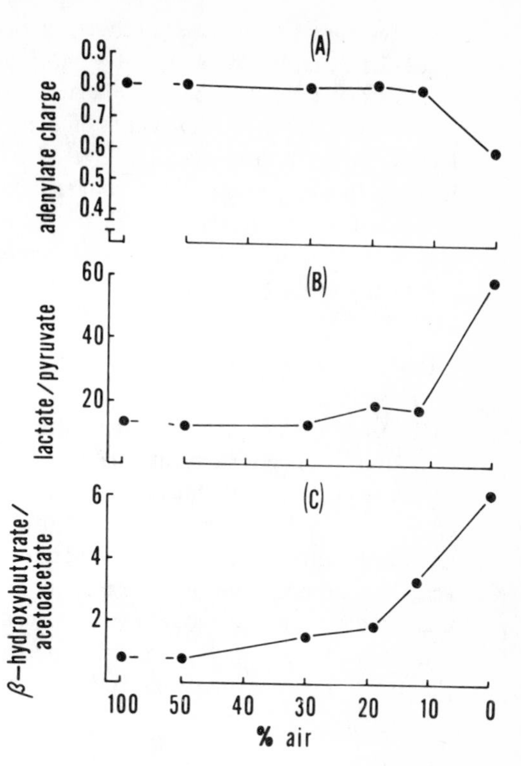

Fig. 3. Livers from animals injected with 20 μmoles L-aspartate were freeze-clamped after 15 minutes exposure to atmospheres which contained different proportions of air and nitrogen. ATP, ADP, AMP, pyruvate, ß-hydroxybutyrate and acetoacetate were determined in neutralized perchlorid acid extracts. (A) adenylate charge, (ATP + 1/2 ADP)/(ATP+ADP + AMP); (B) lactate to pyruvate ratio; (C) ß-hydroxybutyrate to acetoacetate ratio. Each point is the average of determinations on two animals.

total anoxia did the relative proportion of adenine nucleotides change. This result is surprising since the primary effect of oxygen lack will be on adenine nucleotides. A reasonable explanation of this constancy of adenine nucleotides under conditions when gluconeogenesis is severely inhibited would be that the measurements of adenine nucleotides in whole livers are not representative of mitochondrial levels. Furthermore, a fall in mitochondrial ATP may be compensated by an increase in the ATP generated in the cytosol by a glycolytic sequence that could be stimulated by high levels of ADP and inorganic phosphate. These effects might, therefore, result in an unchanged pattern of adenine nucleotides in the cell as a whole.

The cytosol redox as measured by the lactate to pyruvate ratio is relatively constant throughout the varying hypoxic conditions, although with an atmosphere of 12% air, the ratio is increased from a value of 17 to values of 23 and 19 with serine and aspartate, respectively, as precursors. The major change in the lactate-to-pyruvate ratio is between 12% air and 0% air, while gluconeogenesis is already severly depressed in an atmosphere of 12% air. When hepatic gluconeogenesis is increased during fasting or diabetes, there is an increase rather than a decrease in the lactate to pyruvate ratio (Williamson et al. 1967), and it is usually considered therefore that an increase in the ratio will stimulate gluconeogenesis. This is, of course, not consistent with the present hypoxia studies or the measurements in fetal and newborn animals. Krebs has suggested, however, that the increase in the lactate-to-pyruvate ratio is a consequence rather than a cause of increased gluconeogenesis.

He suggests that an increased utilization of ATP to support gluconeogenesis will displace the cytosol redox at the glyceraldehyde-phosphate dehydrogenase reaction, and this redox change would displace the lactate/pyruvate ratio towards a greater proportion of lactate. Our studies on hypoxia are consistent with an increase in the lactate/pyruvate ratio when the level of ATP falls. However, the ATP concentration is not falling as a result of an increase in ATP utilization, but rather as a result of a decrease in ATP formation. Both cytosol redox and ATP change only when the animal is exposed to total anoxia. We consider, therefore, that the increase in the lactate-to-pyruvate ratio is a secondary result of low ATP concentrations and that the high redox under these conditions is not stimulating gluconeogenesis.

Ratios of ß-hydroxybutyrate to acetoacetate with aspartate or serine as gluconeogenic precursors are shown in Fig. 3C, 4C. Unlike the constancy of adenine nucleotides and lactate-to-pyruvate ratios throughout the hypoxic (but not anoxic) conditions, the ß-hydroxybutyrate-to-acetoacetate ratio changes markedly. With aspartate as precursor, this ratio in liver increases 4.1 fold between an atmosphere of 50% air and an atmosphere of 12% air, and 4.6 fold between the same conditions with serine as precursor. We believe that this ß-hydroxybutyrate-to-acetoacetate ratio is a useful indicator of mitochondrial redox in livers from newborn animals since ß-hydroxybutyrate dehydrogenase, the enzyme responsible for the interconversion of these ketone bodies, is present at high specific activity in liver mitochondria. This enzyme is not active in the liver cytosol of rats (Ballard, unpublish-

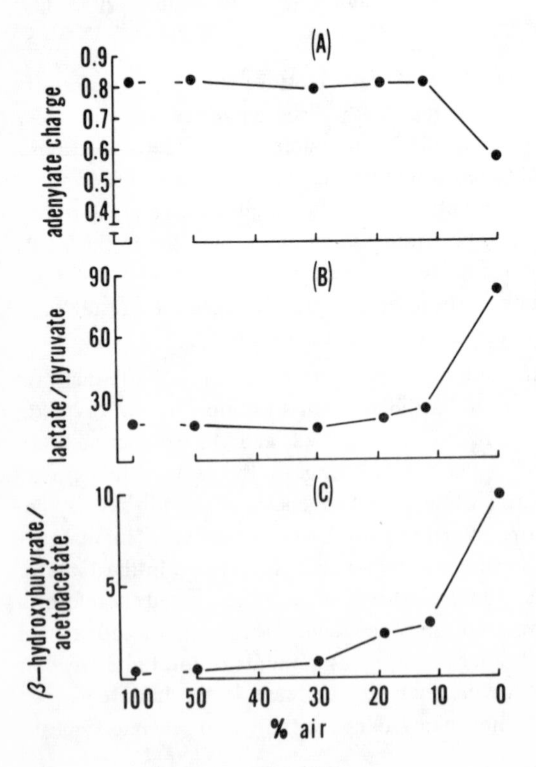

Fig. 4. Effect of graded hypoxia on liver metabolites of animals injected with 20 μmoles L-serine. Other details are described in Fig. 3.

ed). An increase in mitochondrial redox during hypoxia is probably caused by an inhibition of NADH oxidation with oxygen limiting. It does not appear that a highly reduced mitochondrial redox would in itself prevent gluconeogenesis, since this condition would favor malate generation from oxaloacetate via mitochondrial malate dehydrogenase. This would be expected to stimulate rather than depress the overall pathway of glucose synthesis. We assume, therefore, that the mitochondrial redox measurements are also an index of the degree of mitochondrial nucleotide phosphorylation. This would mean a low ATP to ADP ratio in mitochondria which will inhibit gluconeogenesis by a lack of ATP or GTP in the cytosol. These results are summarized in

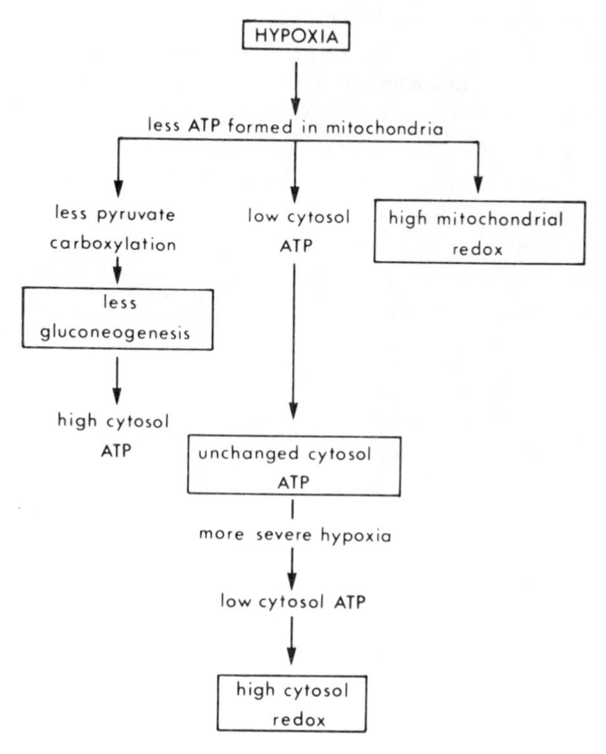

Fig. 5. A possible explanation of the effects of falling oxygen tension or hypoxia on gluconeogenesis, total tissue adenine nucleotides, cytosol redox as indicated by the lactate to pyruvate ratio, and mitochondrial redox as indicated by the ß-hydroxybutyrate-to-acetoacetate ratio. Boxes indicates the measured values.

Fig. 5. The falling oxygen tension during hypoxia causes a lowering in the rate of mitochondrial ATP generation. The effects of less ATP synthesis are an increase in mitochondrial redox since NADH cannot be oxidized, a fall in the cytosol ATP and a lowering of the ATP-dependent pyruvate carboxylation in mitochondria. This last effect will diminish gluconeogenesis which will increase the level of ATP in the cytosol since glucose synthesis requires the utilization of ATP. We presume that this tendency to increase the cytosol ATP and the tendency to decrease ATP that is caused by the mitochondrial-cytosol transphosphorylation system would balance each other, and thus explain the unchanged overall ATP level (Fig. 3A, 4A). Eventually, as the oxygen tension is decreased by an increase in hypoxia, the cytosol will have a lowered ATP concentration, and this could displace the glyceraldehydephosphate dehydrogenase equilibrium to produce a more reduced cytosol (Krebs). This high cytosol redox is indicated by the high lactate-to-pyruvate ratio.

Conclusions

We believe that, although much of the work reported here is preliminary, it is evident that the transition from fetus to newborn offers an interesting model to study the control of gluconeogenesis. During this period in liver, there are marked changes in redox, in the relative proportions of adenine nucleotides, but especially there is the appearance of both the cytosol activity of PEP-carboxykinase and of hepatic gluconeogenesis as an active pathway. We can therefore study the control of gluconeogenesis in the fetus where only the mitochondrial PEP-carboxykinase is present or in the newborn where the principle activity of PEP-carboxykinase is in the cytosol. The results suggest that cytosol redox, mitochondrial redox, the relative proportion of adenine nucleotides as well as the activities of PEP-carboxykinase are important in the regulation of gluconeogenesis.

Acknowledgment

We thank Drs. R. W. Hanson and A. J. Garber for their assistance and encouragement, and acknowledge financial assistance from the Queen Elizabeth Fellowship Committee and Grants HD 02758, CA 10916 and CA 10439 from the National Institutes of Health.

References

Atkinson, D. E.: Science 150 (1965) 851
Ballard, F. J., I. T. Oliver: Nature 195 (1962) 498
Ballard, F. J., I. T. Oliver: Biochim. biophys. Acta. (Amst.) 71 (1963) 578
Ballard, F. J., Ph. D. Thesis, University of Western Australia, 1964
Ballard, F. J., I. T. Oliver: Biochem. J. 95 (1965) 191
Ballard, F. J., R. W. Hanson: Biochem. J. 102 (1967a) 952
Ballard, F. J., R. W. Hanson: Biochem. J. 104 (1967b) 866
Ballard, F. J., R. W. Hanson, D. S. Kronfeld: Fed. Proc. 28 (1969) 218
Ballard, F. J., unpublished observations
Exton, J. H., C. R. Park: Pharmacol. Rev. 18 (1966) 181
Gevers, W.: Biochem. J. 103 (1967) 141
Hanson, R. W., F. J. Ballard: Biochem. J. 108 (1968) 705
Kaiser, I. H., J. N. Cummings: Amer. J. Physiol. 195 (1958) 481
Krebs, H. A., this symposium
Nordlie, R. C., H. A. Lardy: J. biol. Chem. 238 (1963) 2259

Philippidis, H., F. J. Ballard: Biochem. J. 113 (1969) 651
Starck, D.: In Handbuch der Zoologie. Ed. by J. M. Helmcke, H. von Lengerken and D. Starck. De Gruyter, Berlin 1959, p. 1
Stenger, V., D. Eitzman, T. Anderson, J. Cotter, H. Prystowsky: Amer. J. Obstet. Gyn. 93 (1965) 376
Veneziale, C. M., P. Walter, N. Kneer, H. A. Lardy: Biochem. J. 108 (1968) 705
Williamson, D. H., P. Lund, H. A. Krebs: Biochem. J. 103 (1967) 514
Williamson, J. R., E. T. Browning, R. Scholz: J. biol. Chem. 244 (1969) 4607
Yeung, D., I. T. Oliver: Biochem. J. 108 (1968) 325

Discussion to Störmer and Staib, and to Ballard and Philippidis

Wieland: I have a question for Dr. Ballard: Dr. Walter already mentioned the sparing effect of fatty acids on pyruvate oxidation as a possible factor for influencing gluconeogenic rates in the liver. Do you have any information about pyruvate oxidation rates in the newborn? It would be interesting to know whether the interconvertible system of pyruvate dehydrogenase exists already in this early stage of life. This should be examined because this system is possibly important in the regulation of the overall rate of glucose formation. Do you have any information about the oxidation rates for pyruvate in these animals?

Ballard: No, we have not measured pyruvate oxidation. However, the incorporation of 2-^{14}C alanine into specific carbons of glutamate show that most of the pyruvate is decarboxylated in the fetus, while most is carboxylated in the newborn (A. D. Freedman and A. M. Nemeth: J. biol. Chem. 236 (1961) 3085).

Krebs: Dr. Ballard, am I right in thinking that your fetal data were obtained from 21-day-old animals?

Ballard: Yes.

Krebs: But the gestation period is 21 days.

Ballard: No, it is on the twenty-second day that they are born.

Krebs: The point I want to make is that the changes are extremely dramatic.

Ballard: Yes.

Krebs: Could you comment on the mechanisms of these changes? Are they the result of enzyme synthesis or increased enzyme activity?

Ballard: We are using the antibodies to PEP carboxykinase to test this question of enzyme synthesis as against enzyme activation, but we do not yet have satisfactory results.

D. H. Williamson: I would like to congratulate Dr. Ballard on his very interesting experiments. May I ask two questions: Did the one-day-old rats receive the mother's milk, because it may be very important whether these animals received a supply of fat or not?

Ballard: In these experiments, yes.

D. H. Williamson: I think that this should be emphasized. In one of your tables there was a big difference between the ketone body concentrations in fetal and newborn livers; presumably this is due in part to the flux of dietary fat to the liver of the newborn?

Ballard: Yes, that is correct. An additional explanation for the low concentration of ketone bodies in fetal liver would be the high rate of lipogenesia occurring in that tissue.

D. H. Williamson: Is it correct that for a considerable part of the neonatal period the rat exhibits hyperketonemia with high blood sugar concentrations and that this is very unlike the situation in the adult animal?

Ballard: Yes. The rate of ketone body synthesis increases in parallel to gluconeogenesis and reaches a maximum ten days after birth. Glycogen levels in liver fall after birth and then increase to adult levels within two weeks. These changes and the alterations in diet during this period show that the young rat has a different nutritional status to the adult.

D. H. Williamson: Presumably therefore the sites of control are very different in the newborn and the adult.

Exton: May I ask a question of Dr. Ballard? Have you ever withheld the milk from these animals? What happens then?

Ballard: We find no differences between milk-fed and unfed animals in measurements of adenine nucleotide phosphorylation, or in measurements of the lactate to pyruvate ratio.

Söling: Did you conduct your experiments with different nitrogen-oxygen mixtures? And how long did you keep your animals in these atmospheres; also 30 minutes?

Ballard: Animals were kept for five minutes under the controlled atmosphere and for an additional 15 minutes after injection of gluconeogenic precursor.

Söling: My second question is: When you measure the redox state of the 3-hydroxybutyrate/acetoacetate couple and the lactate/pyruvate couple, do you think that you have a steady state or a certain degree of equilibrium between the two systems? I think of the conditions where you have a low percentage of oxygen in the atmosphere but where you already find an increase in the 3-hydroxybutyrate/acetoacetate ratio without an increase in the lactate/pyruvate ratio.

Ballard: We have not followed either the ß-hydroxybutyrate to acetoacetate and the lactate to pyruvate ratios with extended times under the controlled atmosphere. However, we find only minor changes in the degree of adenine nucleotide phosphorylation between five minutes and one hour of exposure to an atmosphere containing 12% air.

J. R. Williamson: Do you have any acetyl-CoA measurements?

Ballard: No, we are unable to do acetyl-CoA assays on the amount of tissue available.

Seubert: I have a question concerning the isotope studies. You observed in the fetus an elevated labeling of malate — as compared with newborn — after injection of C^{14}-aspartate. I missed in the case of C^{14}-pyruvate application similar malate determinations. Have you done any studies on malate with C^{14}-pyruvate?

Ballard: The label in malate after ^{14}C-pyruvate injection is very slight. In the fetus we believe that pyruvate is reduced to lactate rather than carboxylated.

Seubert: Now, when you have a need for ATP, then you would expect in these cases no change with pyruvate?

Ballard: Yes.

J. R. Williamson: Are there any questions concerning the paper of Dr. Störmer?

Hanson: I was somewhat surprised at the effect of aging on P-enolpyruvate carboxykinase as shown in your paper. We have looked at the effect of age on rat liver P-enolpyruvate carboxykinase (Reshef, Hanson, and Ballard: J. biol. Chem. 244 (1970) 1994) and found no change in activity. This situation is different, however, in rat adipose tissue where both activity and the capacity to adapt to starving is markedly age dependent.

Störmer: I just want to mention that this mouse develops a diabetes with increasing age. At three months they have no or a very mild diabetes, but with six months they have a very severe diabetes, and this diabetes does not become more severe up to nine months.

Seubert: You observed after six and nine months if I am right with the development of diabetes a rise of the activities of PEP-carboxykinase and pyruvate cartoxylase. Thereafter no further increase of the enzyme activities could be observed when cortisol was applied. In this connection: have you determined the levels of glucocorticoids in these animals? You are probably aware of the fact that inactivation of these hormones is reduced in the diabetic state? It could be that the rise in the diabetic state already represents a glucocorticoid induced increase of glucogenic enzyme activities. No further increase can then be expected after application of cortisol. Have you done any studies on the levels of glucocorticoids?

Störmer: No.

Seubert: Another question: How long have you treated your animals with insulin?

Störmer: Twenty hours.

Seubert: You will recall we could suppress pyruvate carboxylase (Prinz et al.: Biochem. Biophys. Res. Commun. 16 (1964) 582) in alloxan diabetic rats after insulin treatment. The effects of insulin, however, were observed only after a longer period (14 - 16 hours).

Ruderman: Dr. Like and Mrs. Lauris and their colleagues at Joslin Research Laboratory carried out similar experiments with Bar Harbour Obese mice (ObOb) and with a group of Swiss-Hauschka mice that became diabetic when placed on a diet of Gilford Chow (10% fat) instead of their usual Purina Chow (5% fat). The animals were initially obese and had a marked elevation in plasma insulin with levels of insulin being as high as 50 times normal. Despite the high insulin values, blood glucose levels were elevated. During the hyperinsulinemia period, hepatic glucokinase activity was usually 2-3 times the value found in control fed animals. Liver glycogen also tended to be elevated. Later in their course the animals developed a more severe form of diabetes associated with higher levels of blood glucose, reduced, or at least less elevated, insulin levels and sometimes ketosis. At this stage, glucokinase activity had decreased to normal or subnormal levels. Did you ever observe this hyperinsulinized state of the liver in your obese animals?

Söling: We found exactly the same when we did experiments with New Zealand Obese mice or with Spiny Mice (Acomys cahirimus). In both cases in the stage of hyperinsulinemia, glucokinase and glycogen levels rise. However when they reach a state where insulin secretion drops we found exactly the opposite until finally a situation was reached, similar to that seen in alloxan-diabetic animals. We found the same in human diabetics (Willms, B., et al., Horm., Metab. Res., in press). If you study glucokinase activity in livers from obese maturity onset diabetics you usually find an increase in the activity of glucokinase, whereas in juvenile diabetics you find a strong decrease.

General Discussion

J. R. Williamson: This morning has been devoted to pyruvate carboxylase. We spoke about its possible control by acetyl-CoA or other CoA-derivatives and ADP. Perhaps I could call on Dr. Utter to comment on what his thoughts are on ADP control and possibly calcium control of pyruvate carboxylase.

Utter: Keech and I reported several years ago that ADP was an inhibitor of pyruvate carboxylase but I did not mention it here because I had no new information to add. As Dr. Walter has just indicated, it is quite possible that ADP may be involved in the control of pyruvate carboxylation under some circumstances. Calcium is another story. It can inhibit isolated pyruvate carboxylase but I believe the concentration levels required are rather high.

J. R. Williamson: 100 μM

Utter: Yes, I think you are right. Although mitochondria can concentrate calcium, I wonder if the level of free calcium available to the enzyme ever reaches that level. Perhaps Dr. Seufert could comment.

Seufert: We observed a very strong inhibition of the enzyme by calcium, but the problem is this: what is the intramitochondrial concentration of calcium? I turn the discussion to Dr. Exton because glucagon certainly acts on the distribution of calcium as first has been shown in Dr. Exton's laboratory. Do you have any views on possible calcium control?

Exton: I don't have any clear views about the control of pyruvate carboxylation by calcium or its possible involvement in the action of glucagon. We don t have enough information on this point.

Seubert: As far as I remember, in one of your papers you report a K^+ output on perfusion of liver under glucagon (Adv. Enzyme Regul. 6 (1968) 391). Can you correlate this output in a quantitative manner to changes of the intracellular levels? I mean, is it that far that the intracellular levels of potassium are changed in a wide range?

Exton: Yes. The liver loses up to one third of its potassium.

Seubert: That is very much. As Dr. Weiss has shown in his presentation, at suboptimal levels of acetyl-CoA the activities of pyruvate carboxylase can be controlled by potassium (80-140 mM). The question again arises whether the intracellular acetyl-CoA concentration is suboptimal or whether the enzyme is saturated with acetyl-CoA. What about acetyl-CoA?

J. R. Williamson: Dr. Ruderman has done some very nice experiments. Maybe he will report about his titration curve of acetyl-CoA tomorrow.

Ruderman: I will talk about this tomorrow.

Seubert: I should like to add in this connection, that one should not relate the K_a values which we determined with the purified enzyme to the in vivo situation. As Dr. Utter has shown, the K_a value of pyruvate carboxylase for acetyl-CoA increases with decreasing pH. Dr. Schoner and Dr. Seufert could confirm this for the rat liver enzyme in our laboratory (K_a pH 6.9 = 42.5 μM; K_a pH 7.25 = 27 μM; K_a pH 7.7 = 16.5 μM). In addition, we observed quite different K_a values with a crude and a purified enzyme preparation. I think we need other parameters to correlate acetyl-CoA levels with the activity of pyruvate carboxylase in the intact cell.

Wieland: This raises the basic question whether an enzyme, which can be influenced by various factors in vitro, actually is a rate-limiting factor in vivo. I think it is very hard to say whether pyruvate carboxylase is a rate-limiting enzyme in vivo at all.

J. R. Williamson: Some years ago Prof. Krebs (Krebs, Proc. Roy. Soc. B., 159 (1964) 545) pointed out that rate-limiting steps in a multi-enzyme sequence should be at the disequilibrium steps, particularly at the first disequilibrium step. For gluconeogenesis, this is pyruvate carboxylase. From general considerations, one particular step in a connected enzyme sequence can only be rate-limiting during a transition from one steady state flux to another, and some type of feedback must occur at the first irreversible step which controls the input flux. Once a steady state has been reached, all the enzymes proceed with the same velocity so that metabolite levels are adjusted only in the transition state, and this gives the most information about control sites. Since both pyruvate carboxylase and pyruvate kinase control the input flux to the gluconeogenic pathway, it is not surprising that one should find so many factors controlling their velocity. I doubt that PEP carboxykinase exerts much of a controlling influence on gluconeogenesis, even though the cytoplasmic oxalacetate concentration usually correlates with the rate of gluconeogenesis.

Wieland: Just comparing the activities of pyruvate carboxylase and PEP-carboxykinase the latter enzyme has a lower activity per gram of tissue than pyruvate carboxylase. Therefore one would direct more attention to the enzyme with the lower activity as a possible rate-limiting step.

J. R. Williamson: But does not that point out the non-validity of measurements of enzyme activities in vitro in giving information about rate-controlling steps in vivo?

Krebs: It is true the assays refer to rather artificial conditions, especially in respect to the substrate concentrations. Sometimes we also assay the reaction in the non-physiological direction. I would like to ask Prof. Wieland whether he thinks that pyruvate carboxylase is not a rate limiting step. If it is not, how can the fact be explained that in the well-fed rat the rate of gluconeogenesis is much slower than in the starved animal? There must be a difference in the fate of pyruvate in the well-fed and starved state. In one case pyruvate is decarboxylated and in the other it is converted to oxaloacetate.

Hanson: One thing that has impressed me about pyruvate carboxykinase is its important anaplurotic function in providing oxaloacetate for a number of important pathways. Pyruvate carboxylase is not just the first reaction in gluconeogenesis but it also is involved in lipogenesis and in glyceroneogenesis in adipose tissue. This probably means that P-enolpyruvate carboxykinase is the first committed reaction of gluconeogenesis. Could this be related to your suggestion that the regulation of glucose synthesis may be mediated at this point?

Exton: We have also considered the possibility recently raised by Haynes (Adam, P. J., R. C. Haynes, Jr.: J. Biol. Chem. 244 (1969) 6444) that mitochondrial uptake of pyruvate may be the first rate-limiting step in gluconeogenesis. It is obvious that if one has an acceleration of PEP-carboxykinase with glucagon, some means of increasing flux through pyruvate carboxylase has to be found. This might be increased uptake of pyruvate by the mitochondria perhaps secondary to a reduction in the intramitochondrial concentration of another metabolite. Other possibilities are activation of the enzyme or increased flux because the intramitochondrial concentration of oxaloacetate is lowered. With regard to the latter, I'm not sure whether removal of oxaloacetate can accelerate pyruvate carboxylase. Perhaps Dr. Utter might answer this question.

Utter: Oxalacetate is an inhibitor, at least of the avian liver enzyme, but not a particularly effective one. Our impression is that the levels required are considerably higher than concentrations expected in mitochondria and therefore we would tend to downgrade the potential control significanace of oxalacetate.

Krebs: I have not made myself sufficiently clear. I did not intend to say that pyruvate carboxylase is rate limiting, but that the pyruvate carboxylase reaction is rate limiting. The rate of this reaction depends on many factors including the concentration of pyruvate. When the concentration of lactate is increased the concentration of pyruvate also rises. There are of course other factors which determine the enzyme activity. As the intracellular concentration of pyruvate is in the region of K_M of pyruvate for this enzyme the activity must increase under physiological conditions with the concentration of pyruvate.

J. R. Williamson: I think Dr. Exton earlier touched on a vital point: that one has to consider the small pool of metabolite levels compared with a very much larger flux towards glucose, which is a sink. Therefore if the interaction is solely at the level of PEP-carboxykinase, a depletion of precursors for this enzyme would occur. Therefore one surely has to consider the coordination of control between pyruvate carboxylase and PEP-carboxykinase bearing in mind that the mitochondrial oxaloacetate concentration is very low, and does not come within the inhibitory range for pyruvate carboxylase.

Exton: To clarify what I said earlier: Increased carboxylation of pyruvate may be produced merely by increasing the entry of pyruvate into the mitochondria without direct stimulation of pyruvate carboxylase itself. In other words, control may be simply by substrate.

J. R. Williamson: In answer to Prof. Krebs, I think there is no real correlation between the net rate of PEP formation and pyruvate carboxylase activity, particularly when one considers the three enzymes (pyruvate carboxylase, PEP carboxykinase and pyruvate kinase) as a functional unit.

Hanson: We recently did a rather simple-minded experiment (Tani and Hanson: Biochem. Biophys. Acta. 192 (1969) 420) in which we injected ethionine into starved rats and measured both the level of hepatic P-enolpyruvate carboxykinase and gluconeogenesis from pyruvate. We observed that the rate of gluconeogenesis was enhanced by starving and not effected by ethionine but the normal adaption of P-enolpyruvate carboxykinase was blocked. This of course suggests that changes in the activity, of what may be considered limiting enzymes in a metabolic sequence, are not essential for an increased metabolic flux over that pathway.

Söling: I want to give a comment on this point. We measured the activity of PEP-carboxykinase, pyruvate carboxylase and pyruvate kinase under V_{max}-conditions in fed and 48-hour starved rats. We calculated the ratios of the rate-limiting enzyme, either pyruvate carboxylase or PEP-carboxykinase over pyruvate kinase. This ratio, which is for the rat PEP-carboxykinase over pyruvate kinase, increases to about 400% during 48 hours of starvation. But the type of regulation of gluconeogenesis, for instance by increased fatty acid oxidation, which depends on the pyruvate concentratio does not change very much in spite of the increase of this ratio by a factor of four.

Ballard: I have two comments. Are we not oversimplifying the argument by considering only the rat? This animal is unusual as the liver has a greater activity of pyruvate carboxylase than of PEP-carboxykinase. Secondly, should we not look at precursors of gluconeogenesis other than lactate and pyruvate? These are important pre cursors in exercise or in the newborn, but glucogenic amino acids or propionate may be more important during fasting or in the ruminant. We should be careful not to extrapolate too liberally from the rat to other species. Mammals other than the rat contain a large proportion of the hepatic PEP-carboxykinase in the mitochondria.

J. R. Williamson: I am not too sure how much difference this makes to the overall control of gluconeogenesis.

Ballard: One difference is that the mitochondrial PEP-carboxykinase is not adaptable

Krebs: In ruminants gluconeogenesis from lactate is also of importance, even if it does not arise in the rumen, because lactate is produced within the body by the glycolysis of red cells and of various other tissues.

Ballard: Yes, I agree. I think that lactate contributes up to 30% of the carbon for gluconeogenesis in the ruminant.

Krebs: Yes, propionate is probably the most important single gluconeogenic precursor in ruminants.

Hanson: I agree with Dr. Ballard that the presence of P-enolpyruvate carboxykinase in the mitochondria in various species, such as the guinea pig and human, must mea a very different kind of mitochondrial regulation. The very fact that another reacti (other than NAD-malate dehydrogenase and citrate synthase) competes for oxaloacet would seem important. Also mitochondrial P-enolpyruvate carboxykinase is directly connected to the ATP pool via GTP synthesis. One of the reasons that led to study guinea pig liver mitochondria in the first place is our feeling that the rat may be rather an "a typical" animal in that control of gluconeogenesis may be very differen in this species.

J. R. Williamson: The main difference between an intra- or extramitochondrial location of PEP carboxykinase, aside from differences of enzyme induction, resides in the nature and fluxes of the anions moving across the mitochondrial membrane. Thus, in the guinea pig with lactate as substrate, anion fluxes in and out of the mito chondria should be minimal, i. e. pyruvate enters and PEP leaves. Perhaps this accounts for the lack of an oleate effect with guinea pig liver. On the other hand, wi a substrate more oxidized than glucose, such as pyruvate, there is still a need for re ducing equivalents which are presumably supplied by the malate-aspartate shuttle. I think it remains to be determined whether there is any control by energy level based on the specificity of PEP carboxykinase for GTP, since the ATP to GTP conversion appears to be rapid in liver. Certainly in isolated mitochondria it is possible to see an apparent control of PEP synthesis by substrate level phosphorylation, but how this is concerned with gluconeogenesis in the intact liver remains to be elucidated. Presumably, even in the guinea pig the rate of PEP formation will be determined largely by the rate of pyruvate carboxylation. To what extent any recycling of PEP back to pyruvate occurs in the cytoplasm is unknown.

Ballard: It is not conceivable to me how a highly reduced cytosol redox could depress gluconeogenesis from pyruvate when the PEP is formed within the mitochondria and passed out. However, a reduced cytosol redox could lower glucose synthesis if the PEP is formed in the cytosol since oxaloacetate conversion to malate would be favored. This surely would be on difference in control.

J. R. Williamson: The subject of redox state and the control of gluconeogenesis is a complicated one. Certainly, in rat liver there seems to be little correlation between the cytoplasmic redox state induced by different gluconeogenic precursors (e.g. lactate and pyruvate in the presence and absence of oleate) and the rate of gluconeogenesis. Also, there is not necessarily a correlation between oxalacetate levels and the rate of gluconeogenesis, as revealed by addition of ethanol to livers perfused with alanine in the absence and presence of oleate (Williamson et al.: J. Biol. Chem. 244 (1969) 5044). Ethanol always caused a fall of oxalacetate but increased gluconeogenesis in the absence of oleate and decreased it in the presence of oleate. A high redox state in the cytosol may affect the rate of delivery of reducing equivalents from the mitochondria, but will also affect the levels of intermediates between PEP and FDP due to the altered equilibrium of glyceraldehyde-P-dehydrogenase, causing a fall of PEP and a rise of FDP unless other factors compensate. Since the levels of these two intermediates affect pyruvate kinase activity, the net rate of gluconeogenesis may be affected.

Söling: The question is this: How can you get a reduced redox state in the cytosol when you have pyruvate as a substrate? You discussed the problem of the highly reduced redox state in the cytosol, when pyruvate is a precursor. To me it seems difficult to achieve a reduced redox state in the cytosol under these conditions.

J. R. Williamson: It is clear that pyruvate dehydrogenase is a very powerful supplier of hydrogen equivalents. So in the mitochondria the redox state will get highly reduced as pyruvate is being oxidized.

Ballard: All I wished to illustrate was that different controls may be operative depending on whether PEP is synthesized inside or outside the mitochondria.

Schäfer: I want to comment on Dr. Walter's paper concerning the use of palmitoyl carnitine. This is a very crucial compound, because its effects depend very much on the concentrations which you apply in your experiments. With the long chain fatty acid carnitine esters you easily can uncouple oxidative phosphorylation and damage mitochondria by surface activity. On the other hand, we have found that low concentrations of palmitoyl carnitine obviously change the K_M and the turnover rate of succinate dehydrogenase (R. Portenhauser, G. Schäfer: FEBS-Letters 2 (1969) 281; R. Portenhauser, G. Schäfer, W. Lamprecht: Hoppe-Seylers Z. Physiol. Chem. 350 (1969) 641), and in checking the effect of higher concentrations of palmitoyl carnitine we found that one can reverse the changes in the kinetic behavior of succinate dehydrogenase. This would mean that at low concentrations of palmitoyl carnitine a high rate of malate- or oxaloacetate formation is possible, and that a high flux of acetyl-CoA exists into the cycle. But at higher levels of long chain carnitine esters the enzyme is inhibited, favoring a situation of enhanced ketone body formation and other conversions of acetyl-CoA. My question therefore, is did you check in your experiments the effect of different amounts of palmitoyl carnitine on the rate of gluconeogenesis?

Walter: Yes, we did, but in a limited range. When we added a total of 1.6 μmole of palmitoyl carnitine we found more malate and less citrate than when we added 0.8 μmoles. This palmitoyl carnitine was always added with albumin in ten portions of 0.16 (or 0.08) μmoles every minute. This was to keep mitochondrial damage as low as possible.

J. R. Williamson: I think your point is a very important one, because I think with the fatty acid oxidation you are increasing redox input into the high potential pool of flavines, that is the flavine pool after the rotenone block, and this of course is where succinate feeds in. So one is merely slowing down succinate dehydrogenase, because of the interaction at the flavine step. And the flux through the succinic dehydrogenase will determine the amount of malate you have. And this of course will affect how much dicarbone precursors you have to couple with acetyl-CoA. The amount of malate is very, very critical in determining the changes in acetyl-CoA under the conditions of your experiments, Dr. Walter. Some years ago I did some similar work. With only pyruvate and then giving palmityl carnitine you actually get a decrease of acetyl-CoA levels, and this under conditions where the total carboxylation is not changed by palmityl carnitine. But in the presence of a little bit of malate you can get acetyl-CoA going in the opposite direction. So, in other words, it is a consideration of how much malate you have. This will determine the overall effects in carboxylation and possibly in relation to your ADP-effect.

Schäfer: I would like to add two short comments. I agree with you that palmioyl carnitine is a very important reductant for the mitochondrial flavorproteins. We have compared the reduction of mitochondrial flavoproteins by succinate and by palmitoyl carnitine (R. Portenhauser, G. Schäfer: Abstracts 5th Meeting of the Federation of Europ. Biochem. Soc., Prauge (1968) 142; R. Portenhauser, G. Schäfer, W. Lamprecht: Hoppe-Seylers Z. Physiol. Chem. 350 (1969) 641). It has been shown that one tenth or less of the concentration of palmitoyl carnitine yields about the same degree of reduction as compared to the concentration of succinate required.

The second point I would like to make is that it is critical to claim that the influence on succinate dehydrogenase itself, as mentioned before, must play a role in physiological regulation, because it is difficult to rule out what the surface activity of long chain carnitine esters does with a membrane bound enzyme. However, it should be emphasized that the activating action of low palmitoyl carnitine concentrations on succinate dehydrogenase is specific to the L (-) - compound, which is the physiological one.

Utter: May I ask Dr. Walter a question? I wonder if you have values for magnesium to accompany those for ATP, ADP and AMP. I ask because unchelated ATP is a fairly good inhibitor for pyruvate carboxylase and this factor might help explain some of the difficulties in plotting ATP/ADP ratios against ADP. Do you have measurements of magnesium under the various conditions?

Walter: Unfortunately not.

Effects of Quinolinic Acid on the Free and Total Nicotinamide-Adenine Dinucleotides of Rat Liver+

D. H. Williamson, F. Mayor and Dulce Veloso
Metabolic Research Laboratory, Nuffield Department of Clinical Medicine, Radcliffe Infirmary, Oxford, England

Summary

Quinolinic acid is both an inhibitor of phosphoenolpyruvate carboxykinase and a precursor of NAD. Its administration (50 mg/100 g. body wt.) to fed and starved (48-hour) rats resulted in slight increases (10-15%) in the concentrations of total NAD and NADP in the liver. There was no change in the (total NAD^+)/(total NADH) or (total $NADP^+$)/(total NADPH) ratios in the fed rats, but a decrease in the (total NAD^+)/(total NADH) ratio from 5.1 to 3.6 without significant alteration of the (total $NADP^+$)/(total NADPH) ratio (0.22) in the starved rats. Decrease of the flux of free fatty acids to the liver in the starved rat by injection of 3,5-dimethylpyrazole, an inhibitor of lipolysis, prevented the change in (total NAD^+)/(total NADH) ratio brought about by quinolinate.

The hepatic concentrations of lactate, pyruvate, 3-hydroxybutyrate, acetoacetate, glutamate, 2-oxoglutarate and ammonia were determined in these experiments and used to calculate the (free NAD^+)/(free NADH) ratios at the sites of lactate dehydrogenase (cytosol), 3-hydroxybutyrate dehydrogenase (mitochondrial cristae) and glutamate dehydrogenase (mitochondrial matrix). Quinolinate did not change the cytoplasmic (free NAD^+)/(free NADH) ratio in livers from fed or starved rats, but it decreased the mitochondrial ratio (3-hydroxybutyrate dehydrogenase) from 6.8 to 3.4 in livers from starved rats; there was no change in the mitochondrial ratio in fed livers. Prior administration of 3,5-dimethylpyrazole followed by quinolinate prevented the decrease in the mitochondrial (free NAD^+)/(free NADH) ratio, but increased (two-fold) the cytoplasmic ratio so that it approached that for fed livers. The overall mass action ratio of the glutamate dehydrogenase-3-hydroxybutyrate dehydrogenase system deviated from the equilibrium value by a factor of about 5 after administration of quinolinate to starved rats. Near-equilibrium between the two systems was restored by prior treatment with 3,5-dimethylpyrazole.

These results suggest that the in vivo inhibition of gluconeogenesis at the stage of phosphoenolpyruvate formation results in an accumulation of NADH within the mitochondria which is dependent on the supply of fatty acids to the liver.

\+ Unusual Abbreviations: 3,5-DMP: 3,5-dimethylpyrazole

Enzyme Code Numbers: Phosphoenolpyruvate carboxykinase (EC 4.1.1.32), Lactate dehydrogenase (EC 1.1.1.27), 3-Hydroxybutyrate dehydrogenase (EC 1.1.1.30), Glutamate dehydrogenase (EC 1.4.1.3), Glyceraldehyde-3-phosphate dehydrogenase (EC 1.2.1.12).

Introduction

Quinolinic acid, an intermediate in the catabolism of tryptophan, is both a potential precursor of NAD (Gholson et al. 1964) and an inhibitor of phosphoenolpyruvate carboxykinase in the isolated perfused rat liver (Veneziale et al., 1967). Inhibition of gluconeogenesis at a step below glyceraldehyde-3-phosphate dehydrogenase might be expected to decrease the requirement for reducing equivalents in the cytosol of the liver cell. Consequently, the effects of administration of quinolinic acid on the redox state of the NAD^+-NADH system in livers from fed and starved rats have been studied by freeze-clamping the livers (Wollenberger et al. 1960) and measuring the content of total nicotinamide-adenine dinucleotides. To identify the intracellular site of any changes in redox state the hepatic content of the substrates of lactate, 3-hydroxybutyrate and glutamate dehydrogenases were also determined and from these measurements the cytoplasmic and mitochondrial free NAD^+/free NADH ratios have been calculated (Hohorst et al. 1959, Williamson et al. 1967). The results indicate that in vivo inhibition of hepatic gluconeogenesis in the starved rat at the stage of phosphoenolpyruvate formation causes an accumulation of NADH within the mitochondria which is dependent on the supply of free fatty acids to the liver.

Materials and Methods

Male rats of the Wistar strain weighing 160-190 g were used. Full details of treatment of livers and analytical methods have been described by Williamson et al. (1967) and Williamson et al. (1969).

Results

Evidence for the In Vivo Inhibition of Phosphoenolpyruvate Carboxykinase by Quinolinic Acid

Measurements of phosphoenolpyruvate carboxykinase activity (rate of PEP synthesis) in liver after administration of quinolinate to starved rats showed the same anomalous rise in activity previously described for tryptophan by Foster et al. (1966), but the effect was maintained for a shorter period with quinolinate (Table 1). However, determination of enzyme activity in the direction of oxaloacetate formation (Holton and Nordlie 1965) showed decreased activity (about 40%) in both the quinolinate- and tryptophan-treated rats (Table 2). Picolinate, another metabolite of tryptophan catabolism, had no effect on the activity of the enzyme.

In agreement with findings of the isolated perfused rat liver (Veneziale et al. 1967) administration of quinolinate to fed or starved rats caused large increases in the hepatic content of malate and citrate and, to a lesser extent, of oxoglutarate and aspartate (Fig. 1). The hepatic lactate and pyruvate content only increased in starved rats, while that of glutamate did not change significantly in either fed or starved rats. The increase in the sum of the glucose precursors listed in Fig. 1 was 1.83 μmole/g fresh weight in livers from fed rats treated with quinolinate and 4.93 μmole/g fresh weight in livers from starved rats. These findings are in agreement with the conclusion of

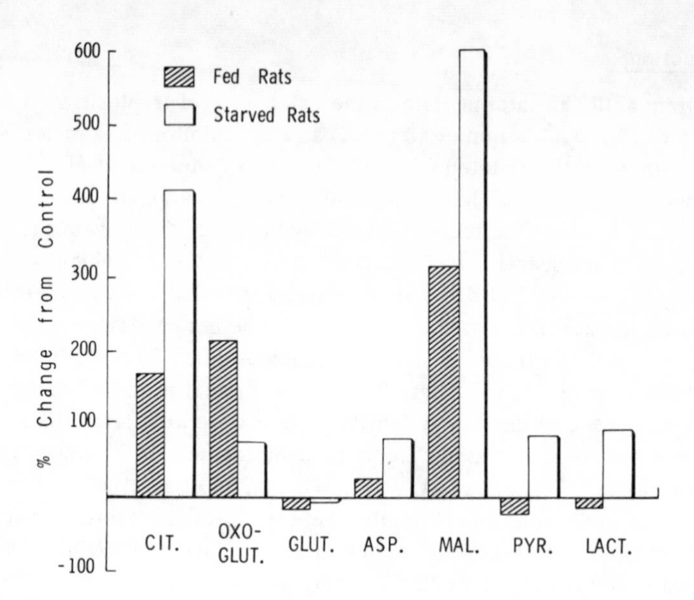

Fig. 1. Comparison of alterations in hepatic metabolites after administration of quinolinate to fed and starved rats. For the dose and method of administration of quinolinic acid, see Table 1. The livers were freeze-clamped (Wollenberger et al. 1960) 30 minutes after injection.

Compound Injected	Activity (μmole/min/g wet wt.): Oxaloacetate → PEP Time after injection			
	30 min	60 min	120 min	210 min
NaCl	3.6	3.7	3.6	3.4
Tryptophan	5.3	7.2	7.5	8.0
Quinolinate	6.0	7.3	5.4	4.6

Table 1. Activity of Phosphoenolpyruvate Carboxykinase in Rat Liver. Rats starved for 48 hours were given an intraperitoneal injection of a neutral solution of tryptophan, quinolinic acid or picolinic acid (50 mg/100 g body wt.). They were killed at timed intervals and the hepatic activity of phosphoenolpyruvate carboxykinase determined in the direction of PEP synthesis (Foster et al. 1966). The results are the mean for two livers at each time.

Compound Injected	Activity (μmole/min/g wet wt.): PEP → Oxaloacetate Time after injection			
	30 min	60 min	120 min	210 min
NaCl	2.8	3.0	2.4	2.3
Tryptophan	1.5	1.5	1.0	1.2
Quinolinate	1.4	1.6	2.6	2.4
Picolinate	2.7	2.8	2.8	2.5

Table 2. Activity of Phosphoenolpyruvate Carboxykinase in Rat Liver. For details of the method of injection and dose, see Table 1. The phosphoenolpyruvate carboxykinase activity was measured in the direction of oxaloacetate synthesis (Holton and Nordlie 1965). The results are the mean for two livers at each time.

Veneziale et al. (1967) that quinolinic acid is an inhibitor of rat liver phosphoenolpyruvate carboxykinase.

Changes in the Hepatic Content of Total Nicotinamide-Adenine Dinucleotides After Administration of Quinolinic Acid

The total nicotinamide-adenine dinucleotide (NAD^+ plus NADP) content of liver only increased slightly (10-15%) after administration of quinolinate and this was not influenced by the nutritional state. The failure of quinolinic acid to act as an effective precursor of NAD may be due to its poor penetration of the liver cell (Ijichi et al. 1966). In the fed rat the total NAD^+/total NADH and total $NADP^+$/total NADPH ratios were not altered by quinolinate (Tables 3, 4), but in the starved rat there was a sustained decrease in the total NAD^+/total NADH ratio with no significant change in the total $NADP^+$/total NADPH ratio; the decrease in the former ratio was mainly due to a 50% increase in total NADH content.

Cytoplasmic and Mitochondrial free NAD^+/free NADH Ratios After Administration of Quinolinic Acid

To determine whether the more reduced state of the hepatic NAD^+-NADH system in livers from starved rats treated with quinolinate applied to the whole cell or a more specific site the cytoplasmic and mitochondrial free NAD^+/free NADH ratios were calculated from the hepatic content of lactate and pyruvate, 3-hydroxybutyrate and acetoacetate, glutamate, oxoglutarate and ammonia (Williamson et al. 1967). Quinolinate did not appreciably alter the cytoplasmic or mitochondrial free NAD^+/free NADH ratio in livers from fed rats, nor did it change the cytoplasmic ratio in the starved rat. There was, however, a marked decrease in the mitochondrial

State of rats	Treatment	Time	(NAD^+)	(NADH)	$(NAD^+)/(NADH)$
		min	nanomole/g wet wt.		
Fed	NaCl	30	760 ± 50	140 ± 30	5.4
	Quinolinate	30	850 ± 70	160 ± 20	5.3
	Quinolinate	60	880 ± 60	160 ± 20	5.5
Starved	NaCl	30	820 ± 80	160 ± 20	5.1
	Quinolinate	30	830 ± 60	230 ± 40	3.6
	Quinolinate	60	930 ± 40	250 ± 30	3.7

Table 3. Effects of Quinolinic Acid on Content of Nicotinamide-Adenine Dinucleotide in Rat Liver. The results are mean values (± S.D.) of at least seven rats in each group. For further details, see Table 1.

free NAD^+/free NADH in livers from starved rats treated with quinolinate (Table 5). This latter result suggests that the observed increase in total NADH in this situation occurred within the mitochondria.

State of rats	Treatment	Time	$(NADP^+)$	(NADPH)	$(NADP^+)/(NADPH)$
		min	nanomole/g wet wt.		
Fed	NaCl	30	67 ± 14	300 ± 60	0.22
	Quinolinate	30	64 ± 6	360 ± 70	0.20
	Quinolinate	60	64 ± 10	350 ± 40	0.18
Starved	NaCl	30	80 ± 21	360 ± 40	0.22
	Quinolinate	30	75 ± 11	380 ± 40	0.20
	Quinolinate	60	74 ± 17	370 ± 50	0.20

Table 4. Effects of Quinolinic Acid on the Content of Nicotinamide-Adenine Dinucleotide Phosphate in Rat Liver. The results are mean values (± S.D.) of at least seven rats in each group. For other details, see Table 1.

State of rats	Treatment	Time	Calculated free NAD^+/free NADH	
			Cytoplasm	Mitochondria
		min		
Fed	Saline	30	875	9.0
	Quinolinate	30	810	7.3
	Quinolinate	60	1170	9.1
Starved	Saline	30	438	6.8
	Quinolinate	30	416	3.4
	Quinolinate	60	416	4.0

Table 5. Values for Ratios of free NAD^+/free NADH in Livers of Normal and Quinolinate-Treated Rats. The ratios have been calculated from the hepatic content of lactate and pyruvate (cytoplasm) and 3-hydroxybutyrate and acetoacetate (mitochondria) by the method of Williamson et al. (1967).

Reversal by 3.5-Dimethylpyrazole of Effects of Quinolinic Acid on the Hepatic Redox State of Starved Livers

The fact that quinolinate only caused a more reduced state of the mitochondrial NAD^+-NADH system in livers from starved rats suggested that this effect might be connected with the increased hepatic oxidation of free fatty acids in the starved state. To test this suggestion, the mobilization of free fatty acids was suppressed in starved rats by prior administration of 3.5-dimethylpyrazole, an inhibitor of lipolysis in adipose tissue (Bizzi et al. 1966). Administration of 3.5-dimethylpyrazole caused a large decrease in plasma free fatty acids and this was accompanied by the expected comparable decrease (about 85%) in hepatic ketone body concentrations. Administration of 3.5-dimethylpyrazole and quinolinate together produced an increase in the sum of glucose precursors comparable to that with quinolinate alone, although the major increase was in aspartate rather than malate and citrate (Table 6). This indicates the importance of fatty acid oxidation in determining the amount of oxalo-acetate diverted to malate or aspartate. The hepatic concentrations of total NAD were similar in both groups of rats, but there was no decrease in the total NAD^+/total NADH ratio in the 3.5-dimethylpyrazole plus quinolinate-treated rats (Table 7). Similarly, in these rats there was no appreciable change in the calculated mitochondrial free NAD^+/free NADH ratio when compared to the rats treated with 3.5-dimethylpyrazole alone (Table 8). Thus free fatty acids appear to be involved in the increased state of reduction of the mitochondrial NAD^+-NADH system in starved rats treated with quinolinate.

Metabolite	Quinolinate	3.5-DMP	Quinolinate + 3.5-DMP
	nanomole/g wet wt.		
Lactate	+ 240	+ 650	+ 1645
Pyruvate	+ 14	+ 23	+ 72
Malate	+ 1740	+ 110	+ 710
Aspartate	+ 690	+ 40	+ 2030
Citrate	+ 2130	+ 250	+ 720
Ketone bodies	- 360	- 1770	- 1645

Table 6. Changes in the Content of Hepatic Metabolites After Administration of Quinolinate, 3.5-dimethylpyrazole or Quinolinate Plus 3.5-dimethylpyrazole to Starved Rats. Rats starved for 48 hours were given an intraperitoneal injection of either 3.5-DMP (10 mg/kg body wt.), 3.5-DMP plus quinolinate (250 mg/kg body wt.) or quinolinate alone and were then killed one hour after injection. The changes in content were calculated from mean values of at least five rats in each group.

Effect of Quinolinic Acid on the Equilibrium Between 3-Hydroxybutyrate and Glutamate Dehydrogenase

For the calculation of the free NAD^+/free NADH ratio in a particular cellular compartment it is essential that at least two enzymes located in that compartment have been shown to be in near-equilibrium in variety situations. Near-equilibrium

Treatment	NAD^+	NADH	NAD^+/NADH
	nanomole/g wet wt.		
NaCl	820 ± 80	160 ± 20	5.1
Quinolinate	830 ± 60	230 ± 40	3.6
3.5-DMP	770 ± 60	140 ± 30	5.5
Quinolinate plus 3.5-DMP	840 ± 70	150 ± 20	5.6

Table 7. Reversal by 3.5-Dimethylpyrazole of Effects of Quinolinate on Content of Nicotinamide-Adenine Dinucleotide in Liver of Starved Rats. The results are mean values (± S.D.) of at least five rats in each group. For other details, see Table 6.

Compound Injected	Calculated free NAD^+/free NADH	
	Cytoplasm	Mitochondria
NaCl	438	6.8
Quinolinate	416	3.4
3.5-DMP	360	7.8
3.5-DMP + Quinolinate	722	6.7

Table 8. Values for Ratios of Free NAD^+/Free NADH in Livers of Starved Rats Treated with 3.5-Dimethylpyrazole or 3.5-Dimethylpyrazole Plus Quinolinic Acid. For details of method of administration of dosage, see Table 6. For other details, see Table 5.

between the reactants of 3-hydroxybutyrate and glutamate dehydrogenase (both mitochondrial enzymes) occurs in vivo in livers from fed, starved and alloxan-diabetic rats (Williamson et al. 1967) and in livers from fed rats during ischemia (Brosnan et al. 1970). Administration of quinolinate to starved rats caused a considerable

State of rats	Compound Injected	$\frac{(\text{Hydroxybutyrate})(\text{oxoglutarate})(NH_4^+)}{(\text{acetoacetate})(\text{glutamate})}$
		mM
Fed	Saline	6.3×10^{-2}
	Quinolinate	10×10^{-2}
Starved	Saline	5.8×10^{-2}
	Quinolinate	23×10^{-2}
	3.5-DMP	5×10^{-2}
	3.5-DMP + Quinolinate	5×10^{-2}

Table 9. The Reactant Ratio of Hydroxybutyrate and Glutamate Dehydrogenase Systems

deviation from the equilibrium position of the 3-hydroxybutyrate-glutamate dehydrogenase system and this effect was reversed by prior administration of 3-5-dimethylpyrazole (Table 9). The precise reasons for this deviation are not known but other experiments (Brosnan and Williamson, unpublished data) suggest that in certain situations oxoglutarate penetration into the mitochondria may be limiting and this results in compartmentation of some of the measured oxoglutarate.

Discussion

It is now generally accepted on the basis of in vitro experiments that the reducing equivalents required in the cytosol of rat liver for the synthesis of glucose from precursors other than glycerol and lactate are provided by the oxidation of fatty acids in the mitochondria (Struck et al. 1965, Williamson et al. 1966). These reducing equivalents are transported in the form of malate and aspartate to the cytosol and are then converted to glucose via oxaloacetate and phosphoenolpyruvate (Lardy et al. 1965b, Lardy 1965a, Krebs et al. 1967). Inhibition of phosphoenolpyruvate formation by quinolinic acid results in a large accumulation of malate and aspartate (Veneziale et al. 1967) with a concomitant decrease in the utilization of reducing equivalents in the cytosol. Hence it can be inferred that the accumulation of NADH within the mitochondrial compartment of the liver cell after administration of quinolinate to starved rats is a direct consequence of this decreased utilization. The fact that quinolinate does not increase mitochondrial NADH in situations where the supply of fatty acids to the liver is decreased (e.g. in fed rats or starved rats treated with 3.5-dimethylpyrazole) confirms that one of their roles in vivo is to provide reducing power for glucose synthesis.

The relatively large increase (about 40%) in total NADH within the mitochondria of livers of starved rats treated with quinolinate suggests that a considerable proportion of total NADH found in the livers of control rats must also be located in this compartment in vivo. The absence of an appreciable change in the amounts of total $NADP^+$ and NADPH after quinolinate administration suggests that either the amounts within the mitochondria are low in relation to the amounts in the cytosol or that the mitochondrial NAD^+-NADH system is not in equilibrium with the $NADP^+$-NADPH system. Further experiments of this type should provide more information on the relationships between total NAD and the free NAD of liver cell compartments.

Acknowledgments

The authors thank Professor H. A. Krebs for advice and encouragement. Dr. D. H. Williamson is a member of the Medical Research Council external staff and Dr. D. Veloso was a Gulbenkian Foundation Research Fellow. This work was supported in part by the Medical Research Council, London.

References

Bizzi, A., M. T. Tacconi, S. Garattini: Experientia, 22 (1966) 664
Brosnan, J. T., H. A. Krebs, D. H. Williamson: Biochem J. 117, (1970) 91
Brosnan, J. T., D. H. Williamson: unpublished data
Foster, D. O., P. D. Ray, H. A. Lardy: Biochemistry 5 (1966) 563
Gholson, R. K., I. Ueda, N. Ogasawara, L. M. Henderson: J. biol. Chem. 239 (1964) 1208
Hohorst, H. J., F. H. Kreutz, Th. Bücher: Biochem. Z. 332 (1959) 18
Holten, D. D., R. C. Nordlie: Biochemistry 4 (1965) 723
Ijichi, H., A. Ichiyama, O. Hayaishi: J. biol. Chem. 241 (1966) 3701
Krebs, H. A., T. Gascoyne, B. M. Notton: Biochem. J. 102 (1967) 275
Lardy, H. A.: Harvey Lectures, Ser. 60 (1965a) 261
Lardy, H. A., V. Paetkau, P. Walter: Proc. nat. Acad. Sci. (Wash.) 53 (1965b) 1410
Struck, E., J. Ashmore, O. Wieland: Biochem. Z. 343 (1965) 107
Veneziale, C. M., P. Walter, N. Kneer, H. A. Lardy: Biochemistry 6 (1967) 2129
Williamson, D. H., P. Lund, H. A. Krebs: Biochem. J. 103 (1967) 514
Williamson, D. H., D. Veloso, E. V. Ellington, H. A. Krebs: Biochem. J 114 (1969) 575
Williamson, J. R., R. A. Kreisberg, P. W. Felts: Proc. nat. Acad. Sci. (Wash.) 56 (1966) 247

Discussion to Williamson, Mayor and Veloso

Hanson: This table shows the response of hepatic and adipose tissue P-enolpyruvate carboxykinase after injection of 5 mmoles per kilo of nicotinamide into starved rats. There is a very sharp drop in adipose tissue P-enolpyruvate carboxykinase to a level about half that of control animals. On the other hand, the liver enzyme is increased about 50%. The same effect can be gotten with nicotinic acid. In adrenalectomized rats we have noted the same effect of nicotinamide as seen in normal animals. This type of experiment is interesting since it shows the very short decay rate of adipose tissue P-enolpyruvate carboxykinase (t 1/2 = 4 1/2 hrs) as well as suggesting that nicotinamide can have a direct effect on tissues without affecting animals.

D. H. Williamson: I imagine that you would not expect an effect within 30 minutes.

Hanson: Not necessarily. I simply wanted to point out a very striking long term effect of nicotinamide (a compound on the same biosynthetic pathway as quinolinate) on an important enzyme in gluconeogenesis.

J. R. Williamson: You mentioned that there must be a compartmentation of α-ketoglutarate in your experiments with quinolinate?

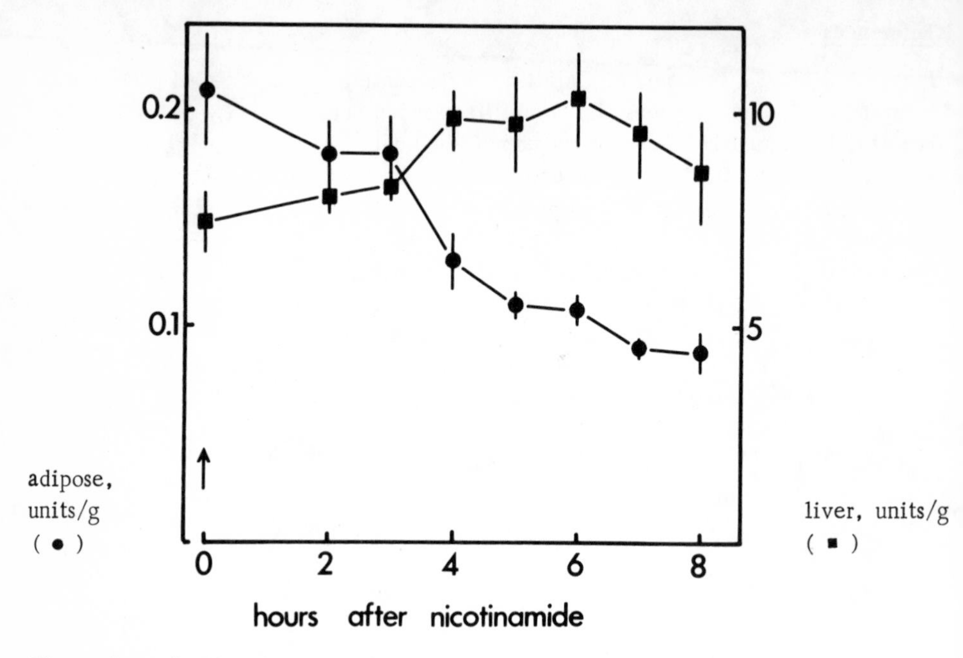

Figure shown by Dr. Hanson in the discussion to Williamson et al.

D. H. Williamson: This is exactly what we think.

Seubert: Have you taken into consideration the role of manganese in the PEP-carboxykinase reaction? The enzyme is activated by this cation in the presence of Mg^{++} by a factor of about two. I do not know if quinolinate forms a complex with manganese. However, it could be a possibility to influence the activity of PEP-carboxykinase in vivo.

D. H. Williamson: I must admit that we hesitated to report the enzyme measurement shown on the first slide, because we have been waiting for Prof. Lardy and his group to come forth with a clear explanation of why in vitro PEP-carboxykinase activities are actually higher after quinolinate treatment although the enzyme is inhibited in vivo. I really would like to hear the comments of the enzymologists here, especially of Prof. Utter on how to explain these apparent anomalies.

Utter: I was very interested to learn of the results obtained by Dr. Williamson with quinolate and the situation reminds me somewhat of some observations made by Dr. Noce in my laboratory several years ago. During studies of the properties of highly purified PEP-carboxykinase from avian liver we were surprised to find that various types of treatment, e. g. $(NH_4)_2SO_4$ precipitation, could alter greatly the ratio of PEP/oxalacetate reactions. Some of these treatments almost abolished the ability of the enzyme to form oxalacetate while maintaining much of the activity in the direction of PEP formation. Such changes seem analogous to those reported here by Dr. Williamson.

On the Redox State of NAD+/NADH Systems in Guinea Pig Liver Under Different Experimental Conditions+

Berend Willms, Jochen Kleineke and Hans-Dieter Söling
Abteilung für klinische Biochemie der Medizinischen Universitätsklinik
Göttingen, FRG

Summary

As fatty acids fail to stimulate gluconeogenesis in guinea pig livers (Söling et al. 1968, Willms and Söling 1968a), changes in the redox state of the substrate pairs lactate/pyruvate, 3-hydroxybutyrate/acetoacetate, glutamate/(α-oxoglutarate) (NH_4^+) and α-glycerophosphate/dihydroxyacetonephosphate as well as changes in the ATP/ADP ratio were studied in guinea pig livers during increased oxidation of fatty acids. The following results were obtained:

1) Under normal conditions the redox state of the lactate/pyruvate system is more negative; under normal conditions that of the 3-hydroxybutyrate/acetoacetate system is more positive in guinea pig liver (in vivo and during isolated perfusion) than in rat liver. The redox state of the α-glycerophosphate/dihydroxyacetonephosphate system is about the same as in rat liver.

2) During increased oxidation of fatty acids, the redox state of the lactate/pyruvate system in isolated perfused guinea pig livers becomes more negative, the redox state of the 3-hydroxybutyrate/acetoacetate system shows a slight increase during more prolonged fatty acid oxidation. In vivo, fat feeding for three days leads to a moderate increase in the lactate/pyruvate ratio, to a dramatic increase in the α-glycerophosphate/dihydroxyacetonephosphate ratio, and to an increase in the 3-hydroxybutyrate/acetoacetate ratio.

3) The ATP/ADP ratio in the guinea pig liver is not altered in vivo by an increased oxidation of fatty acids.

4) Evidence is presented that the activity of the 3-hydroxybutyrate dehydrogenase under the experimental conditions used is not limiting for the 3-hydroxybutyrate/acetoacetate equilibrium.

5) Under normal conditions the 3-hydroxybutyrate dehydrogenase system and the glutamate dehydrogenase system are in equilibrium in rat and in guinea pig livers. Under certain experimental conditions, a disequilibrium is observed as well in guinea pig as in rat liver. This is assumed to be due to different locations of the redox systems within the mitochondria.

+ Unusual Abbreviations:

Enzyme Code Numbers: 3-hydroxybutyrate dehydrogenase (EC 1.1.1.30), glutamate dehydrogenase (EC 1.4.1.3.), triosephosphate dehydrogenase (EC 1.2.1.12).

Introduction

In rat liver, during increased fatty acid oxidation a stimulation of gluconeogenesis and a reduction of the redox state of cytoplasmic and mitochondrial $NAD^+/NADH$ systems is observed (Struck et al. 1966, Williamson et al. 1966, Menahan et al. 1968, Söling et al. 1968, Willms and Söling 1968a). From these results, several authors (Struck et al. 1968, Williamson et al. 1966, Menahan et al. 1968) concluded that the more reduced state of $NAD^+/NADH$ couples would stimulate gluconeogenesis by activation of triosephosphate dehydrogenase reaction.

In guinea pig liver increased fatty acid oxidation fails to stimulate gluconeogenesis (Willms et al. 1968, Söling p. 244 this book). To study the connection between the redox state of NAD couples and the lack of stimulation of gluconeogenesis by fatty acid oxidation, the redox state of several $NAD^+/NADH$ systems in guinea pig liver was examined under various conditions.

Results

The Redox State During Increased Fatty Acid Oxidation In Vivo.

Under steady state conditions in vivo in chow-fed guinea pigs, the lactate/pyruvate ratio was significantly higher, the 3-hydroxybutyrate/acetoacetate ratio significantly lower compared with rat livers (Table 1). Surprisingly, the α-glycerophosphate/dihydroxyacetonephosphate ratio was in the same order as reported for rat liver (Hohorst et al. 1959). During fat feeding for 72 hours, guinea pigs developed a severe ketosis with an increase in the total ketone body concentration in the liver from 0.079 mM to 1.428 mM (Table 1). The ketosis was accompanied by an increase in the lactate/pyruvate ratio from 21.2 to 32.8. Compared with the control animals the α-glycerophosphate/dihydroxyacetonephosphate ratio became considerably more reduced than the lactate/pyruvate system. The 3-hydroxybutyrate/acetoacetate ratio increased from 1.10 to 2.54. The lactate/pyruvate ratios and the 3-hydroxybutyrate/acetoacetate ratios measured in the blood during the in vivo experiments corresponded well with the ratios measured in the liver, showing that at least under steady state conditions these systems in the blood were in equilibrium with the corresponding systems in the liver.

The ATP/ADP ratio was 5.40 in the control animals and 4.90 after fat feeding, ruling out hypoxia as the reason for the high lactate/pyruvate ratio (Table 2).

The redox state of the NAD couple is characterized by the ratio

$$\frac{\text{concentration of free } NAD^+}{\text{concentration of free } NADH}$$

in short: $NAD^+/NADH$ ratio. The $NAD^+/NADH$ ratios calculated from the substrate pairs given in Table 1 are presented in Table 3: Under normal feeding conditions the $NAD^+/NADH$ system being in equilibrium with the α-glycerophosphate/dihydroxyacetonephosphate system was significantly more oxidized than that being in equilibrium with the lactate/pyruvate system. After fat feeding the redox state

	Lact	Pyr	L/P	GlycP.	DHAP.	α-GlycP./DHAP.	3-OHB	AcAc	3-OHB/AcAc
	(nmoles/g)		(ratio)	(nmoles/g)		(ratio)	(nmoles/g)		(ratio)
Chow-fed animals (n=6)	1,656 ± 153	86 ± 9	21.2 ± 1.5	336 ± 16	92 ± 9	3.8 ± 0.3	41.1 ± 5.3	37.7 ± 2.4	1.10 ± 0.12
Fat-fed animals (n=6)	1,286 ± 131	40 ± 4	32.8 ± 2.7	681 ± 35	32 ± 3	22.0 ± 1.5	1,025 ± 150	403 ± 36	2.54 ± 0.51

Table 1. Steady state concentrations of L-lactate (Lact.), pyruvate (Pyr), α-glycerophosphate (α-GlyP.), dihydroxy-acetonephosphate (DHAP), 3-hydroxybutyrate (3-OHB) and acetoacetate (AcAc) and ratios of these substrate pairs in livers of chow-fed and fat-fed guinea pigs in vivo. The fat fed animals were starved for 24 hours and then received 3 ml of olive oil per animal every eight hours for three days. The animals were anesthetized and liver tissue was obtained by freeze stop. Results are mean values ± s.e.m.

	ATP (nmoles/g)	ADP (nmoles/g)	ATP/ADP (ratio)
Chow-fed (n = 17)	2948 ± 124	550 ± 14	5.40 ± 0.30
Fat-fed (n = 11)	3117 ± 138	693 ± 55	4.90 ± 0.40

Table 2. Steady state concentrations of ATP and ADP, and ATP/ADP ratios in guinea pig livers in vivo. The fat-fed animals were starved for 24 hours and then received 3 ml of olive oil every eight hours for three days. Liver tissue was obtained by freeze stop. Results are ± s.e.m.

of the $NAD^+/NADH$ system being in equilibrium with the α-glycerophosphate/dihydroxyacetonephosphate system was considerably more reduced and therefore both cytoplasmic $NAD^+/NADH$ systems became nearly equilibrated. The mitochondrial NAD^+/NA ratio calculated from the 3-hydroxybutyrate dehydrogenase system was more reduced than the cytoplasmic systems before and after fat feeding.

The Redox State During Increased Fatty Acid Oxidation in the Isolated Perfused Liver

In the isolated perfused guinea pig liver the concentrations of lactate and pyruvate and those of 3-hydroxybutyrate and acetoacetate were in the same range in tissue and perfusion medium, as was reported for rat liver (Schimassek 1963, Krebs and Gascoyne 1968). In control experiments the lactate/pyruvate ratio in the perfusate was similar to that observed in vivo and again significantly higher than in the perfused rat liver under the same conditions (Fig. 1). This higher lactate/pyruvate ratio remained constant during the whole experiment. When 0.72 mmole/h of sodium hexanoate was infused intraportally, the lactate/pyruvate ratio in the medium increased significantly. This increase was much higher than the increase seen in experiments with isolated perfused rat livers receiving the same amount of sodium hexanoate (Söling et al. 1968, Willms and Söling 1968a) in spite of the fact that ketogenesis of the isolated perfused rat liver was even higher.

The 3-hydroxybutyrate/acetoacetate ratio, on the other hand, was significantly lower in the isolated perfused guinea pig liver than in the isolated perfused rat liver and also lower than in vivo (Table 4). During the perfusion experiment, the ratio remained constant as well as in control experiments in experiments with intraportal infusion of hexanoate, even when the hexanoate infusion was increased to 2.16 mmole/h. In similar experiments with rat livers the 3-hydroxybutyrate/acetoacetate ratio increased (Söling et al. 1968).

In summary we find during increased fatty acid oxidation in guinea pig liver in vivo and in vitro an increased reduction of cytoplasmic and — during more prolonged fatty acid oxidation — also of the mitochondrial redox states without any stimulation of gluconeogenesis.

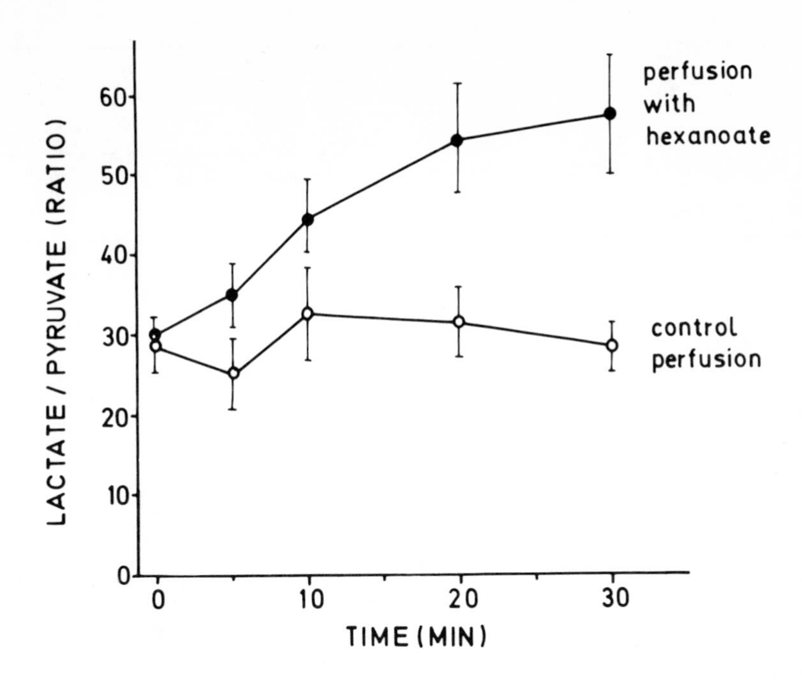

Fig. 1. Changes of the lactate/pyruvate ratio in the medium during perfusion of isolated guinea pig livers with and without intraportal infusion of 0.72 mmole/h of sodium hexanoate. Points and vertical bars represent the mean values and s.e.m.

	$NAD^+/NADH$ (L/P) (ratio)	$NAD^+/NADH$ (α-Glycp./DHAP.) (ratio)	$NAD^+/NADH$ (3-OHB/AcAc.) (ratio)
Chow-fed animals	333	1047	18.4
Fat-fed animals	216	199	8.0

Table 3. $NAD^+/NADH$ ratios calculated from the ratios of the substrate pairs lactate-pyruvate, α-glycerophosphate-dihydroxyacetonephosphate and 3-hydroxybutyrate-acetoacetate. For the lactate-pyruvate system and the α-glycerophosphate-dihydroxyacetonephosphate system the equilibrium constants given by Hohorst et al. (1959) were used. For the 3-hydroxybutyrate-acetoacetate system, the equilibrium constant given by Krebs et al. (1967) was used. All equilibrium constants were corrected for a temperature of 37 °C. For details, see Table 1.

	3-Hydroxybutyrate/acetoacetate (ratio)		
Duration of perfusion (min)	No hexanoate	0.72 mmoles/h hexanoate	2.16 mmoles/h hexanoate
	(n = 11)	(n = 12)	(n = 4)
0	0.58 ± 0.11	0.51 ± 0.08	0.43 ± 0.11
20	0.41 ± 0.08	0.48 ± 0.06	0.46 ± 0.08
30	0.45 ± 0.09	0.51 ± 0.06	0.48 ± 0.08

Table 4. The 3-hydroxybutyrate/acetoacetate ratio in the medium during perfusion of isolated guinea pig livers without and with intraportal infusion of 0.72 or 2.16 mmoles/h of sodium hexanoate. Results are ± s.e.m.

3-Hydroxybutyrate Dehydrogenase in Rat and Guinea Pig Livers				
	rat µmole/g/min	µmole/g/h	guinea pig µmole/g/min	µmole/g/h
V_{max} (D-3-OHB)	29.94	1796	2.65	159
V_{max} (AcAc)	12.80	768	0.86	56
	mM		mM	
K_M (D-3-OHB)	1.02		0.72	
K_M (AcAc)	0.29		0.18	

Table 5. The 3-hydroxybutyrate dehydrogenase in rat and guinea pig livers. The assay was carried out according to Lehninger et al. (1960) at 30^0 C. The reaction was started with substrate.

Validity of the Calculation of the Mitochondrial NAD^+/NADH System

The determination of the NAD^+/NADH ratio depends on a sufficiently high activity of the dehydrogenase to establish equilibrium (Krebs 1967). Its activity must be greater than that of the enzymes that cause the formation and removal of the substrate. As in guinea pig liver a low activity of the 3-hydroxybutyrate dehydrogenase has been reported by Lehninger et al. (1960) (only 7% of the rat liver activity), we had to exclude the possibility that the lower 3-hydroxybutyrate/acetoacetate ratio resulted only from a disequilibrium of the dehydrogenase reaction.

By re-examining the values of Lehninger et al. (1960), we found higher activities of the 3-hydroxybutyrate dehydrogenase activity, especially in guinea pig liver (Table 5). Possibly this is due to different strains of guinea pigs. The K_M value for D-3-hydroxybutyrate in our experiments was slightly higher than given by Lehninger et al. (1960) or Sekuzu et al. (1963), but were similar in rat liver and guinea pig liver. K_M values for acetoacetate as substrate were somewhat lower, but again did not differ significantly from rat to guinea pig liver (Table 5).

In case of a disequilibrium of the 3-hydroxybutyrate dehydrogenase reaction, one would expect the ratio 3-hydroxybutyrate/acetoacetate to become lower when the hexanoate infusion would be extended over a longer time. But when the intraportal infusion (Fig. 2) of 0.72 mmole/h of sodium hexanoate in isolated perfused guinea pig livers was extended from 60 to 180 minutes, the 3-hydroxybutyrate/acetoacetate ratio started to increase to about the same degree as in the isolated perfused rat liver, though on a lower level.

Moreover, when 0.430 mmole/h of sodium acetoacetate was infused intraportally into isolated perfused livers, a constant 3-hydroxybutyrate/acetoacetate ratio in the perfusate was reached in experiments with guinea pig livers as fast as in those with rat livers (Fig. 3). The ketone body concentrations in the liver were identical with those in the perfusate. A constant decrease of the 3-hydroxybutyrate/acetoacetate ratio would have been expected in case of limiting activity of the 3-hydroxybutyrate dehydrogenase.

From these experiments we assume that the 3-hydroxybutyrate dehydrogenase activity is not rate limiting and that their substrate pairs are in equilibrium with the mitochondrial NAD^+/NADH system.

Comparison of the Redox State of the Glutamate Dehydrogenase System with that of the 3-Hydroxybutyrate Dehydrogenase System.

In another attempt to assess the validity of the calculation of the redox state of the mitochondrial NAD^+/NADH system from the redox state of the 3-hydroxybutyrate dehydrogenase system, we determined the substrates of the glutamate dehydrogenase system. This system is located in the mitochondrial matrix, whereas the 3-hydroxybutyrate dehydrogenase is located in the cristae. According to Krebs (1967) both mitochondrial dehydrogenases share a common NAD^+/NADH pool. Thus corresponding NAD^+/NADH ratios should be expected. In livers of normal, well-fed guinea pigs, both mitochondrial redox systems are more oxidized than in rat liver and give the same values which

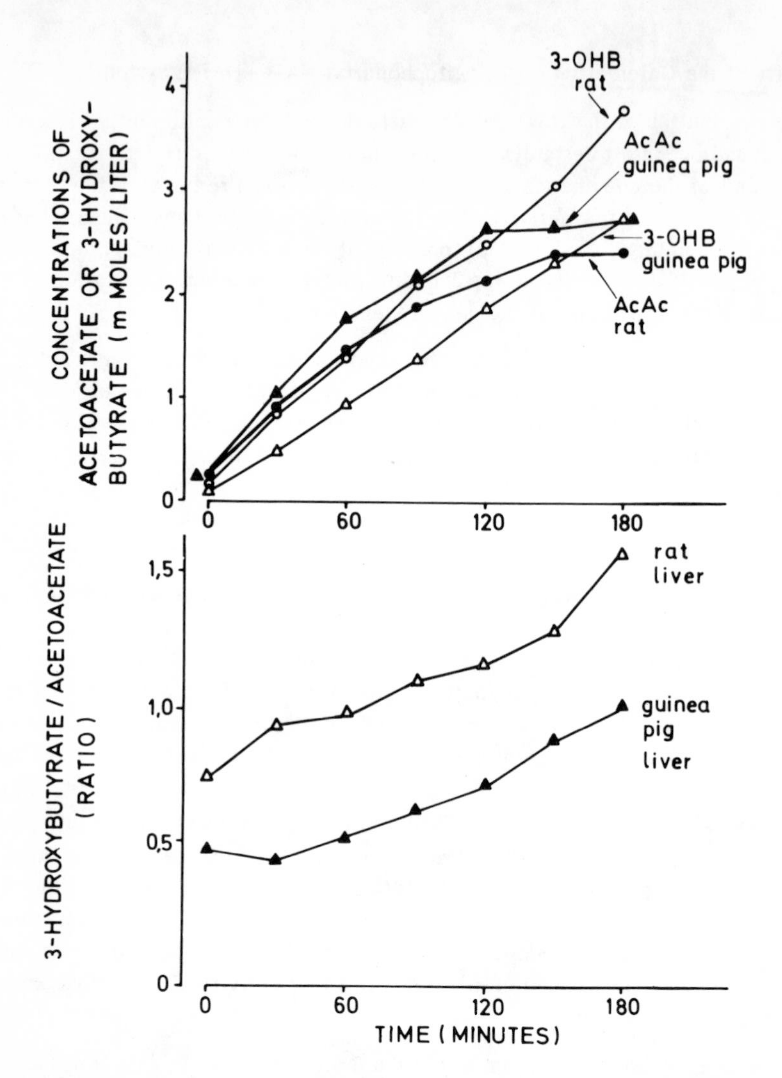

Fig. 2. Changes of the medium concentrations of acetoacetate and 3-hydroxybutyrate (upper part) and of the 3-hydroxybutyrate/acetoacetate ratio (lower part) during intraportal infusion of 0.72 mmole/h of sodium hexanoate in experiments with isolated perfused rat and guinea pig livers.

means that they are in equilibrium (Table 6). After fat feeding for 72 hours, both systems are more reduced, but again an equilibrium is observed. In the isolated perfused guinea pig liver both systems differ from the in vivo values: The 3-hydroxybutyrate dehydrogenase system is more oxidized, but the glutamate dehydrogenase

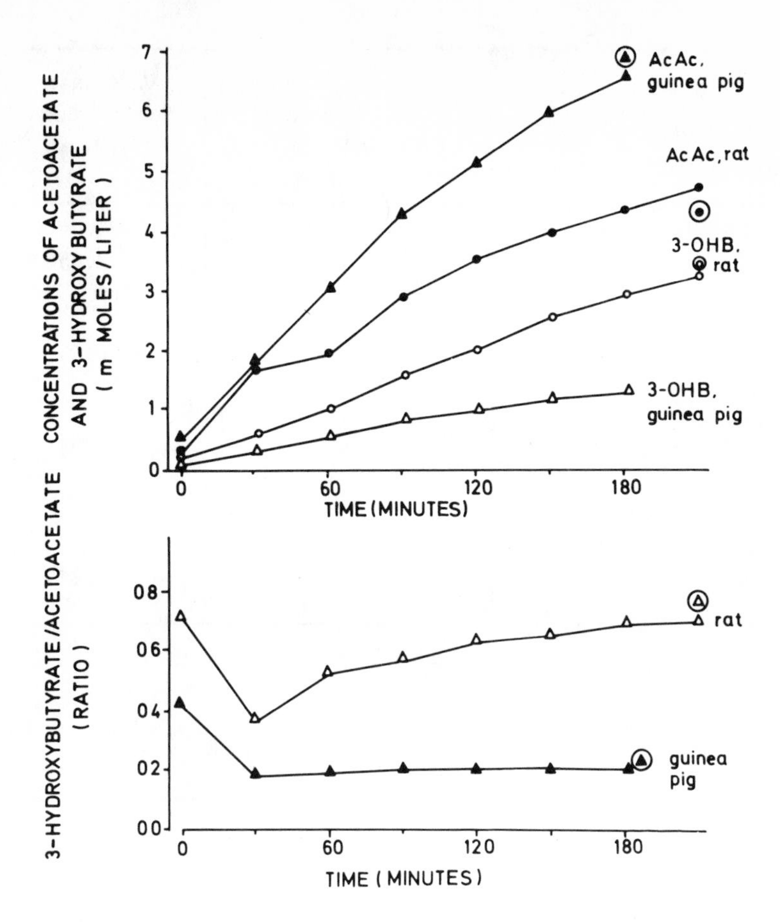

Fig. 3. Changes of the medium concentrations of acetoacetate and 3-hydroxybutyrate (upper part) and of the 3-hydroxybutyrate/acetoacetate ratio (lower part) during intraportal infusion of 0.43 mmole/h of sodium acetoacetate in experiments with isolated perfused rat and guinea pig livers. The encircled symbols correspond to the values measured at the end of the perfusion in the freeze-stopped liver tissue.

system is more reduced than in vivo, thus resulting in a disequilibrium of both systems. This disequilibrium is not influenced when fatty acid oxidation and ketogenesis are increased by infusing 0.72 mmole/h of sodium hexanoate. When 4 mM NH_4Cl is added to the perfusion medium 15 minutes prior to the freeze stop, the glutamate dehydrogenase system is considerably more oxidized than the 3-hydroxybutyrate dehydrogenase system.

From these experiments we conclude that the disequilibrium of the two mitochondrial $NAD^+/NADH$ systems is more due to a location in different mitochondrial compartments than to a limiting activity of the 3-hydroxybutyrate dehydrogenase.

	3-Hydroxybutyrate dehydrogenase system (cristae)	Glutamate dehydrogenase system (matrix)
Chow-fed in vivo	18.27	21.9
72 h fat fed in vivo	7.99	6.55
Chow-fed, control perfusion	42.94	6.80
Chow-fed, perfusion with hexanoate	41.40	6.84
Chow-fed, perfusion with hexanoate, addition of NH_4Cl	58.22	27.87

Table 6. Mitochondrial NAD^+/NADH ratios in guinea pig liver. The equilibrium constants given by Krebs (Krebs 1967, Krebs et al. 1962) were used.

	3-Hydroxybutyrate dehydrogenase system (cristae)	Glutamate dehydrogenase system (matrix)
Chow fed, in vivo	6.95	7.35
48 h starved, perfusion with 10 mM alanine	21.08	2.07

Table 7. Mitochondrial NAD^+/NADH ratios in rat liver. The equilibrium constants given by Krebs (Krebs 1967, Krebs et al. 1962) were used.

This idea is further supported by the fact that similar disequilibria occur in rat liver which has the highest activity of 3-hydroxybutyrate dehydrogenase (Table 7). In the well-fed rat in vivo we find the same NAD^+/NADH ratios as reported by Williamson et al. (1967), but in livers from starved rats perfused with 10 mM alanine the NAD^+/NADH ratios diverge in the same directions as in the isolated, perfused guinea pig liver.

Acknowledgments

This work was supported by the Deutsche Forschungsgemeinschaft. The authors are grateful to Mrs. G. Janson and H. Tabatowski for skilled and conscientious technical assistance.

References

Hohorst, H. J., F. Kreutz, Th. Bücher: Biochem. Z. 332 (1959) 18

Krebs, H. A., J. Mellanby, D. H. Williamson: Biochem. J. 82 (1962) 96

Krebs, H. A., In: Advances in Enzyme Regulation. Ed. by G. Weber, Pergamon, New York. Vol. 5. 1967, 409

Krebs, H. A., T. Gascoyne: Biochem. J. 108 (1968) 513

Lehninger, A. L., H. C. Sudduth, J. B. Wise: J. biol. Chem. 235 (1960) 2450

Menahan, L. A., B. D. Ross, O. Wieland, In: Stoffwechsel der isoliert perfundierten Leber. Ed. by W. Staib and R. Scholz. Springer, Berlin 1968, p. 142

Schimassek, H.: Biochem. Z. 336 (1963) 460

Sekuzu, I., P. Jurtschuk, D. E. Green: J. biol. Chem. 238 (1963) 975

Söling, H. D., B. Willms, D. Friedrichs, J. Kleineke: Europ. J. Biochem. 4 (1968) 364

Söling, H. D., B. Willms, J. Kleineke, this symposium

Struck, E., J. Ashmore, O. Wieland, In: Advances in Enzyme Regulation. Ed. by G. Weber, Pergamon, New York. Vol. 4. 1966, p. 219

Williamson, D. H., P. Lund, H. A. Krebs: Biochem. J. 103 (1967) 514

Williamson, J. R., R. A. Kreisberg, P. W. Felts: Proc. nat. Acad. Sci. (Wash.) 56 (1966) 247

Willms, B., H. D. Söling, In: Stoffwechsel der isoliert perfundierten Leber. Ed. by W. Staib and R. Scholz. Springer, Berlin 1968a, p. 118

Willms, B., M. Jellinghaus, J. Kleineke, D. Friedrichs, A. Janson, H. D. Söling, In: 14. Symposion Dtsch. Ges. Endokrin., Springer, Berlin 1968b, p. 268

Gluconeogenesis and Redox State

H. A. Krebs
Metabolic Research Laboratory, Nuffield Department of Clinical Medicine, Radcliffe Infirmary, Oxford, England

Summary

It has been known for some time that the redox state of the NAD-couple in the liver, as measured by the [lactate]/[pyruvate] ratio, moves in the direction of reduction when the rate of gluconeogenesis increases. It was not evident however whether the redox state is of regulatory importance; in other words, whether the change in the redox state is the primary factor causing an increased rate of gluconeogenesis or whether it is the consequence of gluconeogenesis. The evidence favors the latter alternative. The fact that maximum rates of gluconeogenesis from lactate or pyruvate occur over a wide range of [pyruvate]/[lactate] ratios, i.e. [free NAD]/[free $NADH_2$] ratios indicate that the cytoplasmic redox state is not rate-controlling. The primary cause for the greater reduction of the NAD-couple is the ATP requirement of gluconeogenesis (6 ATP for each glucose molecule formed from lactate). This results in a rapid conversion of ATP into ADP and Pi in the cytoplasm. The increased concentration of Pi shifts the equilibrium in the glyceraldehyde phosphate dehydrogenase system in the direction of reduction because in this system Pi is equivalent to a reductant:

$$\frac{[\mathrm{NAD}]}{[\mathrm{NADH_2}]} = K \times \frac{[\text{1, 3-Di-phosphoglycerate}]}{[\text{Glyceraldehyde-P}]\ [\mathrm{Pi}]}$$

Any change in the redox state of the glyceraldehyde-P dehydrogenase system is paralleled by changes of other cytoplasmic NAD-linked dehydrogenase systems which are near-equilibrium, such as the lactate dehydrogenase system.

The [lactate]/[pyruvate] ratio in the liver is known to rise in diabetes and in starvation (Hohorst et al. 1967, Wieland and Löffler 1963). This implies that the redox state of the cytoplasmic NAD couple moves in the direction of reduction. In both starvation and diabetes the rate of gluconeogenesis is higher than in the normal liver and this raises the question of whether the redox state of the NAD couple plays a role in the regulation of gluconeogenesis. In other words, is the change in the redox state the primary factor causing the increased rate of gluconeogenesis or is it a consequence of gluconeogenesis?

Experiments on the rate of gluconeogenesis in the perfused rat liver (Ross et al. 1967) show that the rates of glucose formation are almost identical when 10 mM lactate or when 10 mM pyruvate are added to the perfusion medium (1.06 and 1.02 μmol glucose synthesized min/g respectively). Addition of 10 mM lactate or pyruvate is bound to cause major changes in the lactate/pyruvate ratio within the tissue (although the redox state of the NAD couple may, under such conditions, not be directly proportional to the lactate/pyruvate ratio and the two substrates are liable to produce changes in opposite directions. For example, 30 minutes after addition of 10mM L-lactate to the perfusion medium, the ratio within the freeze-clamped liver was 12.1 and 20 minutes after addition of 10 mM pyruvate it was 0.9. Thus, maximum rates of gluconeogenesis can occur over a considerable range of cytoplasmic redox states. This suggests that the redox state does not directly regulate the rate of gluconeogenesis. Indirectly it may do so by controlling the pyruvate concentrations when these are within a low range. However this critical range does not obtain upon addition of pyruvate or lactate.

On the other hand, there is evidence that the change in the redox state of the pyridine nucleotides is the consequence of the raised rate of gluconeogenesis. Gluconeogenesis requires six molecules of ATP for one molecule of glucose and at least four of these react in the cytoplasm. The consumption of ATP causes a fall in the steady state concentration of ATP, as shown in Table 1. The fall in the concentration of ATP must be accompanied by an equivalent rise in the concentration of Pi. These changes in the phosphorylation state of the adenine nucleotide system necessarily causes changes in the redox state of the pyridine nucleotides because the two systems are interlinked by equilibrium reactions. These involve the glyceraldehyde phosphate dehydrogenase, 3-phosphoglycerate kinase and lactate dehydrogenase reactions. The activity of the three enzymes concerned is high enough to maintain near equilibrium in the liver under normal conditions. At equilibrium the following relation holds (Veech et al. 1970):

$$\frac{[\text{Pyruvate}]}{[\text{Lactate}]} = \frac{1}{K} \times \frac{[\text{3-P-G}]}{[\text{GAP}]} \times \frac{[\text{ATP}]}{[\text{ADP}][\text{Pi}]} \qquad (1)$$

where

$$K = \frac{K_{\text{GAP-dehydrogenase}} \times K_{\text{3-PG-kinase}}}{K_{\text{Lactate dehydrogenase}}}$$

The experimental demonstration that the three enzyme systems maintain near equilibrium is shown in Table 2. In these experiments the concentrations of ATP, ADP and Pi were measured directly in the freeze-clamped liver. At the same time, lactate, pyruvate, 3-phosphoglycerate and glyceraldehyde phosphate were determined so that the phosphorylation state of the adenine nucleotides could be calculated from equation (1). The agreement between the calculated and measured values must be regarded as satisfactory and this implies that the three enzyme systems were at near equilibrium. Exact agreement cannot be expected because the measurements of the adenine nucleotides and of Pi were made on the whole tissue whereas the required values are those for the cytoplasm. In view of the fact that the distribution of ATP, ADP and Pi between mitochondria and cytoplasm is not even, it is surprising that the discrepancies are not greater. This is probably connected with the fact that by far the greater part of the liver tissue is made up by the cytoplasm and the compartments which readily communicate with the cytoplasm (nuclei and outer mitochondrial space).

Substrates	None		10 mM L-Lactate	
Period of perfusion (min)	15	85	15	85
[ATP] [mM]	2.17	2.05	1.70	1.70
[ADP] [mM]	0.86	0.66	1.00	0.86
[ATP]/[ADP]	2.54	3.11	1.70	1.99

Table 1. Effect of Lactate on the Concentration of ATP and ADP in the Perfused Rat Liver. The values represent concentrations in the freeze-clamped perfused liver. For experimental details, see (Hems et al. 1966).

State of rat	Phosphorylation state of adenine nucleotides $\frac{[ATP]}{[ADP][Pi]}$	
	Calculated	Measured
Normal, well fed	648	551
Starved, 48 hrs.	199	282
High sucrose diet, 3 days	1,290	874
High glucose diet, 3 days	738	660
High-fat, low-carbohydrate diet, 3 days	612	427

Table 2. Measured and Calculated Values for the Phosphorylation State of Adenine Nucleotides in Rat Liver Cytoplasm.

The further analysis of the relations between the redox state of the pyridine nucleotides and the phosphorylation state of the adenine nucleotides indicates that the key substance responsible for the linking is Pi which is equivalent to a reductant in the glyceraldehyde phosphate dehydrogenase system. The equilibrium equation of the glyceraldehyde phosphate dehydrogenase system is defined by this relation:

$$\frac{[NAD]}{[NADH_2]} = K \times \frac{[1,3\text{-di-PG}]}{[GAP]\ [Pi]}$$

Thus increasing the concentration of Pi shifts the redox state of the NAD system in the direction of reduction. Combination of the equilibrium equations for the lactate and GAP dehydrogenases gives:

$$\frac{[\text{Lactate}]}{[\text{Pyruvate}]} = K \times \frac{[\text{GAP}]\,[\text{Pi}]}{[1,3\text{-di-PG}]}$$

This equation makes it clear why a rise of Pi increases the value of the [lactate]/[pyruvate] ratio and that the ratios of [reductant]/[oxidant] of redox couples at equilibrium are not necessarily the same if the electron donor and acceptor are only considered. Previous work (Hohorst et al. 1959, Bücher and Rüssmann 1963) has shown that under many conditions there is a constant relationship between the [lactate]/[pyruvate], [malate]/[oxaloacetate] and [α-GAP]/[DHAP] ratios. An exception however is the glyceraldehyde phosphate system because of the involvement of a third component, Pi.

Since the concentration of Pi depends on the phosphorylation state of the adenine nucleotide system it may be said that the phosphorylation state of the adenine nucleotides regulates the redox state of the cytoplasmic NAD couple. The present treatment makes it clear that this regulation is mediated by Pi and by the phosphate dependance of the redox state of the glyceraldehyde phosphate dehydrogenase system.

The direct experimental proof is hampered by the difficulty of determining the concentration of di-phosphoglycerate in tissues or tissue preparations. The difficulty arises from the exceedingly low concentration of this intermediate from the rapid spontaneous decomposition of di-phosphoglycerate. Its half life at pH 7.2 and 38^0C is 27 minutes (Negelein 1963).

Another experimental difficulty arises from the compartmentation of Pi. It is known that Pi tends to be unevenly distributed between cytoplasm and mitochondria. Its concentration can be much higher in the mitochondria than in the cytoplasm. There is no reliable procedure for determining (Pi) in the cytoplasm which is the value required for the present considerations. So the main experimental support for the time being remains the demonstration of the existence of approximate equilibrium between the adenine and pyridine nucleotide systems.

References

Bücher, Th., W. Rüssmann: Angew. Chem. 75 (1963) 881

Hems, R., B. D. Ross, M. N. Berry, H. A. Krebs: Biochem. J. 101 (1966) 284

Hohorst, H. J. F. H. Kreutz, Th. Bücher: Biochem. Z. 332 (1959) 18

Hohorst, H. J. F. H. Kreutz, M. Reim, H. J. Hübener: Biochem. biophs. Res. Commun. 4 (1961) 163

Negelein, E., In: Methods of Enzymatic Analysis. Ed. by H.-U. Bergmeyer. Verlag Chemie, Weinheim 1963, p. 234

Ross, B. D., R. Hems, H. A. Krebs: Biochem. J. 102 (1967) 942

Veech, R. L., L. Raijman, H. A. Krebs: Biochem. J. (1970) in press

Wieland, O. G. Löffler: Biochem. Z. 339 (1963) 204

Discussion to Willms, Kleineke and Söling, and Krebs

J. R. Williamson: I have a question to Dr. Krebs: Do I understand you right, that you are suggesting a regulatory relationship between the fall in the ATP/ADP ratio and an increase in gluconeogenesis?

Krebs: No, definitely not. I was merely concerned with the question why the redox state moves in the direction of reduction when gluconeogenesis occurs. This does not imply that the phosphorylation state of the adenine nucleotides directly controls gluconeogenesis. Indirectly the adenine nucleotides, of course, play a role by affecting the activities of FDP-ase and phosphofructokinase.

Utter: Did you include the two cases in which you started with pyruvate and lactate in the five states for which you performed calculations?

Krebs: No lactate or pyruvate was given to the rats because we intended to examine livers which were at near-equilibrium as far as possible. Giving lactate would upset the equilibrium.

Utter: I was wondering how far away from equilibrium these were.

Krebs: We have not carried out any experiments on the perfused liver, bearing on this question.

Wieland: I think this would also help to explain the stimulatory effect of fatty acids on gluconeogenesis from pyruvate and from lactate as well. In the case of pyruvate the role of fatty acid oxidation can be explained by providing the reducing equivalents for glucose synthesis, but this would hardly apply to lactate which would yield the hydrogen for gluconeogenesis itself. I wonder whether changes in the energy potential as discussed by Prof. Krebs could perhaps give a uniform explanation for the stimulatory effect of fatty acids. Another point: How would the ratio ATP/AMP + inorganic pyrophosphate fit in your equation?

Krebs: Concerning the stimulating effect of fatty acids on gluconeogenesis from lactate I think that the increased rate of supply of acetyl-CoA is an important factor. Since lactate is almost quantitatively converted to glucose it is not likely to be an effective source of acetyl-CoA. As for your second question, I have no information on the pyrophosphate content of the liver. It appears to be exceedingly low because the pyrophosphatase activity is high. We have measured AMP as well as ADP and ATP in order to calculate the mass action ratio in the adenylate kinase system. Like other workers, we found considerable variations, between 0.6 and 2 for the ratio $[ATP][AMP]/[ADP]^2$. This may be connected with compartmentation as adenylate kinase is not active within the mitochondrial matrix.

Wieland: As far as I remember, Dr. Williamson has measured AMP levels during increased fatty acid oxidation. Didn't you find a striking increase in the AMP level?

J. R. Williamson: Yes, it increases two-to three-fold. Before conclusions can be reached whether the fall of phosphate potential is the cause of a change of the redox state towards reduction in gluconeogenesis, it is important to establish the kinetics of the processes involved. Addition of lactate or pyruvate to substrate-free perfused livers produces an immediate reduction of the pyridine nucleotides: in both cytoplasmic and mitochondrial spaces with lactate, and in the mitochondrial space with only pyruvate as it is converted to acetyl-CoA by pyruvate dehydrogenase. I would say that the observed NAD-reduction represents a substrate effect on the dehydrogenases and is unrelated to the altered energy demands imposed by gluconeogenesis. I would like to reiterate a point concerning possible redox control at the glyceraldehyde-P dehydrogenase step. In my first studies with rat livers perfused with alanine, we found that the stimulation of gluconeogenesis induced by oleate was associated with a crossover at the glyceraldehyde-P dehydrogenase step (Williamson et al.: Proc. Nat. Acad. Sci. 56 (1966) 247). This information was incorrectly interpreted as indicating a control site at the dehydrogenase step due to the increased NADH/NAD ratio resulting from the enhanced rate of ß-oxidation. I have since become convinced, largely from isotope exchange data, that glyceraldehyde-P dehydrogenase is a non-equilibrium enzyme step, and as such is unlikely to function as a rate controlling site. In other words, it is unlikely that a change of cytoplasmic NADH/NAD ratio per se could directly result in a change of gluconeogenic flux by its interaction with glyceraldehyde-P dehydrogenase. The question arises, therefore, concerning the significance of a crossover when one is observed between 3PGA and GAP. A crossover always indicates the location of site of interaction in an enzyme sequence. It is immaterial whether or not the pathway contains branch points. However, the significance each crossover requires individual interpretation to determine its relationship to the cause of a flux change, as the system moves from one steady state to another. A useful initial approach is to analyze the crossover by the Fault Theorem (Williamson, in Red Cell Metabolism and Function, G. J. Brewer (ed.), Adv. in Exp. Med. and Biol. 6, Plenum Press, (1970) 117). By this means it is possible to ascertain whether the crossover of the carbon intermediates is consistent with changes of the other reactants and products, if the reaction is bimolecular, or whether it is caused by an activation or inhibition not involved in the particular enzyme reaction. Special considerations relating to the tissue concentration of intermediates, K_m and K_i values have to be taken into account when the intermediate acts as both a substrate or product and an inhibitor or activator. With regard to the glyceraldehyde-P dehydrogenase reaction, it is evident that an increased NADH/NAD ratio in the cytoplasm will tend to decrease levels of 1,3PGA relative to those of GAP, possibly causing a crossover. In the absence of any large changes in the ATP/ADP ratio, and assuming near-equilibrium for the enzymes involved, this redox change will result in a relative decrease of intermediates up to PEP and increase of intermediates up to FDP, although the increase of GAP, DAP and FDP may be attenuated by the NADH/NAD interaction at α-glycerophosphate dehydrogenase. However, since pyruvate carboxylase and possibly also PEP-carboxykinase are essentially irreversible enzymes under physiological conditions, it is clear that the fall of PEP cannot directly influence flux through pyruvate carboxylase, and therefore the rate of substrate input into the sequence. In fact, because of the redox interaction at lactate dehydrogenase, flux through pyruvate carboxylase may be expected to decrease as a result of the fall of pyruvate if no other factors change to affect its activity, as pointed out by Prof. Krebs with regard to the effect of ethanol on gluconeogenesis.

However, it is apparent that a number of feedback interactions can occur as a result of the redox change which affect the net rate of PEP formation. Thus, the fall of PEP would be expected to decrease pyruvate kinase flux, while the rise of FDP would have the opposite effect. Additionally, as shown by Dr. Utter, an increase of the NADH/NAD ratio would be expected to decrease the acetoacetyl-CoA concentration, thereby resulting in an increased flux through pyruvate carboxylase. Other feedback interactions resulting from changes of the adenine nucleotides are possible, which further complicate the picture, but these are generally of less importance since the concentration of adenine nucleotides in the cytoplasm is usually much greater than their K_m_ and K_i_ effects on the enzymes involved. It follows, therefore, that a considerable amount of information, in addition to knowledge of changes of the major gluconeogenic intermediates, is needed for an interpretation of crossover points, particularly in relation to attempts to locate rate-controlling sites in the gluconeogenic sequence. In summary, a crossover point locates sites of interaction or feedback sites, but does not necessarily give direct information concerning which step was initially rate-limiting for the enzyme sequence, nor which step was primarily responsible for the flux change during a metabolic perturbation. To return to my starting point, namely, the effect of oleate on livers perfused with alanine, it is now clear that the major factor determining the increased rate of glucose production is the rise of acetyl-CoA which increases flux through pyruvate carboxylase. The net effect of this activation, however, is modified by a number of factors which include feedback effects to pyruvate kinase and the control of anion fluxes across the mitochondrial membrane (Williamson et al.: J. Biol. Chem. 244 (1969) 4607, 4617, 5044).

Krebs: There now seems to be a good deal of agreement between us regarding control at the glyceraldehyde phosphate dehydrogenase step. If the triosephosphate dehydrogenase system is at near-equilibrium as our recent work indicates (R. L. Veech, L. Raijman, H. A. Krebs: Biochem. J. 117 (1970) 499), then the normal tendency is just towards equilibrium, but there is no controlling directive concerning the flux rate. The crossover theorem has of course proved very valuable in analyzing events in the respiratory chain which is an essentially irreversible process even though under special conditions several steps can be reversed. But then the respiratory chain is in many ways simpler than glycolysis and gluconeogenesis. All steps in the respiratory chain are oxidoreductions. Glycolysis and gluconeogenesis include many types of reactions (oxidoreductions, isomerizations, phosphorylations and dephosphorylations, condensation and fission), and moreover there are branching points.

Haeckel: I have a question to Prof. Krebs: What were your concentrations of lactate and pyruvate in the medium? Was it saturated?

Krebs: The initial concentrations of lactate or pyruvate in the medium were 10 mM. In the case of lactate the perfusion was continued for 30 minutes and in the case of pyruvate for ten minutes before the liver was freeze-clamped.

Haeckel: In the guinea pig liver, I suppose, we have a different situation. If you add lactate or pyruvate in a concentration of 10 mM, you don't get the same rate of gluconeogenesis. With pyruvate gluconeogenesis is only 70% of that with lactate. If you determine all the metabolites you get a crossover between 3PGA and GAP. The 3PGA is increased by about 400%. It is a very marked crossover. In the presence of ethanol this crossover disappears, and then you get the same rate of gluconeogenesis with pyruvate and lactate . So I think in the case of the guinea pig liver we can explain these changes only by alterations of the NAD/NADH ratio. (Biochemistry 7 (1968) 3803).

Krebs: Before I can comment I have to study the data in detail. Comparative investigations are certainly very interesting. You refer to 3-hydroxybutyrate dehydrogenase in guinea pig liver. The activity of this enzyme is relatively low. Is it sufficient to establish equilibrium? As for Dr. Williamson's question the rates of gluconeogenesis are not seriously affected by very large variations in the hepatic (lactate)/(pyruvate) ratios.

Söling: I have a question about Dr. Williamson's concept, his present concept. You just mentioned that you cannot calculate from a crossover that the flux is increased at this step, and I agree with this, but then I have a somewhat naive question: You have an increased net synthesis of glucose, and you are not able to conclude that this takes place at the point of crossover. What would you propose to do to find the step where the increase in flux was started?

J. R. Williamson: Dr. J. Higgins at the Johnson Research Foundation was responsible for my interest in crossover theorems, and their use in the analysis of glycolytic or gluconeogenic flux changes. Unfortunately, his full treatment of multi-enzyme sequences modeled after these pathways has not yet been published. These are sophisticated theoretical treatments based on generalized rate laws and generalized rate equations, which so far are not completely applicable to the multi-enzyme sequences of glycolysis and gluconeogenesis. However, the Fault Theorem already referred to is applicable, and this theorem is what is actually used implicitly or explicitly by most people when attempts are made to interpret relative changes of metabolic intermediates in a sequence. In general, however, much less information is obtained by comparing two steady state conditions which differ in flux, than the transition from one steady state to other. Only in the transition state does the concept of a rate controlling step have any meaning, since in the steady state, the flux through each enzyme step is the same. I will attempt to answer your question briefly. If the flux through a metabolic pathway increases, this must result in a decrease in concentration of the metabolite feeding into the pathway and an increase in the concentration of the end product. Substrate and product changes, therefore, provide a basic crossover, which must be a result of imposing an external stimulus on the system (e.g. addition of fatty acids, a hormone or an inhibitor), which directly or indirectly affects flux through a particular enzyme step or permeability barrier. The purpose of further measurements of changes in metabolite levels is to identify specific sites of interaction and control sites. An important aspect of crossover theory is the fact that for most biochemical sequences, a crossover point does not necessarily identify a control site, but a control site must appear as a crossover. For technical reasons (e.g.

difficulty of measuring mitochondrial oxalacetate) a crossover site may be missed, particularly if it is immediately followed by a reverse crossover (i. e. change from an accumulation to depletion of intermediates for a flux increase). Therefore, the analytical data available in each experiment need to be evaluated in relation to the known topology of the sequence. Thus, the kinetic and allosteric properties of individual enzymes, the existence of feedback or feedforward loops, flux changes in branched chains and possible compartmentation of intermediates, have all to be taken into account in data interpretation. Neglect of some of these considerations is largely responsible for the fact that many workers tend to misinterpret data of metabolite changes in tissues. However, it is clear that more detailed knowledge of the intimate functioning of metabolic control can only be achieved from measurements of cellular constituents.

D. H. Williamson:

Concerning Dr. John Williamson's comments on control at the triose phosphate dehydrogenase step, I though it was now accepted, at least since the excellent review of Newsholme and Gevers (Newsholme, E. A. and Gevers, W.: Vitamins & Hormones, 25 (1967) 1) that an equilibrium step is unlikely to be a control step. I would like to ask Dr. Willms two points: Do you have any idea why the guinea pig liver mitochondrial compartment appears to be more oxidized than that of the rat liver?

Willms: I have no idea why.

D. H. Williamson: Have you measured total NAD/NADH in the whole liver?

Willms: No, we haven't measured it so far.

D. H. Williamson: Dr. R. A. Hawkins in Prof. Kreb's laboratory has done some preliminary experiments which suggest that the total NAD/NADH ratio is higher in guinea pig liver than in rat liver and this may have some bearing on your findings.

Fröhlich: I have a question to Prof. Krebs: Did you measure the ß-hydroxybutyrate/acetoacetate ratio in these experiments? Did it change as the lactate/pyruvate ratio?

Krebs: In these experiments we did not measure the [ß-hydroxybutyrate]/[acetoacetate] ratio, but we did this in many other experiments. In short-term experiments the redox state of the mitochondria does not necessarily change in the same way as that of the cytoplasm, but over a longer period of time changes in the cytoplasm are transmitted to the mitochondria. As J. R. Williamson, R. Scholz, E. T. Browning, R. G. Thurman and M. H. Fukami have shown (J. biol. Chem. 244 (1969) 5044) this occurs for example after dosing with ethanol.

Utter: Could we go back to the first part of your talk where you concluded from the same rates of gluconeogenesis from pyruvate and lactate that the redox state was not a control factor? This is certainly the simplest interpretation but I wonder if an alternate explanation could be offered. For example, if gluconeogenesis is limited by the available concentration of pyruvate, increasing the concentration of that substrate would speed the reaction. However, if at the same time a redox-mediated control operated in the other direction, that is, a decrease in the NADH/NAD ratio caused a decrease in the rate of gluconeogenesis, the net effect of increasing pyruvate concentration might be no change in rate. Therefore, a sort of balance of two opposing controls might exist. The only relatively simple question I can ask about this concerns the pyruvate levels in the experiments starting with lactate. In this case do you think the pyruvate levels are high enough to be saturating?

Krebs: What you suggest is certainly a possibility.

Utter: I wonder whether you do not really need a buffering system to keep the process going at the same rate?

Krebs: It is a striking phenomenon that the rate of glucose formation from pyruvate and lactate, at different concentrations, and different concentration ratios of the substrates, is constant. This suggests that the steps which limit the rate of glucose formation under the test conditions do not depend on the concentration of lactate or pyruvate, or on the redox state of the NAD couple.

Exton: I want to make a comment on Dr. Krebs'talk. I certainly agree with Dr. Krebs that the cytoplasmic redox state is unlikely to control per se, except under extreme circumstances. But to show what happens to intermediate levels in extreme circumstances I am showing you the concentrations of gluconeogenic intermediates in livers perfused with 10 mM pyruvate expressed as percentages of those in livers perfused with 10 mM lactate. The interesting observation is that all the redox changes seem to balance out, resulting in levels of dihydrooxyacetone phosphate, fructose-1,6-diphosphate, glucose-6-phosphate and glucose which are the same under the two circumstances. There are very large differences in lactate and pyruvate. We don't have oxalacetate, but presumably it was increased. There is a very large effect at the triosephosphate dehydrogenase step which causes an enormous increase in 3-phosphoglycerate. The dihydroxyacetone phosphate level is the same, showing that all these changes in the equilibrium reactions finally adjust to give the same rate of gluconeogenesis.

One thing we have been puzzled by is the fact that we find initially a very large lactate/pyruvate ratio in livers perfused with lactate which gradually returns to a normal value of about 10/1, but we have never been able to find changes in the ATP/ADP ratio. We expected to find these in the earlier stages of perfusion where the lactate/pyruvate ratios were so greatly displaced. The measurements were made at 20 minutes. If you analyze after one hour, the patterns with lactate and pyruvate are very similar.

Krebs: You did not maintain 10 mM pyruvate?

Exton: No, this was the initial concentration.

Krebs: It declines very quickly. Within 20 minutes. 10 mM pyruvate will come down to 1 or 2 mM.

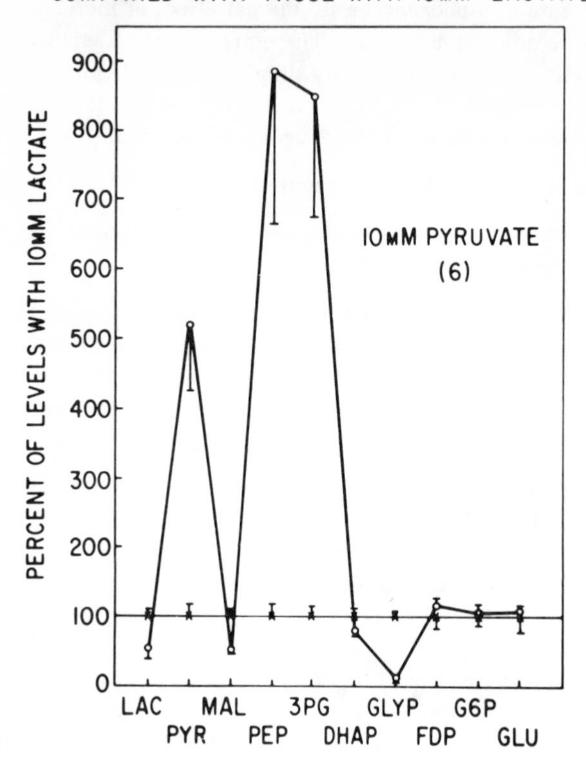

Slide shown by Dr. Exton.

J. R. Williamson: I would like to come back to a point made earlier by Prof. Krebs; namely, whether some of the phenomena observed in tissues are due to a balance of controls. This is undoubtedly so, and our problem is to sort out their nature and relative strengths. A problem that has interested me for some time is the question whether the rate of delivery of reducing equivalents from mitochondria to cytosol can provide a control step for gluconeogenesis. In a broader context, can the rate of anion transport across the mitochondrial membrane control metabolic pathways? Possibly, it may be important to distinguish this factor as a rate-controlling step or as a control responsible for adjusting metabolite levels and their concentrations in the mitochondrial and cytoplasmic spaces. With lactate as substrate, reducing equivalents are generated directly in the cytoplasm but when pyruvate is the principal substrate for gluconeogenesis in the perfused liver, reducing equivalents are thought to be carried to the cytoplasm as malate. We have recently suggested that a high malate efflux from the mitochondria is only possible if there is a large malate concentration gradient (Williamson et al.: J. biol. Chem. 244 (1969) 4617). Addition of fatty acids to the liver apparently increases this malate gradient as a result of a stimulation of pyruvate carboxylase and an increased state of reduction of the mitochondrial pyridine nucleotides. In comparing the results of liver perfusions using rats and guinea pigs, I wonder if some of the differences observed, especially the lack of oleate effect with guinea pig livers, can be explained by differences in the requirements for anion transport resulting from the mitochondrial location of PEP-carboxykinase in the guinea pig.

Söling: I want to ask Prof. Krebs: Did you ever do experiments where you added to your perfusion medium high amounts of lactate and then measured the ATP/ADP ratio at different times after the addition of lactate? Under these conditions you still have a linear relationship between time and rate of gluconeogenesis, but your lactate/pyruvate ratio changes very rapidly, and if the systems are in equilibrium, you should get a change in the ATP/ADP ratio during the change of the lactate/pyruvate ratio. Is this the case?

Krebs: Experiments bearing on your question were reported by us in Biochem. J. 101, 291 (Table 7). The concentrations of ADP and ATP remained fairly constant during perfusion. The equilibrium studies were not carried out on the perfused liver, but on the freeze-clamped organ removed as quickly as possible from the animal after death (Veech, Raijman, Krebs: Biochem. J. 117 (1970) 499).

Söling: That means it was balanced by the inorganic phosphate?

Krebs: In our earlier perfusion experiments we did not carry out all the necessary determinations because at that time the theoretical concepts were not sufficiently developed to indicate which intermediates must be measured.

Hanson: I have a question for Dr. D.H. Williamson: You were talking about malate gradients but in one of your recent papers (Williamson et al.: Biochem. J. 114 (1969) 575) you calculate the intramitochondrial oxaloacetate concentration based on the assumption that malate is equally distributed in the mitochondria and the cytosol. The approach is quite different from that of Dr. J. R. Williamson who calculates the distribution of malate and oxaloacetate based on redox couplets. I wonder if you would care to discuss this difference?

D. H. Williamson: I think that we stated this was pure hypothesis.

Krebs: I must emphsize that we calculated the redox state on the assumption that metabolite determinations in the whole liver reflect their concentration in the cytoplasm. The agreement between the results obtained suggests that this assumption is justified, even though there may be major concentration differences between mitochondria and cytoplasm, for example in respect to oxoglutarate, glutamate and malate, but I doubt whether these occur in vivo.

To return to the question of whether the supply of reducing equivalents may be rate limiting, this would seem unlikely when lactate is the precursor. With pyruvate the reducing equivalents have to be provided by the mitochondria and in this connection it is relevant that the rate of gluconeogenesis from pyruvate in rat kidney is very much higher, by a factor of two or three, than gluconeogenesis from lactate (Biochem. J. 86 (1963) 22; 102 (1967) 275). This indicates that the potential flow of reducing equivalents into the cytoplasm is very high, and the fact that in the liver the rate of gluconeogenesis from pyruvate and lactate is the same would suggest that the transport of reducing equivalents is not a rate limiting factor.

J. R. Williamson: Just to illustrate the possibility that one might find a control by mitochondrial permease systems is the fact that when lactate is the substrate, the carbon precursor has to be carried out of the mitochondria as aspartate, while with pyruvate as substrate, malate is transported. So I would interprete Prof. Krebs' data as indicating that aspartate transport in the kidney has a much lower capacity than malate transport. Much work has been done recently in isolated mitochondria on the control of anions transport. Of course, one does not know how much of this work is relevant to the whole tissue. Certainly isolated mitochondria can maintain very high concentration gradients across the mitochondrial membrane. The method of calculation of mitochondrial and cytoplasmic oxaloacetate is basically very simple: One has four unknowns; two are the total concentrations of malate and oxaloacetate in the tissue. Moreover one has the coupling in the cytoplasm between the lactate dehydrogenase and the malate dehydrogenase systems, so that one has another equation based on the equilibrium constants of LDH and MDH and measured lactate/pyruvate ratios. And a similar phenomenon occurs in the mitochondria where the 3-hydroxybutyrate/acetoacetate ratio is coupled to the malate/oxaloacetate ratio in the mitochondria. So one can solve the four equations with the four unknowns for calculating the distribution. This is merely a method of calculation, but the conclusions are valid. It is only with pyruvate perfusion that you do have this high mitochondrial malate, and this just fits in with a necessity for a high malate transport. With other substrates you have a more or less equal concentration of malate.

D. H. Williamson: Have you made your calculations on the data of Lardy and his group (Ray, P. D., D. O. Foster, H. A. Lardy: J. biol. Chem. 241 (1966) 3904) where there is a large accumulation of hepatic malate after tryptophan injection?

J. R. Williamson: No, we have done the calculations from your data in vivo. They agree with the perfused liver. Thus we come to the conclusion that α-ketoglutarate is largely cytoplasmic, and glutamate and aspartate largely within the mitochondria. There is a possibility, therefore, that the apparent similarity of the ratios of free NAD/ free NADH calculated from the glutamate dehydrogenase system and those calculated from the 3-hydroxybutyrate dehydrogenase system are based on coincidence resulting from the opposite distributions of α-ketoglutarate and glutamate. Under many conditions, the imbalances of distributions may be cancelled from the equations. One has to find experiments where they are in real disagreement, as you already have.

D. H. Williamson: I agree that the apparent similarity between the ratios of free NAD/free NADH calculated from the two systems could be coincidental, but we do not think this is so. However, we are now examining as many situations as possible to check this point. Expecially situations where the flux rates are changing rapidly and already we have shown that the agreement holds in short-term ischemia (Bronsnan, J. T., H. A. Krebs, D. H. Williamson: Biochem J. 117 (1970) 91).

On the Inhibition of Gluconeogenesis by 1-β-Phenyl-Ethylbiguanide in the Perfused Guinea Pig Liver

R. Haeckel and H. Haeckel
Medizinische Universitätsklinik Heidelberg, FRG, and Institute for Enzyme Research, University of Wisconsin, Madison, U.S.A.

Summary

Altschuld and Kruger (1967) reported that gluconeogenesis from lactate was inhibited by phenethylbiguanide (DBI) in the perfused guinea pig liver. The authors attributed this effect to a concomitant suppression of the ATP/ADP ratio. We found that glucose formation from pyruvate, alanine and fructose was also lowered by biguanides in the isolated perfused guinea pig liver. Under our experimental conditions 2×10^{-5} M DBI reduced gluconeogenesis from lactate about 20%.

The pattern of hepatic metabolites indicated a crossover phenomenon between 3-phosphoglycerate and glyerinaldehyde-phosphate. The pyruvate metabolism was also affected, however, no clear-cut results were obtained with regard to pyruvate carboxylation.

The glutamate/2-oxoglutarate x NH_4 ratio rose almost 300% in the presence of 2×10^{-5} M DBI, whereas the ATP/ADP-, lactate/pyruvate- and 3-hydroxybutyrate/acetoacetate ratio were not changed. We therefore assume that small variations of the intramitochondrial redox state are probably not reflected by the lactate/pyruvate- and the 3-hydroxybutyrate/acetoacetate ratio in the perfused guinea pig liver.

As the hepatic citrate and 2-oxoglutarate concentrations were decreased by DBI, the activity of the citric acid cycle seemed to be affected. Whether the crossover phenomenon observed in the gluconeogenic pathway can be attributed to an inhibitory effect of the biguanides on cell respiration could not be decided from the metabolite pattern. A decrease of the ATP/ADP ratio was not the cause of the crossover phenomenon, however, it may have played an additive role if it occurred with increasing doses of the biguanides.

The influence of the biguanides on hepatic glucose output was dependent on the time the animals had been starved prior to the liver perfusion and on the hydrogen concentration in the perfusion medium. Alkalinization of the perfusate increased the action of DBI.

Introduction

Investigations from different laboratories have shown that the biguanides usually accumulate in the liver (Wick et al. 1960, Beckman 1965) and decrease hepatic glucose output (Nielsen et al. 1958, Beringer 1958). Since Hollunger (1955) discovered the inhibition of cell respiration by guanidine, there has been a long controversy whether this effect also causes the decrease of hepatic gluconeogenesis. Inhibition of mitochondrial respiration could reduce the hepatic ATP/ADP ratio and the AMP concentration which could influence the pyruvate carboxylase (Walter et al. 1966) and the fructosediphosphatase reactions (Patrick 1966). An increase of the intra-mitochondrial NADH/NAD ratio as another consequence expected from a lowered respiration rate could diminish the acetyl-coenzyme A level, an allosteric activator of the pyruvate carboxylation to oxaloacetate.

In the last several years some criticism against this hypothesis as a possible in vivo mechanism arose from the fact that usually in diabetic patients without concomitant renal insufficiency the lactate/pyruvate ratio is not elevated during the treatment with biguanides (Tranquada et al. 1968).

This argumentation persumes that the blood lactate/pyruvate ratio always reflects the mitochondrial redox state. However, Williamson et al. (1967) have shown that this is not the case in rat liver. We started to reinvestigate this problem at the isolated perfused liver and tried to find out the following:

1) Are there crossover phenomenons in the hepatic metabolite concentrations at reactions which are in equilibrium with the adenosine phosphates?
2) Does the inhibition of gluconeogenesis by DBI correlate with a decrease of the ATP/ADP ratio and is it possible to reduce the DBI concentration to a dosis which is still suppressing glucose formation without affecting the hepatic adenosine phosphate concentrations?
3) Does the lactate/pyruvate ratio correlate with variations of mitochondrial redox couples and, therefore, can it be used as a sensitive indicator for the mitochondrial respiration state in the isolated perfused guinea pig liver?

As the biguanides did not reduce gluconeogenesis in the perfused rat liver (Altschuld and Kruger 1967, Söling 1967), we used the isolated guinea pig liver which seemed to be more sensitive to these compounds with regard to gluconeogenesis.

Our modified perfusion procedure was that of Miller et al. (1951) as described by Haeckel and Haeckel (1968). The perfusion medium contained 3 g% bovine albumin, 1 mg of sodium ampicillin and bovine erythrocytes, washed three times and taken up in Krebs-Ringer bicarbonate buffer (aerated with 95% oxygen and 5% CO_2). The hemoglobin concentration was 5 g%. Substrates and effectors of gluconeogenesis were added to the perfusion medium as indicated in Fig. 1. The animals were starved 24 or 48 hours before the experiments since we had found that under this condition the action of the biguanides was more pronounced. This relationship is demonstrated in Fig. 2 with fructose as substrate.

Results and Discussion

In livers from animals fasted 24 hours 0.08 mM DBI decreased gluconeogenesis from pyruvate by about 30% (Fig. 3). In the presence of ethanol (10 mM) the action of DBI was more marked under our experimental conditions. DBI inhibited the glucose formation from both pyruvate and lactate by about 50% when ethanol was added to the perfusate. The conversion of pyruvate to glucose was stimulated whereas the glucose formation from lactate was impaired by ethanol as previously reported by

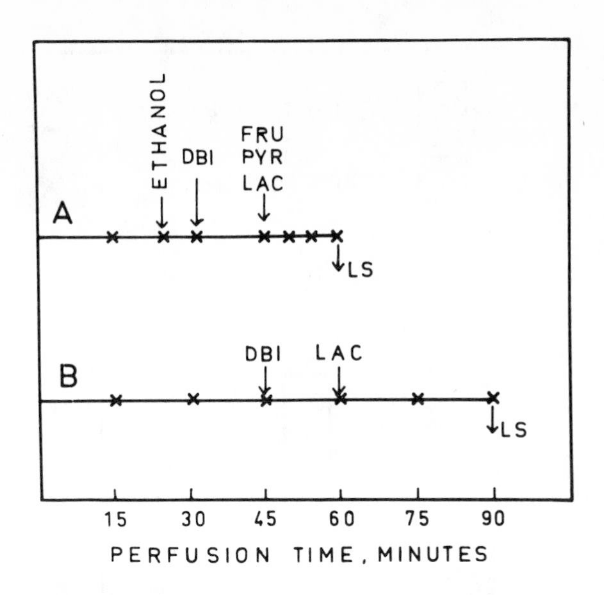

Fig. 1. Time schedule of the perfusion experiments. Liver samples (LS) were taken at the end of the perfusion by Wollenberger's method (1958). Blood samples for glucose determinations were withdrawn as marked by an "X". FRU = fructose, PYR = pyruvate, LAC = lactate.

Haeckel and Haeckel (1968). In the perfused guinea pig liver ethanol did not alter the intrahepatic redox potential as extremely as has been found in perfused rat livers (Kreisberg 1967, Williamson 1969). In the presence of alanine, ethanol increased the lactate/pyruvate ratio only from 5.9 to 8.8 (Haeckel and Haeckel 1968)

From the pattern of hepatic metabolite concentrations (Fig. 4), no distinct cross-over phenomenon could be found with DBI when pyruvate was used as substrate. The inhibition of glucose formation appeared to have occurred very early in the gluconeogenetic pathway. In the presence of ethanol phosphoenolpyruvate, 2-phosphoglycerate and 3-phosphoglycerate appeared to be elevated, whereas the triosephosphates were lowered, however, without mathematical significance. With lactate instead of pyruvate this crossover phenomenon between 3-phosphoglycerate and glycerinaldehydephosphate was more marked (Fig. 5). The pyruvate carboxylation could also have been reduced since malate, a metabolite usually considered to take part in the gluconeogenetic pathway, was not elevated as expected from the increased mitochondrial redox state (as will be shown later) and the elevated phosphoenolpyruvate concentration.

The increased 3-phosphoglycerate/glycerinaldehydephosphate ratio indicated that the phosphoglyceratekinase or the glycerinaldehydephosphatedehydrogenase reaction was inhibited by the guanidine compound. DBI concentrations up to 8 mM did not effect the conversion of 3-phosphoglycerate to glycerinaldehydephosphate, if added

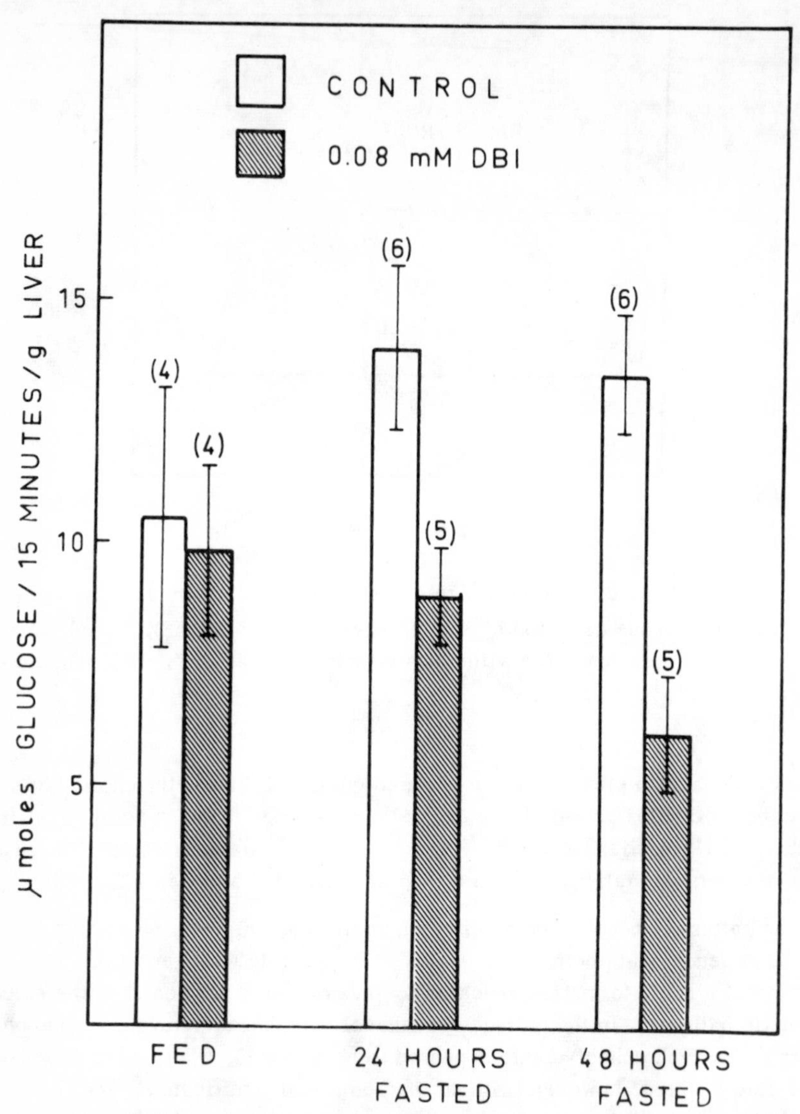

Fig. 2. Glucose production from fructose (10 mM) by perfused livers from guinea pigs starved at different times, in the absence and presence of the same dosis of DBI (0.08 mM). The vertical bars represent standard deviations. The number of contributing values are given in parentheses.

to a NADH-coupled assay system using the 100,000 g x h supernatant fraction of liver homogenate, 0.5 - 6.0 mM $MgSO_4$ and 40 mM triethanolamine buffer (pH 7.4). Therefore, we assume that the biguanides affected these reactions indirectly. This could be achieved by an alteration of the ATP/ADP ratio since the phosphoglyceratekinase activity is influenced by this ratio (Parker and Hoffman 1967).

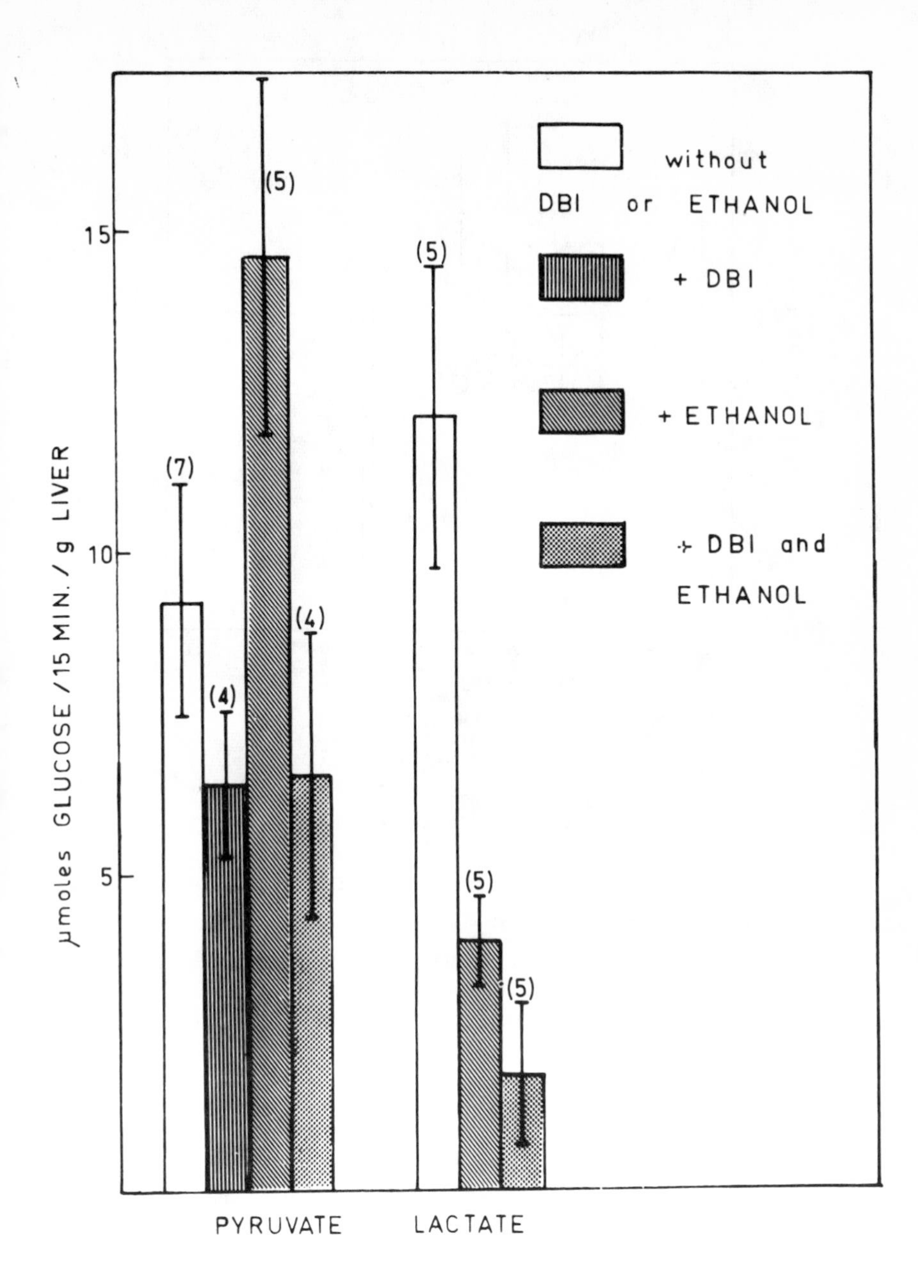

Fig. 3. The influence of 10 mM ethanol and 0.08 mM DBI on the glucose formation from lactate and pyruvate. For further explanation, see legend to Fig. 3 or Scheme A in Fig. 1.

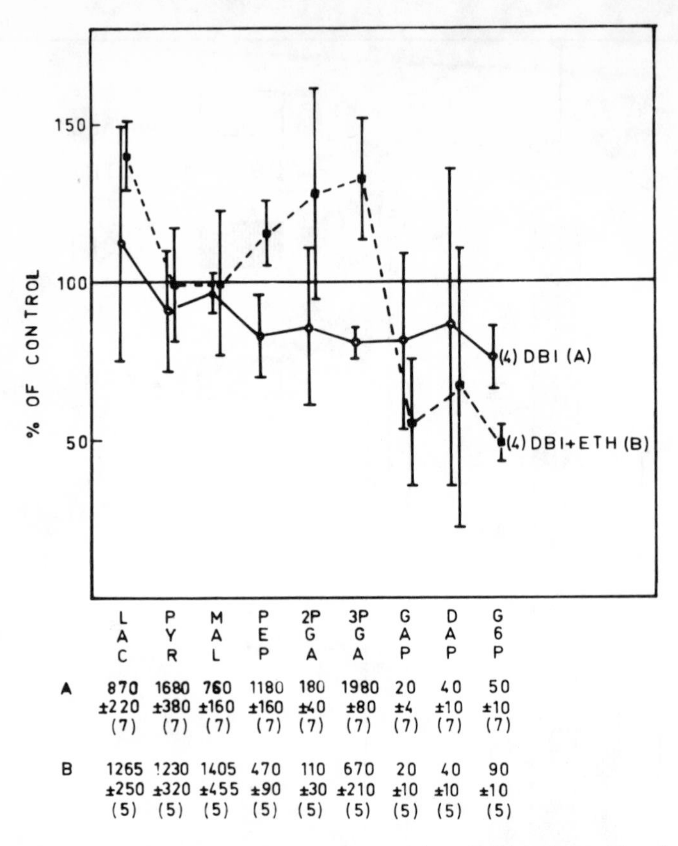

Fig. 4. . The effect of 0.08 mM DBI alone and in combination with ethanol (curve B) on the pattern of hepatic metabolites in the presence of pyruvate (scheme A in Fig. 1.) In the control experiments to curve A, only pyruvate was applied. In the controls for curve B both pyruvate and ethanol were added to the perfusion medium. The control values on the bottom of the figure are given in nmoles with standard deviations and the number of the contributing experiments in parentheses.

Table 1 shows that under the conditions of the experiments just reported, the ATP/ADP ratio was decreased by 0.08 mM DBI. The 3-hydroxybutyrate/acetoacetate ratio significantly rose in the presence of 0.08 mM DBI only with lactate as substrate (Table 2). It should be mentioned that this ratio was 0.4 with both pyruvate and lactate. That means no equilibrium between the lactate/pyruvate- and the 3-hydroxybutyrate/acetoacetate ratio was established during these experiments. DBI had no effect on the total concentration of the ketone bodies.

From the data on the adenosine phosphate level (Table 1), a causal correlation between a primary suppression of the ATP/ADP ratio and a secondary inhibition of gluconeogenesis by the biguanides could be assumed (Hollunger 1955, Altshuld and Kruger 1967, Jangaard 1968). An increase of the 3-phosphoglycerate/glycerinaldehydephosphate ratio as a consequence of a lowered ATP/ADP ratio would fit into this hypothesis. However, experimental conditions were found under which this interrelationship could not be demonstrated.

In livers from 48 hours starved animals, even 0.02 mM DBI decreased the glucose formation from lactate (Fig. 6) by about 20% if the pH value of the perfusion medium was kept between 7.40 and 7.45 with the addition of sodium bicarbonate.

The ATP/ADP ratio, the lactate/pyruvate- and the 3-hydroxybutyrate/acetoacetate ratio were not altered by 0.02 mM DBI (Table 3), whereas the ratio coupled with the glutamatedehydrogenase reaction was elevated almost 300%. According to Williamson et al. (1967), the last ratio is in equilibrium with the intramitochondrial NAD^+/NADH system. Since the glutamatedehydrogenase system is located in the matrix and the 3-hydroxybutyratedehydrogenase system in the cristae of the mitochondria, perhaps they do not share a common NAD^+/NADH system.

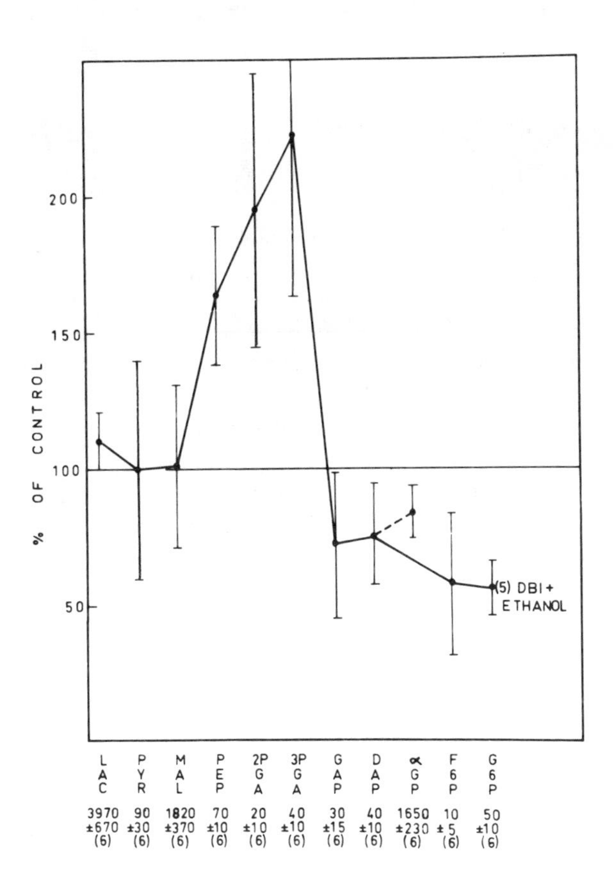

Fig. 5. The influence of 0.08 mM DBI on the pattern of hepatic metabolites in the presence of lactate and ethanol. For further explanation of symbols, see legend to Fig. 4.

Substrate DBI, 0.08 mM Ethanol, 10 mM	PYR - -	PYR + -	PYR - +	PYR + +	LAC - -	LAC - +	LAC + +
ATP	2650	2225	2490	2190	2580	2560	2145
	± 190	± 160	± 170	± 90	± 315	± 190	± 150
	(7)	(4)	(6)	(4)	(6)	(6)	(5)
ADP	570	610	580	905	650	650	840
	± 60	± 85	± 120	± 170	± 60	± 115	± 110
	(7)	(4)	(6)	(4)	(6)	(6)	(5)
ATP/ADP	4.6	3.7	4.3	2.4	4.0	4.0	2.5
	± 0.8	± 0.5	± 0.6	± 0.5	± 0.8	± 0.6	± 0.3
	(7)	(4)	(6)	(4)	(6)	(6)	(5)
Citrate	860	540	830	415	590	780	485
	± 180	± 80	± 100	± 90	± 60	± 110	± 40
	(7)	(4)	(6)	(4)	(6)	(6)	(5)

Table 1. Influence of DBI and Ethanol on the Hepatic Concentrations of Citrate and Adenosine Phosphates. All Concentrations are given in nmoles/g wet weight with standard deviations and the number of contributing values in parentheses.

The 3-phosphoglycerate and the 2-phosphoglycerate concentrations (Table 3) were slightly but significantly ($P < 0.05$) elevated, although the ATP/ADP ratio was not changed. Both citrate and 2-oxoglutarate were lowered which led to the conclusion that the activity of the citric acid cycle was decreased even by 0.02 mM DBI.

We assume that the mitochondrial respiration was inhibited by the biguanide leading to an intramitochondrial accumulation of free NADH as indicated by the increased glutamate/ 2-oxoglutarate x NH_4 ratio. However, under these conditions the reduction of the mitochondrial respiration was probably not strong enough to cause a response of the intracellular ATP/ADP ratio. Therefore, we suggest that the biguandes do not suppress gluconeogenesis primarily by reducing the ATP/ADP ratio.

In our system the lactate/pyruvate- and the 3-hydroxybutyrate/acetoacetate ratios appeared to be unsuitable as sensible indicators for the intramitochondrial redox state. We suggest that also in diabetic patients an unchanged lactate/pyruvate ratio should not be used alone to prove that the biguanides do not affect intracellular respiration.

Substrate	PYR	PYR	PYR	PYR	LAC	LAC	LAC
DBI, 0.08 mM	-	+	-	+	-	-	+
Ethanol	-	-	+	+	-	+	+
3-Hydroxybutyrate	430	450	460	635	440	680	750
	± 90	± 100	± 140	± 100	± 110	± 110	± 225
	(7)	(4)	(5)	(4)	(6)	(6)	(5)
Acetoacetate	1070	1110	980	880	1110	1070	750
	± 260	± 250	± 310	± 240	± 290	± 300	± 135
	(7)	(4)	(5)	(4)	(6)	(6)	(5)
3-Hydroxybut./ Acetoacetate	0.40 ± 0.06	0.41 ± 0.07	0.47 ± 0.05	0.72 ± 0.09	0.40 ± 0.08	0.64 ± 0.22	1.01 ± 0.19
	(7)	(4)	(5)	(4)	(6)	(6)	(5)
Ketone Bodies	1490	1570	1430	1510	1530	1750	1500
	± 330	± 300	± 440	± 340	± 370	± 340	± 325
	(7)	(4)	(5)	(4)	(6)	(6)	(5)

Table 2. Influence of DBI and ethanol on the hepatic concentrations of ketone bodies. All concentrations are given in nmoles/g wet weight with standard deviations and the number of contributing values in parentheses. These data are from the experiments illustrated in Fig. 5,6.

In the last experiments the guinea pig liver usually took up potassium ions from the medium (Fig. 7) during a perfusion period of two hours — quite contrary to the rat liver which releases potassium under similiar conditions according to several authors (Creutzfeldt et al. 1967, Exton and Parks 1967). DBI did not significantly influence this influx of potassium ions as long as the ATP/ADP ratio was not changed. These results will be reported in detail by Haeckel et al. (in press).

In agreement with experiments reported by Jangaard et al. (1968), DBI suppressed the pyruvate oxidation as indicated by a decreased citrate concentration. Some evidence was presented that the pyruvate carboxylation was also influenced by the biguanides. In the presence of lower concentrations of pyruvate, as they are received when lactate or ethanol in combination with various substrates were used, a crossover phenomenon was observed between 3-phosphoglycerate and glycerinaldehydephosphate. So, two steps of the gluconeogenetic pathway appeared to be affected by the biguanide:

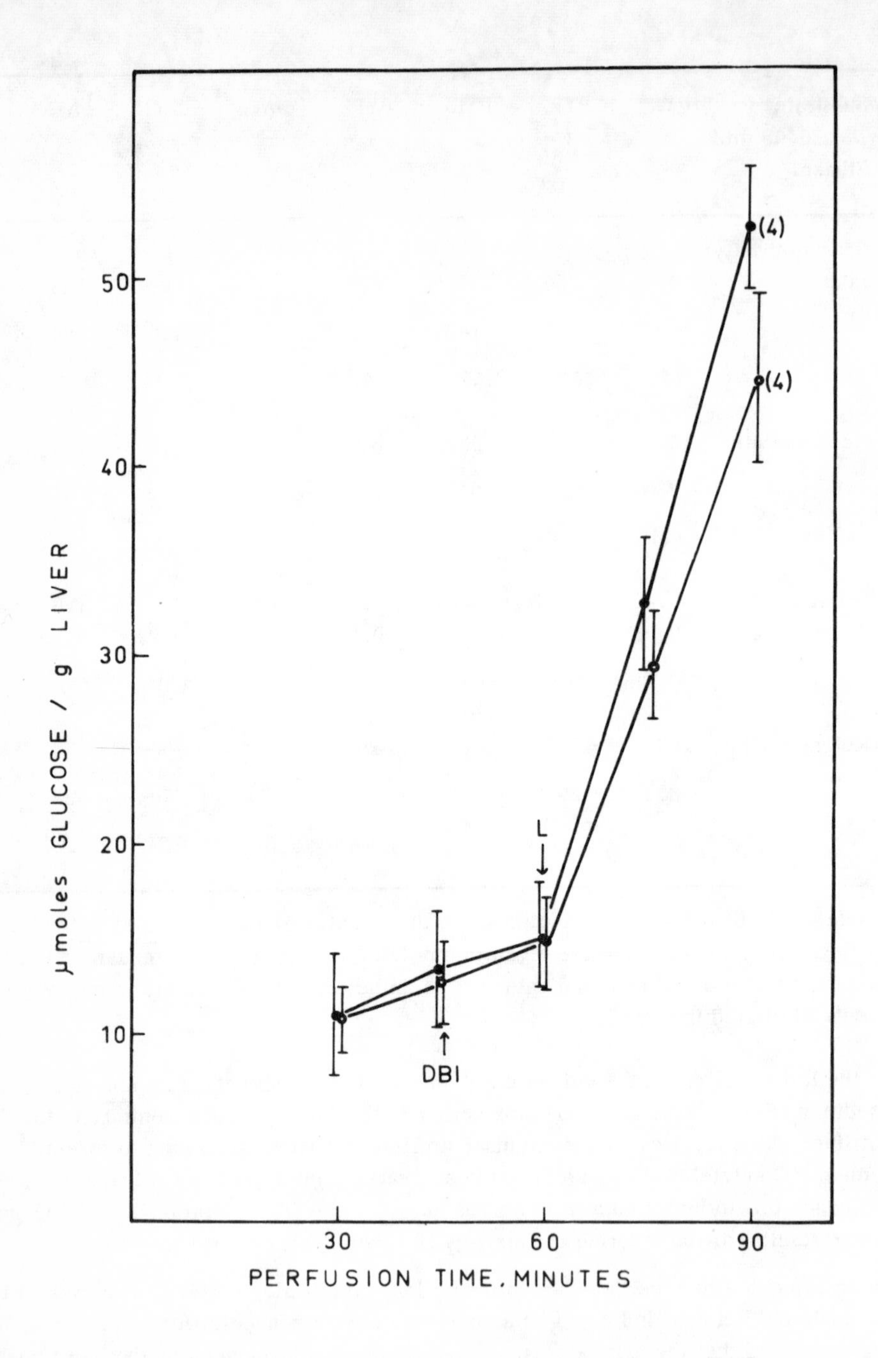

Fig. 6. The influence of 0.02 mM DBI on glucose formation from lactate as substrate. The experimental conditions are described by Scheme B in Fig. 1. The livers were from animals starved for 48 hours.

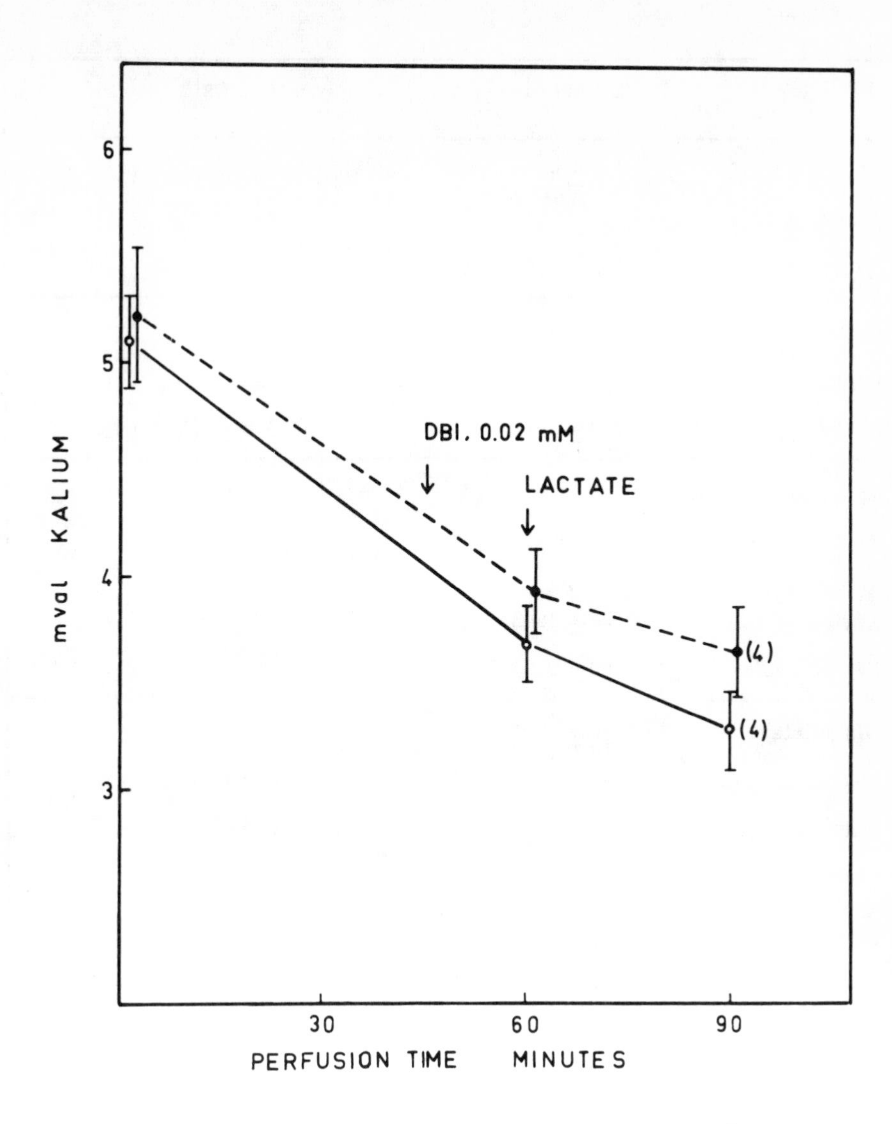

Fig. 7. The potassium concentration of the medium during the perfusion. Erythrocytes were centrifugated prior to the determination of the ions by flame photometry. These data are from the experiments which are represented in Fig. 7.

the conversion of pyruvate to malate and the conversion of 3-phosphoglycerate to glycerinaldehydephosphate.

	Control		DBI	
ATP	2,810 ± 490	(5)	2,580 ± 330	(4)
ADP	870 ± 260	(5)	830 ± 70	(4)
ATP/ADP	3.20 ± 0.35	(5)	3.15 ± 0.25	(4)
Lactate	2,370 ± 950	(4)	2,130 ± 602	(4)
Pyruvate	140 ± 60	(4)	130 ± 40	(4)
LAC /PYR	17.3 ± 5.1	(4)	16.8 ± 1.0	(4)
3-Hydroxybutyrate	260 ± 35	(4)	260 ± 85	(4)
Acetoacetate	1,260 ± 290	(4)	1,090 ± 170	(4)
3-Hydroxybut./ Acetoacetate	0.21 ± 0.03	(4)	0.23 ± 0.06	(4)
Ketone Bodies	1,520 ± 320		1,340 ± 250	(4)
Glutamate	270 ± 69	(4)	240 ± 60	(4)
2-Oxoglutarate	1,120 ± 390	(4)	500 ± 110	(4)
Ammonium	1,070 ± 280	(4)	815 ± 310	(4)
Glutamate/ 2-Oxoglutarate x NH_4	2.2 ± 1.0	(4)	6.0 ± 1.6	(4)
Citrate	510 ± 110	(4)	350 ± 50	(4)
Malate	400 ± 130	(4)	400 ± 130	(4)
PEP	200 ± 120	(4)	270 ± 50	(4)
2-PGA	20 ± 10	(4)	55 ± 4	(4)
3-PGA	400 ± 100	(4)	590 ± 120	(4)

Table 3. The Influence of 0.02 mM DBI on the concentrations of some hepatic metabolites. Liver samples were taken for analysis after 90 minutes of perfusion (Scheme B, Fig. 1). All metabolite concentrations are given in nmoles/g wet weight with standard deviations and the number of contributing values in parentheses.

Acknowledgment

We would like to thank Prof. Dr. H. A. Lardy for encouraging this work and for his very helpful suggestions and discussions. Financial support was offered by the Deutsche Forschungsgemeinschaft.

References

Altschuld, R. A., F. A. Kruger, New York Acad. Sci. Symposium: Diabetes, Obesity and Phenformin as a Research Tool. February 27, 28, March 1, 1967

Beckmann, R.: Drug Research 15 (1965) 761

Beringer, A., K. Hupka, K. Mösslacher, K. Moser, R. Wenger: Wien. med. Wschr. 108 (1958) 639

Creutzfeldt, W., E. Skutella, D. Moshagen, P. Kneer, H. D. Söling: Diabetologia 3 (1967) 9

Exton, J. H. C. R. Parks: J. biol. Chem. 242 (1967) 2622

Haeckel, R., H. Haeckel: Biochem. 7 (1968) 3803

Haeckel, R., H. Haeckel, M. Anderer, in press

Hollunger, G.: Acta pharm. Toxicol. 11, suppl. 1 (1955) 1

Jangaard, N. O., J. N. Pereira, R. Pinson: Diabetes 17 (1968) 96

Kreisberg, R. A.: Diabetes 16 (1967) 784

Miller, L. L., C. G. Bly, M. L. Watson, W. F. Bale: J. exp. Med. 94 (1951) 431

Nielsen, R. L., H. E. Swanson, D. C. Tanner, R. H. Williams, M. O'Connell: Arch. intern. Med. 101 (1958) 211

Parker, J. C., J. F. Hoffmann: J. gen. Physiol. 50 (1967) 893

Patrick, S. J.: Canad. J. Biochem. 44 (1966) 27

Söling, H. D., D. Moshagen, E. Skutella, P. Kneer, W. Creutzfeldt: Diabetologia 3 (1967) 318Tranquada, R. E., Ch. R. Kleeman, J. S. Brown: Diabetes 17 (1968) 96

Walter, P., V. Paetkau, H. A. Lardy: J. biol. Chem. 241 (1966) 2523

Williamson, D. H., P. Lund, H. A. Krebs: Biochem. J. 103 (1967) 514

Williamson, J. R., R. Scholz, E. T. Browning, R. G. Thurman, M. H. Fukami: J. biol. Chem. 244 (1969) 5044

Wick, A. N., Ch. J. Stewart, G. S. Serif: Diabetes 9 (1960) 163

Wollenberger, A., E. G. Krause, B. E. Mahler: Naturwissenschaften 45 (1958) 294

Announced Discussion

Effects of Biguanides on Gluconeogenesis in Rats and Guinea Pigs

H. Stork and F. H. Schmidt
Forschungslaboratorien der Boehringer Mannheim GmbH., Mannheim

In guinea pigs doses of 10 mg/kg lower blood-glucose. In rabbits and rats no effect or only at very high doses could be demonstrated. Söling has already shown in guinea pigs that the fall in blood-glucose is accompanied by an increase in blood-lactate. Our own experiments demonstrated that 10 mg/kg Phenformin (i.p.) in guinea pigs caused a nearly statistical significant fall in blood-glucose and a diminished concentration of liver-glycogen. The increase of lactate at this dose is in contrast highly significant. In our experiments with isolated kidney and liver cells the effects of pretreatment with Phenformin on gluconeogenesis and pyruvate-consumption in normal rats, alloxandiabetic rats and guinea pigs were studied (Fig. 1a and 1b). In normal rats, no influence of Phenformin on glucose-formation could be seen. In contrast to this a strong inhibition of gluconeogenesis could be demonstrated in the tissue of guinea pigs (Fig. 1b). There are also differences in liver and kidneys in pretreated alloxan-diabetic rats; the glucose-formation rate is higher than in normal animals. The effects observed in the guinea pigs could be related to the compartmentation of the gluconeogenetic key enzymes which differs from that of the rat.

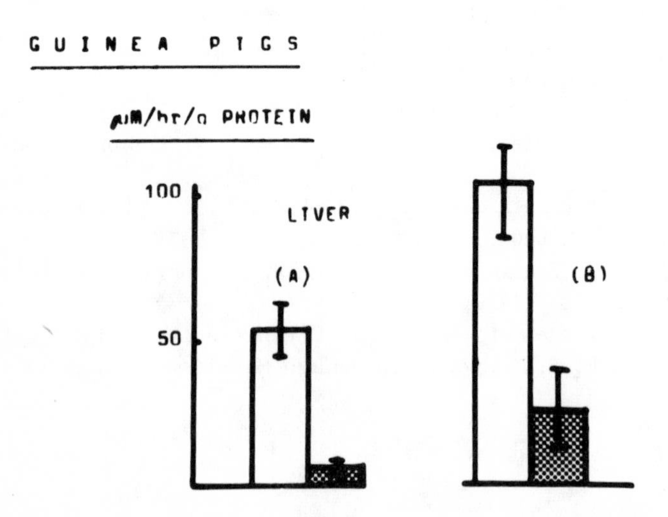

Fig. 1a. Net formation of glucose by isolated liver cells from guinea pigs. White columns: controls, black columns: 20 mg/kg Phenformin i.p. (A) or 10 mg/kg Phenformin i.p. for 10 days (B).

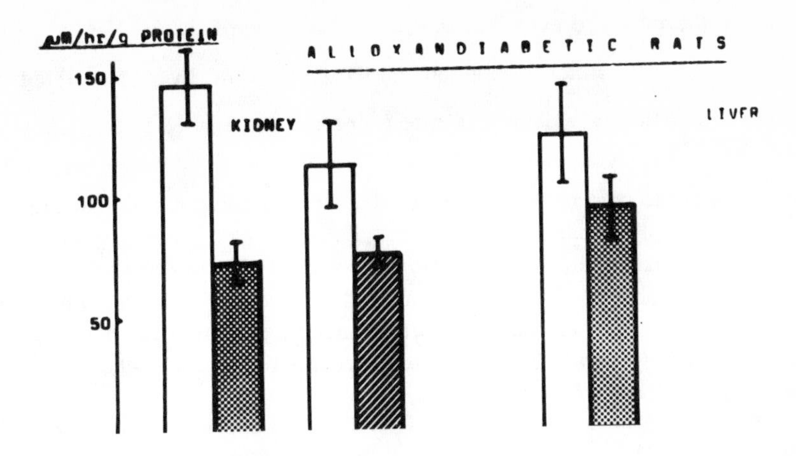

Fig. 1b. Net formation of glucose (left column pair) and net utilization on pyruvate (right column pair) by kidney cells, and net formation of glucose by liver cells from untreated (white columns) and Phenformin treated (dark columns) alloxandiabetic rats.

Discussion to Haeckel and Haeckel

Söling: I want to make a comment concerning the effect of biguanides in rat and guinea pig livers. There are biguanide derivatives available — at least we had the opportunity to test one of those derivatives — which act very strongly also on gluconeogenesis in the rat liver. The second point refers to that what Dr. Krebs reported this afternoon: Did you ever check for changes in the concentrations of inorganic phosphate during your experiments, because this would be very interesting with respect to the changes in the concentrations of 3-phosphoglycerate and glyceraldehydphosphate you observed in your crossover studies.

Haeckel: We measured the concentration of inorganic phosphate in the liver and we found an increase of the phosphate concentration only when the ATP/ADP ratio was decreased. If the ATP/ADP ratio was unaltered the concentration of inorganic phosphate remained constant.

Söling: Was there any relationship between crossover phenomena and the concentration of inorganic phosphate?

Haeckel: I have data on inorganic phosphate only from the first experiments, where the ATP/ADP ratio was decreased and the concentration of inorganic phosphate was increased. I have no data on inorganic phosphate concentrations from the last experiments in which the ATP/ADP ratio was not changed, and 3-P-GA elevated.

Krebs: Dr. Woods at Oxford has also carried out experiments with biguanides. At 1 or 2 mM phenethylbiguanide caused a production of lactate in the well-fed perfused rat liver, but this concentration is much higher than those encountered clinically. Is there a similar effect in the guinea pig liver?

Haeckel: I did not make experiments with the whole animal.

Krebs: Yes, our experiments were done on the perfused rat liver of well-fed rats.

Haeckel: We do not have data of lactate formation by perfused livers from well-fed guinea pigs.

Schäfer: I want to comment on what Dr. Haeckel has pointed out, that the ATP/ADP ratio has not any significance in respect to the action of biguanides. Can you differentiate in perfusion experiments by estimating ATP/ADP ratios between the pools of ATP in the mitochondria and in the cytoplasm? There might well be some change of ATP concentration in the mitochondrial compartment which you could not detect at this low concentration of biguanides. It is first within the mitochondria that ATP interacts with gluconeogenesis.
The other point I want to make is that biguanides, or in general guanidine derivatives mainly are taken up by the mitochondria in form of cations. They do not merely accumulate in the liver, they especially seem to accumulate within the mitochondria (G. Schäfer, D. Bojanowski, unpublished observations (1970)). At the membrane these compounds compete with cations for translocation (B. C. Pressman, J. K. Park: BBRC 11 (1963) 182; F. Davidoff: Diabetes 18, Suppl. 1 (1969) 331). The exchange of cations is strongly related to the exchange of anions through the mitochondrial membrane according to Palmieri (E. Quagliariello, F. Palmieri, The energy level and metabolic control in mitochondria; S. Papa, J. M. Tager, E. Quagliariello, E. C. Slater (Eds.) Adriatica Editrice, Bari (1969) 45) and Pressman (B. C. Pressman, E. Quagliariello, F. Palmieri (1969) 87). Since in gluconeogenesis we have a highly active shuttling of substrates across the mitochondrial membrane, it should be considered whether there is a diminished rate of provision of reducing equivalents for gluconeogenesis. Actually, I am not yet convinced that the ATP/ADP level within the mitochondria is not changed under the influence of biguanides during rat liver perfusion, because the mitochondrial redox state is known to be changed, as revealed with isolated liver mitochondria (G. Schäfer: BBA 93 (1964) 279; B. C. Pressman in B. Chance Energy linked functions of mitochondria, Acad. Press, New York (1963) 181).

Haeckel: First, I want to point out that I do not want to turn down the established meaning that the ATP/ADP ratio has no influence on gluconeogenesis. I also think if you add biguanides in high concentrations to the liver, then you get a decrease of the ATP/ADP ratio, which might have an effect on gluconeogenesis. If you raise the concentration of biguanides up to 0.2 mM, then gluconeogenesis is almost completely blocked in the perfused liver, but this has no great physiological significance. We were interested in the effects of very low concentrations of biguanides. We wanted to see what happens first: The decrease of the ATP/ADP ratio or the inhibition of gluconeogenesis during the influence of the biguanides. Concerning your question I do not know whether the measurement is exact enough to make final statements on the ATP/ADP ratios within the mitochondria. But when the increase of the 3-PGA/GAP ratio would correlate with a change of the ATP/ADP ratio in the cytosol, I think, this change would be measurable.

Wieland: Dr. Haeckel, if I understood you correctly, you gave an explanation for the effect of biguanides on gluconeogenesis and on pyruvate metabolism by two mechanisms: First, inhibition of pyruvate oxidation, and second: inhibition of pyruvate carboxylation. I may have missed it on your slides, but if this was true, one would expect highly accumulating amounts of pyruvate or lactate under these conditions. Did you see this?

Haeckel: Yes, that is true. However there must be an early block, especially in the experiments with pyruvate.

Wieland: I just refer to your crossover plots, and there was obviously no increase of pyruvate. This puzzles me: Where does pyruvate go, if it can neither be oxidized nor be transformed into glucose? It must remain somewhere!

Haeckel: Another explanation would be that the uptake of pyruvate by the liver cell could be inhibited by the biguanides.

Wieland: Perhaps one could get more insight in this by isotopic experiments with labeled pyruvate, studying the oxidation rate. Has anyone done this?

Haeckel: Yes, Jangaard et al. (Diabetes 17, (1968) 96), had done this and he found an increase of pyruvate in the guinea pig liver homogenate. But I do not know why pyruvate did not accumulate under our experimental conditions.

Söling: I think we are dealing here with very different ATP/ADP ratios. We started this morning with Dr. Ballard's ratio of about 7, then that of Dr. Krebs with 3.11 and your values are about 4.5. Therefore I think there are still lots of experimental problems in the determination of ATP/ADP ratios. Therefore we are not really able to correlate the ATP/ADP ratio to rate-controlling steps.

Papenberg: Are these compounds, the biguanides, metabolized or converted?

Haeckel: Yes, they are hydroxylated. The guinea pig liver which is more sensitive to biguanides does less hydroxylate the biguanides than the rat liver. The rat liver has a higher rate of hydroxylation. After addition of the hydroxylated compound to the perfused liver, no effect on the carbohydrate metabolism was observed.

Hanson: I would like to ask Dr. Söling what his thinking is concerning the differences in cytosol redox state between the rat and guinea pig? These differences are so marked and you did not discuss in your paper how this might affect the metabolism of the liver. Also, could the more highly reduced cytosol of guinea pig liver be due to technical reasons such as anoxia during the freeze-clamping? Dr. D. H. Williamson may also care to comment.

Söling: I do not think that the differences in the redox state are due to technical reasons. The values are obtained in both species under the same experimental conditions. The lactate/pyruvate ratios were in the rat about 10 to 11, and in the guinea pig about 20 to 25. The ATP/ADP ratio, as Dr. Willms has shown, was in the order of 5.4 in the guinea pig, and this ratio is usually in the rat between 3 and 4. The lactate dehydrogenase activity is about four times lower in the guinea pig liver than in the rat liver, but this is still sufficient to establish an equilibrium under normal conditions. Triosephosphate dehydrogenase activity is also much lower in the guinea pig liver than in the rat liver. But both these differences are in my opionion not responsible for the differences in redox state. I would like very much to hear comments from the audience whether these differences in redox state could be correlated to the differences in regulation of gluconeogenesis.

Krebs: What is the maximal rate of gluconeogenesis under optimal conditions in the guinea pig liver? In starved rat liver we find a rate of about 1 μmole/g/min glucose formed when no oleate has been added.

Haeckel: I got the same value for lactate, but for pyruvate the rate of gluconeogenesis is only 70% that of lactate.

Söling: We get higher values: The rate of gluconeogenesis from lactate is 1.5 μmoles/g/min in the guinea pig liver. The rate of gluconeogenesis from pyruvate is somewhat low in both species, but the difference is greater in the guinea pig liver than in the rat liver.

Haeckel: The glucose formation rate from fructose is only 50% that of rat liver and I think you could confirm this!

Söling: Yes, this results from the low activity of fructokinase.

Haeckel: I have a question to Dr. Stork: You have shown that 10 mg/kg body weight of phenformin increase the lactate/pyruvate ratio. If I remember the paper of Dr. Söling and his co-workers (Arch. Exp. Path. Pharmak. 244 (1963) 290) correctly, they did not get a change of the lactate/pyruvate ratio with this dosis, they had to apply higher amounts of phenformin to get an increase of the lactate/pyruvate ratio. Ten mg/kg lowered the blood glucose concentration but did not change the lactate/pyruvate ratio. Can you explain this difference?

Stork: As it was shown in Fig. 1, the increase in blood lactate is highly significant after 10 mg/kg Phenformin. The increase in the blood-pyruvate concentration by contrast is not. In accordance with Söling's findings we also do not see a statistically significant increase in the lactate-pyruvate ratio.

Regulation of Gluconeogenesis with Ethanol and Fructose by the Isolated Perfused Rat Liver

Joachim Papenberg
Medizinische Universitäts-Klinik, Heidelberg, FRG

Summary

1) Ethanol and fructose metabolism of isolated perfused livers from fed rats is discussed. In four groups of experiments the livers are perfused with no additions to the perfusate and with loads of ethanol, fructose, and ethanol-fructose. The perfusate concentrations of ethanol, fructose, glucose, lactate, pyruvate, beta-hydroxybutyrate, acetoacetate as well as the glycogen content of the liver tissue are used as parameters.

2) In comparison to the control experiments, ethanol causes gluconeogenesis, formation of glycogen and an increase of the lactate/pyruvate ratio.
3) Addition of fructose to the perfusate is followed by a high release of glucose and an increase of the glycogen content of the liver. There is increased formation of lactate, pyruvate and of the ketone bodies beta-hydroxybutyrate and acetoacetate. Simultaneously the ratios of lactate/pyruvate and beta-hydroxybutyrate/acetoacetate rise reflecting a shift to a reduced state of the liver cell.
4) If ethanol and fructose are added to the perfusate there is a distinct decrease of gluconeogenesis, fructose conversion and formation of glycogen in comparison to the fructose experiments. The oxidation of ethanol is also inhibited. As a result of both ethanol and fructose using oxidative pathways for their conversions the shift of the redox ratio to a reduced state is more pronounced as compared to the ethanol and fructose experiments. The lactate/pyruvate ratio is increased as high as 673. This result possibly reflects the mutual inhibitions of the oxidative pathways of ethanol and fructose.

Introduction

Several authors (Pletscher et al. 1952, Stuhlfauth et al. 1955, Tygstrup et al. 1965), have observed an increased oxidation of ethanol under the influence of fructose. Therefore they propose a fructose treatment for ethanol oxidation. Other authors (Hassinen 1964, Lundquist et al. 1963, Vitale et al. 1954) cannot confirm an enhancement of ethanol oxidation by fructose. In a study on the fructose effect on ethanol metabolism of the human liver with catheters in the hepatic veins, Tygstrup et al. (1965) found that fructose doubles ethanol conversion to acetate and vice versa and that ethanol increases the uptake of fructose into the liver. These authors can also demonstrate that the output of glucose from the liver is greater during ethanol-fructose infusion than during addition of fructose alone.

In studies with isolated livers from fed rats, ethanol causes gluconeogenesis (Forsander et al. 1965, Papenberg 1970). It is assumed that this release of glucose comes from the glycogen depots of the liver. Since it has been demonstrated by Forsander et al. (1965) that, by contrast, livers of fasted rats which are depleted of glycogen ethanol causes glucose uptake.

Field et al. (1963) have shown that glycogen formation from fructose is inhibited by ethanol in livers from rats starved for 18 hours. They also conclude from these experiments that ethanol causes glycogen depletion of the liver. On the other hand, they have demonstrated that glucose formation from fructose is not inhibited by ethanol. Kaden et al. (1969) have observed with isolated perfused rat livers that gluconeogenesis from fructose is unchanged under the influence of ethanol while fructose conversion is even enhanced by ethanol.

With the isolated rat liver preparation, we have found (Papenberg et al. 1970) that fructose conversion is inhibited by ethanol and vice versa. Simultaneously ethanol decreases the rate of glucose formation from fructose.

In the present study we have further tried to clarify the way of ethanol and fructose utilization of the isolated perfused rat liver in respect to glucose, lactate, pyruvate, beta-hydroxybutyrate and acetoacetate formation. Furthermore, glycogen kinetics are demonstrated during perfusions under the various conditions.

Methods

The method of perfusion of isolated livers from fed male Wistar rats (body weight 200-250 g) is described in a previous report (Papenberg et al. 1970). The rat livers are perfused over a period of three hours. The semisynthetic perfusate consists of 60 ml ringer bicarbonate buffer, 40 ml washed erythrocytes, 1.5 g/100 ml bovine albumine, 150 mg/100 ml glucose and 1 mg/100 ml ampicillin. The perfusate content of glucose appears to be important in respect to gluconeogesis and glycogen metabolism (Hems et al. 1966, McCraw 1968). During an initial period of 60 minutes the livers are allowed to recover from hypoxemia during the operation for isolation. Initial metabolite concentrations are determined after this time. Alterations of metabolite concentrations and glycogen contents are calculated on the basis of these control values. Ethanol and fructose are then added to start the two-hour experimental period. Perfusate samples for measurements of glucose, ethanol, fructose, lactate, pyruvate, beta-hydroxybutyrate and acetoacetate are taken at 60, 90, 120, 150 and 180 minutes of perfusion. Small liver lobes for determinations of glycogen are ligated and cut at 60, 120 and 180 minutes. All tissue contents are related to one gram of liver wet weight. The rates of formation and degradation of metabolites are referred to liver wet weight. At the end of the perfusions, the livers weigh on the average 11.0 g.

Four groups of experiments are carried out in which livers are perfused without any additions to the perfusate and in which livers are loaded with ethanol, fructose, and ethanol-fructose.

Results and Discussion

Control Experiments

If only glucose is added to the perfusate at a concentration of 150 mg/100 ml, it is utilized by the rat liver to a small extent (Fig. 1). According to Table 1, during the first hour -8.4 micromoles/g of glucose are taken up by the liver; during the second hour, -5.9 micromoles/g of glucose are taken up. This consumption of glucose by the liver agrees with the results of McCraw (1968). Glucose in converted in the glucokinase and glucose-6-phosphatase systems (Haas and Byrme 1960). Apparently the isolated rat liver regulates a glucose concentration in the perfusate which is about double the concentration in normal rat blood (Schimassek 1963b). Over the experimental period of two hours, the glycogen content of the liver disappears at a rate of -13.1 micromoles/g and -15.1 micromoles/g during the first and second hour, respectively (Fig. 1, Table 1). Schimassek has also found a decrease of the glycogen content during the perfusion (Schimassek 1963a). The lactate/pyruvate ratios as indicators of the redox state in the cytoplasma remain at 10 (Fig. 2), agreeing with other reports (Schimassek 1963a, 1963b, Papenberg et al. 1970).

Experiments with Ethanol

Addition of 200 mg/100 ml of ethanol to the perfusate causes formation of glucose (Fig. 1). As can be seen on Table 1, glucose amounting to +13.6 micromoles/g during the first hour and +12.1 micromoles/g during the second hour are newly formed by the liver. Likewise, there is a formation of +4.9 micromoles/g of glycogen during the first 60 minutes of the experiment and a disappearance of glycogen of -8.7 micromoles/g during the second 60 minutes (Table 1, Fig. 1). Ethanol is oxidized independently of its concentration in the perfusate with a rate of -67.0 micromoles/g during the first and -83.0 micromoles/g during the second hour (Fig. 1, Table 1). As a result of ethanol oxidation to acetic aldehyde and acetate the lactate/pyruvate ratio rises as high as 143 (Fig. 2).

Ethanol leads to a formation of glucose in livers of fed rats (Forsander et al. 1965, Papenberg et al. 1970). These authors postulate that ethanol-induced gluconeogenesis is derived from the glycogen depots of the livers. This assumption agrees with findings by Akabane et al. (1964) who demonstrated depletion of liver glycogen storage of fed rats by ethanol. In contrast to these results, we find an increase of the glycogen content of the perfused rat liver during the first hour of ethanol oxidation. Even during the second hour of perfusion with ethanol, the glycogen decrease is less than in the control experiments (Table 1). The difference of these results to those of Akabane et al. (1964) may be explained by the fact that Akabane used whole animals. Glycogen phosphorylase acts dependently on the $NADH_2$/NAD ratio in the perfused liver. The enzyme activity increases if this redox ratio is augmented in the cytoplasmatic compartment (Spiess and Meyer 1968) as it is during ethanol oxidation (Forsander et al. 1965, Forsander 1966, Papenberg et al. 1970). Haeckel and Haeckel (1968) have demonstrated that increased gluconeogenesis of the perfused

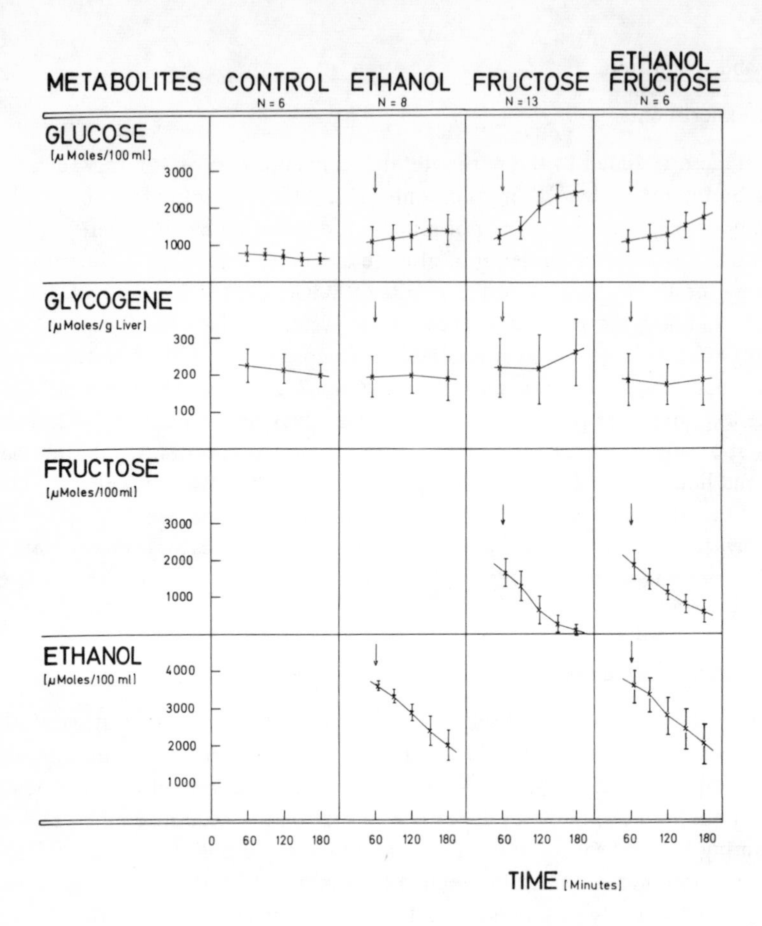

Fig. 1. Effect of ethanol and fructose on the rates of formation and conversion of glucose, glycogen, fructose and ethanol. The figures show means and standard deviations. The arrows indicate additions of substrates. Glucose, fructose and ethanol are calculated as micromoles/100 ml perfusate, glycogen as micromoles/g liver tissue.

fasted guinea pig liver during ethanol oxidation is caused by the increased ratio of $NADH_2/NAD$, since the conversion of 3-phosphoglycerate to 1.3-diphosphoglycerate and 3-phosphoglycerolaldehyde from pyruvate is promoted by an increased level of $NADH_2$ in the liver cell.

Thus glucose newly formed under the influence of ethanol is not derived from glycogen but from endogenous precursors like pyruvate of which concentration decreases during ethanol oxidation (Fig. 2).

Experiments with Fructose

An amount of 360 mg/100 ml of fructose is converted to glucose and glycogen (Fig.

1). As shown in Table 1, +72.5 micromoles/g and 35.3 micromoles/g glucose are newly formed after the first and second hours of perfusion respectively. At the same time, it comes to a glycogen depletion of -4.4 micromoles/g during the first hour, while during the second hour, +46.7 micromoles/g glycogen are formed. Fructose is metabolized exponentially (Fig. 1, Table 1), during the first 60 minutes with a rate of -99.3 micromoles/g and during the second hour with -48.2 micromoles/g. Simultaneously in the cytoplasma the concentrations of lactate and pyruvate rise (Fig. 2). To confirm that fructose increases the reduced state of the liver, the mitochondrial redox ratio of beta-hydroxybutyrate and acetoacetate is also studied during fructose metabolism (Fig. 3). While the concentrations of beta-hydroxybutyrate rise from 235 nanomoles/ml to 916 nanomoles/ml, the concentrations of acetoacetate show only slight changes (from 312 to 365 nanomoles/ml) during the perfusion time. The ratios of lactate/pyruvate rise from 13 to 32, while those of beta-hydroxybutyrate/acetoacetate increase from 0.8 to 2.5 when fructose is metabolized.

According to Leuthart and Stuhlfauth (1960), Heinz and Lamprecht (1967), Heinz (1968), Heinz et al. (1968) and Heinz and Junghäanel (in press), after phosphorylation to D-fructose-1-phosphate, D-fructose is metabolized to D-glyceraldehyde and dihydroxyacetonphosphate. The latter enters into the glycolytic pathway. D-glyceraldehyde

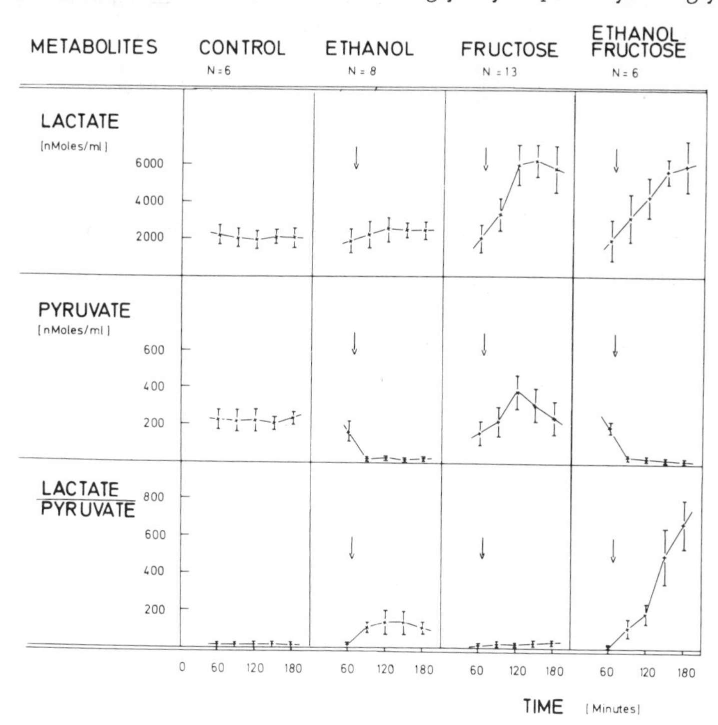

Fig. 2. Effect of ethanol and fructose on lactate and pyruvate concentrations in the perfusing medium. The figures show means and standard deviations. The arrows indicate additions of substrates. Lactate and pyruvate are calculated as nanomoles/ml perfusate.

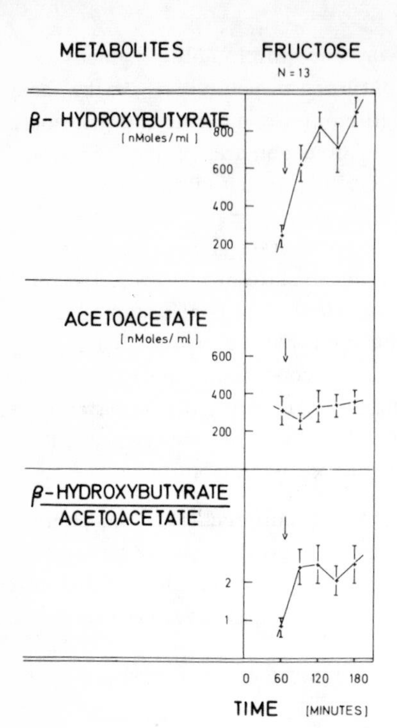

Fig. 3. Effect of fructose on the formation of beta-hydroxybutyrate and acetoacetate. The figures show means and standard deviations. The arrows indicate addition of fructose. Both metabolites are calculated as nanomoles/ml perfusate.

METABOLITES	GLUCOSE [μMoles/g Liver]		ETHANOL [μMoles/g Liver]		FRUCTOSE [μMoles/g Liver]		GLYCOGENE [μMoles/g Liver]	
TIME [Minutes]	60 - 120	120 - 180	60 - 120	120 - 180	60 - 120	120 - 180	60 - 120	120 - 180
CONTROL N = 6	- 8.40 ± 2.42	- 5.95 ± 5.44					- 13 1 ± 5 6	- 15.1 ± 10.7
ETHANOL N = 8	+ 13.60 ± 1.81	+ 12.10 ± 1.40	- 67.0 ± 10.1	- 83.0 ± 10.4			+ 4 9 ± 0.3	- 8 7 ± 4 3
FRUCTOSE N = 13	+ 72.5 ± 18.7	+ 35 30 ± 2 50			- 99 3 ± 0 5	- 48 2 ± 11 2	- 4 4 ± 3 7	+ 46 7 ± 1 2
ETHANOL FRUCTOSE N = 6	+ 12 2 ± 6 4	+ 37 90 ± 0 6	- 63 0 ± 4 6	- 60 1 ± 8 3	- 57 0 ± 13 4	- 42 0 ± 7 7	- 11 6 ± 3 9	+ 12 8 ± 7 8

Table 1. Effect of ethanol and fructose on the rates of formation and conversion of glucose, fructose and ethanol (micromoles/g liver) in the perfusate and of glycogen (micromoles/g liver) in liver tissue. The figures represent means and standard deviations of substrate conversion rates during 60-120 and 120-180 minutes of perfusion time; + indicates a new formation, - means disappearance of a substrate.

is oxidized to D-glycerate in a NAD dependent reaction. Further on D-glycerate is converted to 3-phosphoglycerate. On the other hand, D-glyceraldehyde can be reduced to glycerol by alcoholdehydrogenase which yields after phosphorylation glycerol-1-phosphate and by further oxidation dihydroxyacetonphosphate.

Since the redox ratios of lactate/pyruvate (Fig. 2) and beta-hydroxybutyrate/acetoacetate (Fig. 3) as well as alphaglycerophosphate/dihydroxyacetonphosphate in the liver tissue (Papenberg et al. 1970) are increased during fructose metabolism it can be assumed that a major part of fructose is oxidized via D-glyceraldehyde to D-glycerate. The data by Exton and Park (1969) confirm this pathway of fructose. So far it is not clear why it comes to an increased formation of the ketone-bodies beta-hydroxybutyrate and acetoacetate during fructose metabolism. It may be caused by a possible lack of oxaloacetate or by the increased redox ratio $NADH_2/NAD$.

Cori and Cori (1926) have demonstrated the conversion of fructose to glycogen in rat liver.

Experiments with Ethanol and Fructose

If ethanol (200 mg/100 ml) and fructose (360 mg/100 ml) are added to the perfusate simultaneously, gluconeogenesis is decreased as compared to the experiments with fructose alone (Fig. 1). As demonstrated by Table 1, the inhibition of gluconeogenesis from fructose by ethanol is about six fold during the first hour (+12.2 micromoles/g), while it is comparable to the fructose experiments during the second hour.

Fructose conversion and glycogen formation are also inhibited in comparison to the fructose experiments (Fig. 1, Table 1). Thus during the first hour, fructose degradation (-57.0 micromoles/g) is inhibited 1.7 fold by ethanol, while it is similar to the fructose experiments (-42.0 micromoles/g) during the second hour of perfusion. The glycogen kinetics differ with -11.6 micromoles/g during the first hour and with +12.8 micromoles/g during the second hour from those of the fructose experiments. Thus glycogen formation from fructose is diminished 3.6 fold by ethanol during the second hour of perfusion. The rate of ethanol degradation is comparable to the ethanol experiments during the first hour of perfusion, while it is inhibited 1.4 fold during the second hour (Table 1).

With ethanol and fructose in the perfusate it comes to a great shift of the lactate/pyruvate ratios (Fig. 2) to a value of 673 after 180 minutes of perfusion. This considerable change of the redox state is possibly caused by the fact that both compounds ethanol and fructose are converted through oxidative pathways exhausting the NAD which is available.

As we have already reported (Papenberg et al. 1970) the inhibition of fructose conversion by ethanol and vice versa with a simultaneous decrease of gluconeogenesis and glycogen formation is possibly due to mutual inhibitions of the oxidative pathways of ethanol and fructose. Inhibition of glycogen formation from fructose by ethanol is reported by Field et al. (1963). However, in contrast to our data, they find that fructose conversion is similar with and without ethanol. Kaden et al.

(1969) observed even an increased fructose utilization rate and an unchanged gluconeogenesis from fructose by ethanol in the perfused rat liver.

Acknowledgment

This work was supported by the Deutsche Forschungsgemeinschaft. The author is indebted to Mrs. D. Kocklemus and Mr. H. Koch for their technical assistance.

References

Akabane, J., S. Nakanishi, H. Kohei, R. Matsumara, H. Ogaba: Med. J. Shinshu University 9 (1964 25
Cori, C. F., G. T. Cori: J. biol. Chem. 70 (1926) 577
Exton, J. H., C. R. Park: J. biol. Chem. 244 (1969) 1424
Field, J. B., H. E. Williams, G. F. Mortimore: J. Clin. Invest. 42 (1963) 497
Forsander, O. A., N. Raiha, M. Salaspuro, P. Maenpää: Biochem. J. 94 (1965) 259
Forsander, O. A.: Biochem. J. 98 (1966) 244
Haas, L. F., W. L. Byrme: Science 131 (1960) 991
Hems, R., B. D. Ross, M. N. Berry, H. A. Krebs: Biochem. J. 101 (1966) 284
Haeckel, R., H. Haeckel: Biochemistry 7 (1968) 3803
Hassinen, J: Ann. Med. exp. Fenn. 42 (1964) 76
Heinz, F., W. Lamprecht: Hoppe-Seylers Z. physiol. Chem. 348 (1967) 855
Heinz, F.: Hoppe-Seylers Z. physiol. Chem. 349 (1968) 399
Heinz, F., W. Lamprecht, J. Kirsch: J. Clin. Invest. 47 (1968) 1826
Heinz, F., J. Junghänel: in press
Kaden, M., N. Oakley, J. B. Field: Amer. J. Physiol. 216 (1969) 756
Leuthardt, F., K. Stuhlfauth: In: Medizinische Grundlagenforschung, Vol. III, Thieme, Stuttgart (1960), p.416
Lundquist, F., J. Svendsen, P. H. Petersen: Biochem. J. 86 (1963) 119
McCraw, E. F.: Metabolism 17 (1968) 833
Papenberg, J., J. P. v. Wartburg, H. Aebi: Enzym. Biol. Clin. 11 (1970) 237
Pletscher, A., A. Bernstein, H. Staub: Helv. Physiol. Acta 10 (1952) 74
Schimassek, H.: Biochem. Z. 336 (1963a) 460
Schimassek, H.: Biochem. Z. 336 (1963b) 468
Schimassek, H.: Life Sciences 11 (1962) 635
Spiess, J. E. Meyer: Hoppe-Seylers Z. physiol. Chem. 349 (1968) 1261
Stuhlfauth, K., A. Engelhardt-Golkel, J. Schaffry: Klin. Wschr. 33 (1955) 888
Tygstrup, N., K. Winkler, F. Lundquist: J. Clin. Invest. 44 (1965) 817
Vitale, J. J., D. M. Hegsted, H. McGrath, E. Grable, N. Zamchek: J. biol. Chem. 210 (1954) 753

Announced Discussion

Fructose as a Precursor for Ketogenesis+

H. D. Söling and B. Willms
Abteilung für Klinische Biochemie, Medizinische Universitätsklinik
Göttingen, FRG

Experiments with isolated perfused guinea pig livers revealed an increased rate of ketogenesis when fructose was used as a gluconeogenic precursor, whereas pyruvate and glycerol diminished the rate of ketogenesis and L-lactate had no effect (Fig. 1). Similar results had been obtained in former experiments with isolated perfused rat livers (unpublished observations). We therefore performed experiments with isolated perfused rat livers in order to clarify where the carbon for the higher rate of ketone body formation might come from. Livers from rats starved for 48 hours were perfused with 2 mM (1-^{14}C) oleate bound to albumin with and without 20 mM fructose in the medium and the incorporation of label and the net formation of ketone bodies were measured. The incorporation of label into ketone bodies was not affected by fructose and the rate of ketogenesis was depressed rather than stimulated.

In further experiments the incorporation of (^{14}C) activity from (U-^{14}C) fructose into ketone bodies was determined. There was a considerable incorporation of label into ketone bodies. The specific radioactivity (dpm/ugatom carbon) of the ketone bodies formed was about the same as that of fructose.

This made it likely that the ketone bodies derived directly from fructose. Therefore the effect of fructose on the rate of pyruvate oxidation was studied (Fig. 2).

In these experiments, (1-^{14}C) pyruvate was infused at a rate of 1 mmole/hr. after a priming dose to give an initial pyruvate concentration of 5 mM. After 60 minutes, cold fructose was added to the medium (20 mM final concentration). In spite of the fact that the labeled pyruvate pool became diluted by the cold pyruvate derived from fructose, the formation of $^{14}CO_2$ increased immediately by about 200%.

This clearly demonstrates a stimulation of pyruvate oxidation by fructose.

This observation is paralleled by a significant increase in the concentration of acetyl-S-CoA (Table 1) in the isolated perfused rat liver in presence of fructose.

When this work was in progress, we learned about the studies of Raivio et al. (1969), who found that fructose led to a very rapid fall in the levels of adenine nucleotides in rat liver. When ATP, ADP and AMP were measured under our conditions, the adenine nucleotides had dropped dramatically indeed, especially at the expense of ATP (Table 1).

+ Enzyme Code Number: Pyruvate dehydrogenase (EC 1.2.4.1).

ATP is known to inhibit pyruvate oxidation (Wieland and v. Jagow-Westermann 1969) as it is the substrate for an enzyme which inactivates pyruvate dehydrogenase by phosphorylation (Linn et al. 1968).

We therefore believe that the sequence of events is as follows:

Fructose utilizes ATP for phosphorylation and stimulates the breakdown of adenine nucleotides to uric acid and allantoin. As a result, the rate of pyruvate oxidation increases as pyruvate dehydrogenase becomes deinhibited. This in turn leads to an increased formation of acetyl-S-CoA which then is converted to ketone bodies or oxidized.

Work is in progress to find out whether a decreased activity of the Krebs cycle contributes additionally to the formation of ketone bodies.

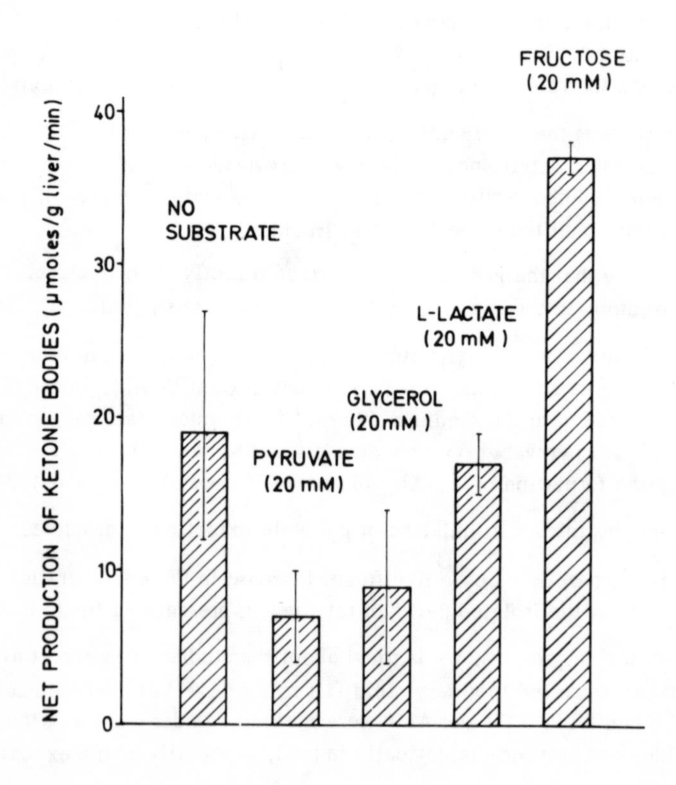

Fig. 1. Net formation of total ketone bodies by isolated perfused livers from 48-hour-starved guinea pigs in the absence of and in the presence of various gluconeogenic precursors.

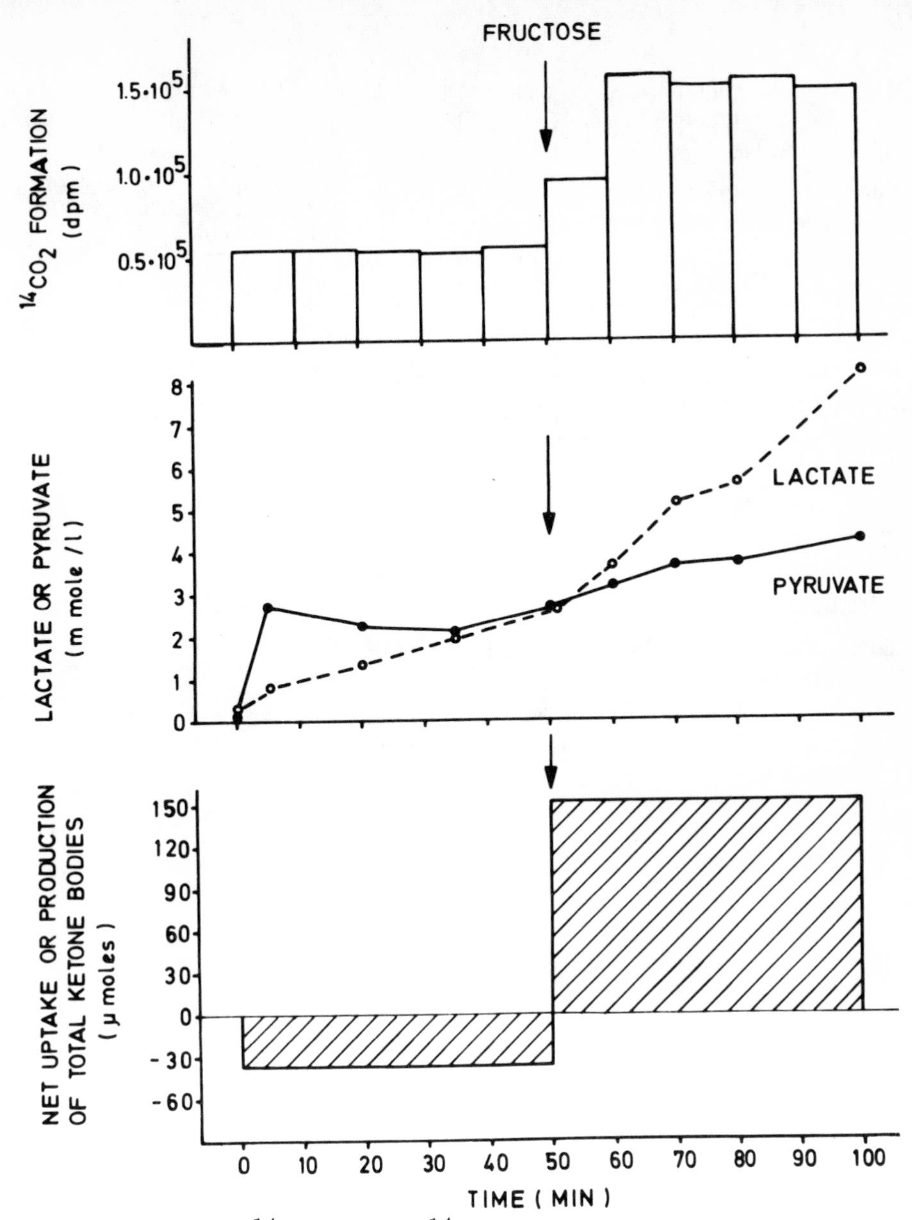

Fig. 2. Formation of $^{14}CO_2$ from (1-^{14}C) pyruvate, concentrations of lactate and pyruvate in the perfusion medium, and net balance of ketone bodies (cumulative) in experiments with isolated perfused rat livers.

A priming dose of (1-^{14}C) pyruvate was given at zero time to give an initial concentration of 5 mM. This was followed by a constant intraportal infusion of 1 mmole/hr of (1-^{14}C) pyruvate of the same specific activity throughout the experiment. After 50 minutes (arrow) cold fructose (20 mM final concentration) was added. The ketone body balance refers to the total amount of ketone bodies taken up within the first 50 minutes or being produced during the second 50 minutes.

	Acetyl-S-CoA (nmole/g)	ATP	ADP (μmole/g)	AMP	ATP+ADP+AMP (μmole/g)	ATP/ADP (ratio)
20 mM glucose	45.6	2.24	0.75	0.095	3.09	2.98
20 mM fructose	76.7	1.19	0.53	0.104	1.82	2.28

Table 1. Levels of acetyl-S-CoA, ATP, ADP and AMP in isolated livers from 48-hour-starved rats perfused with either 20 mM glücose or 20 mM fructose. Glucose or fructose were added 60 minutes prior to the freeze-stop.

References

Raivio, K. O., M. P. Kekomäki, P. H. Mäenpää: Biochem. Pharmacol. 18 (1969) 2615

Wieland, O., B. v. Jagow-Westermann: FEBS Letters 3, (1969) 271

Linn, T. C., F. H. Pettit, L. J. Reed: Proc. nat. Acad. Sci. (Wash.) 62 (1968) 234

Discussion to Papenberg

Krebs: We had experiences very similar to those of Dr. Söling. Dr. Woods also noted a large fall in the concentrations of adenine nucleotides on perfusing the isolated rat liver with fructose, but we found that Finnish biochemists had already published this observation a year before we made it (Mäenpää, P. H., K. O. Raivio and M. P. Kekomäki: Science, N. Y. 161 (1968) 1253). According to the Finnish authors, the purine ring of the adenine nucleotides appears as uric acid and allantoin, but after a relatively short time — one hour -- the adenine nucleotides are replaced. It is surprising how quickly the liver recovers from this disturbance. The reason for the adenine nucleotide loss is as follows: Fructose is rapidly converted to fructose-1-phosphate which accumulates. This causes a fall in the concentration of ATP and inorganic phosphate and the inhibitory effects of ATP and P_i on the enzymes which can cause a hydrolytic degradation of AMP (AMP-deaminase and 5-nucleotidase) no longer operate. So the factors which normally protect AMP are lost.

Ruderman: I would like to describe briefly some studies in which we observed a diphasic effect of ethanol on hepatic gluconeogenesis. In these experiments, livers were perfused with a medium containing 10 mM alanine and linoleate. During the first 15 minutes of perfusion in the presence of ethanol, gluconeogenesis from alanine was diminished by about 75%. After this, the rate of gluconeogenesis progressively increased, until at 30 minutes and beyond, there tended to be an even higher rate of gluconeogenesis than was observed in the control livers.
In light of these observations, Dr. Toews and Dr. Lowy measured metabolite levels in liver biopsies taken after 15 minutes of perfusion, when gluconeogenesis was inhibited, and at 45 minutes when it appeared to be restored. At 15 minutes the lactate/pyruvate ratio was 200% of the control value. Pyruvate levels were significantly diminished as were the levels of 3-phosphoglycerate, 2-phosphoglycerate, phosphoenolpyruvate, aspartate and oxaloacetate. In biopsies taken at 45 minutes, the concentration of pyruvate and the lactate/pyruvate ratio were approximately the same as they were at 15 minutes. On the other hand, the concentration of malate, which was 180% of the control level at 15 minutes, had gone up even further and oxaloacetate, which was depressed at 15 minutes, was now elevated to 150% of normal. In addition aspartate, phosphoenolpyruvate and 3-phosphoglycerate levels had returned to normal.
We feel that the data suggest that the concentration of oxaloacetate was rate limiting for gluconeogenesis in livers perfused with ethanol. The basis for the secondary increase in oxaloacetate, aspartate and malate, and with this the restoration of gluconeogenesis remains to be established.

Krebs: Dr. Papenberg, much glucose and glycogen seems to have disappeared in your control experiments. Can you account for this disappearance?

Papenberg: I think, McCraw (Metabolism 17 (1968) 833) has published the same phenomenon. This is glucose uptake and I do not know where this glucose goes. And a glycogen depletion is also observed over these two hours of the experiment. Schimassek (Biochem. Z. 336 (1963) 460) has published exactly the same results.

Krebs: So there is a deficit. Can this be explained by lactate formation? I suppose the rate of oxygen consumption is too low to explain the deficit by an oxidation of glucose?

Hanson: I would like to ask Dr. Ruderman if he measured ethanol consumption? I wonder whether this diphasic effect of ethanol has something to do with decreasing concentration of ethanol in the perfusion medium? Did you maintain the ethanol concentration?

Ruderman: Yes we did. Ethanol concentration was 2 mM at the end of the experiment. We also carried out studies in which ethanol was infused, in order to maintain a relatively constant concentration throughout the experiment. The results were essentially the same.

Güder: I want to ask Dr. Papenberg if he knows the source of his carbon he gets into the glucose in the presence of ethanol alone?

Papenberg: I do not know. But experiments of Dr. Haeckel (Biochem. 7 (1968) 3803) suggest that it might come from lactate or pyruvate. Only in perfusion experiments with livers of well-fed rats, is glucose formed with ethanol. If you perfuse livers from starved rats you get glucose uptake by the liver.

Guder: Did you find a decrease of these metabolites?

Papenberg: Not of these metabolites.

Krebs: Did you add lactate or pyruvate, or do you think it was derived from endogenous precursors?

Papenberg: I think, in the case of livers of fed rats the lactate must be endogenous.

Krebs: But the endogenous store of lactate and pyruvate is so small that it could hardly account for the phenomenon.

Papenberg: It is very hard to know where this glucose is coming from. At least in the first hour it does not come from glycogen. I think it cannot come from fatty acids or amino acids because the oxidative deamination of amino acids and oxidation of fatty acids are inhibited during ethanol administration. It is very hard for me to imagine where this glucose is coming from.

J. R. Williamson: You mentioned that fructose increased the ß-hydroxybutyrate/acetoacetate ratio. I did not understand your explanation for this.

Papenberg: Fructose metabolsim goes through the oxidative pathway from glyceraldehyde to blycerate, at least partly. I think, therefore the ratios are increasing.

J. R. Williamson: Didn't you mention something about α-glycerol- phosphate?

Papenberg: The α-glycerol-phosphate increases under fructose, as does the α-glycerol-phosphate/dehydroxy-acetonephosphate ratio. That depends on the way the oxidative pathway of fructose is going.

Krebs: I am puzzled by your remark that fructose inhibits ethanol utilization because it has long been established that under many conditions, including in man in vivo, fructose stimulates ethanol degradation (K. Stuhlfauth and H. Neumaier, Med.

Klin. 46 (1951) 591; F. Lundquist and H. Wolthers: Acta Pharmacol. Toxicol. 14 (1958) 290; N. Tygstrup, K. Winkler, F. Lundquist: J. clin. Invest. 44 (1965) 817; H. I. D. Thieden, F. Lundquist: Biochem. J. 102 (1967) 177; J. Merry, V. Marks: The Lancet II (1967) 1328; G. L. S. Pawan: Nature 220 (1968) 374).

Papenberg: This is not always the case. There are reports that fructose will not do this. (Hassinen: Ann. Med. exp. Fenn. 42 (1964) 76; Lundquist, F.: Biochem. J. 86 (1963) 119; Vitale, J. J.: J. biol. Chem. 210 (1954) 753; Pfleiderer: Klin. Wschr. 32 (1954) 560).

Wieland: One could call your effect, Dr. Söling, a paradoxical effect. Fructose has been well known as an antiketogenic agent either clinically or experimentally. In our studies with fructose in liver perfusion many years ago with Dr. Matschinsky (O. Wieland: Excerpta Medica, Internat. Congr. Ser. No. 84 (1965) 533) we could demonstrate an antiketogenic effect of fructose. We also found an increase of lactate, pyruvate and α-glycerol-phosphate. It could be that the ketogenic effect of fructose reported today had been masked in these experiments, because we produced high rates of ketogenesis by infusing fatty acids.

Söling: Yes, this is true. If you add fructose to isolated perfused livers in the presence of oleate, then you do not get an additional effect of fructose on ketogenesis. On the contrary, ketogenesis becomes even inhibited by fructose.

Ballard: If fructose was metabolized to sorbitol this could alter the NAD to NADH ratio. Did you measure sorbitol?

Söling: No, we didn't.

D. H. Williamson: Two questions on Dr. Söling's contribution. Were these fed or starved rats and what were the rates of ketogenesis in your experiments?

Söling: There were starved rats. With 20 mM fructose added, the rate of ketogenesis was in the range of 0.5 μmoles/g/min.

D. H. Williamson: If we are quoting references, Lowry and his group have also reported this fall of ATP concentration on fructose loading (Burch, H. G., P. Max, K. Chyu, O. H. Lowry: Biochem. biophys. Res. Commun. 1967).

Exton: I have not had any experience with fructose in the fed animal, but it is quite clear that there is a great difference in the effect of lactate on ketone-body production in the fed rat compared with the fasted or diabetic animal. In the fasted or diabetic animal, lactate reduces ketogenesis to about one third. On the other hand, in fed animals, lactate is ketogenic. The effect is not large. The ketone bodies are partly coming from lactate as demonstrated by isotope studies. So the nutritional state of the animal has an influence on the metabolism of lactate. This could be partly due to a greater metabolism of pyruvate by way of pyruvate-dehydrogenase in the fed animal.

Wieland: In this context I would like to recall a paper on fructose metabolism in liver by Helmreich et al. (Hopp-Seylers Z. physiol. Chem. 292 (1953) 184) which appeared more than 15 years ago. These workers claimed an increased formation of acetyl-CoA from fructose, and I am pleased to see this substantiated now by direct measurements of acetyl-CoA levels. These authors could not quantitate acetyl-CoA directly but they used sulfanilamide as an acceptor, and measured the acetylated product which was markedly increased.

Mechanism of Glucagon Activation of Gluconeogenesis++

J. H. Exton[+], M. Ui, S. B. Lewis, and C. R. Park
Department of Physiology Vanderbilt University School of Medicine,
Nashville, Tennessee, U.S.A.

Summary

The gluconeogenic action of glucagon was studied in rat livers perfused with bicarbonate buffer containing bovine albumin and red cells. In livers from fed rat perfused with 20 mM ^{14}C-lactate, glucagon increased the formation of ^{14}C-glucose plus ^{14}C-glycogen approximately two-fold but did not alter the production of $^{14}CO_2$ or ^{14}C-ketone bodies. The ^{14}C-lactate uptake was increased by 28% and the labeling of lipids and proteins was decreased by approximately 40%. Oxygen consumption, ketogenesis, ureogenesis and glycogenolysis were also increased. The data are consistent with the activation of gluconeogenesis from lactate and endogenous sources (probably protein). There was no evidence of inhibition of lactate oxidation. Acetyl-CoA production from endogenous substrate (probably lipid) appeared to be increased resulting in increased ketogenesis and a reduction in the specific activity of ^{14}C-ketone bodies. The radioactivity of cholesterol and fatty acid esters was decreased to an extent corresponding to the reduction in acetyl-CoA specific activity, suggesting that glucagon did not inhibit cholesterol and fatty acid synthesis. The increase in respiration was greater than that attributable to increased ketogenesis and protein catabolism indicating that the Krebs cycle was activated. The extra substrate utilized in the cycle was apparently derived from endogenous sources.

In livers perfused with 1 mM lactate and 0.1 mM pyruvate and tracer amounts of 1-^{14}C-glutamate, glucagon rapidly and markedly stimulated ^{14}C-glucose and $^{14}CO_2$ production. Tissue levels of pyruvate, citrate, α-oxoglutarate and glutamate were decreased and aspartate and P-pyruvate were increased. Measured 2.5 minutes after glucagon treatment, the radioactivities of pyruvate, oxalacetate, α-oxoglutarate and glutamate were decreased while those of malate and citrate were unchanged and those of succinate, fumarate, aspartate and P-pyruvate were increased. Very similar changes were seen in livers perfused with ^{14}C-lactate plus ^{14}C-pyruvate. With glucagon treatment for 8.5 minutes, the concentrations of intermediates were not further changed but the specific radioactivity of glutamate de-

---

+ Investigator of the Howard Hughes Medical Institute.

++ Unusual Abbreviations: P-pyruvate = phosphoenolpyruvate

Enzyme Code Numbers: Citrate synthase (EC 4.1.3.7), Glutamate oxalacetic transaminase (EC 2.6.1.1), Nucleoside diphosphate kinase (EC 2.7.4.6), P-pyruvate carboxykinase (phosphoenol-pyruvate carboxykinase) (EC 4.1.1.32), Pyruvate carboxylase (EC 6.4.1.1), Pyruvate kinase (EC 2.7.1.40).

clined causing corresponding changes in the radioactivity of the other intermediates.

Glucagon also increased the labeling of aspartate, P-pyruvate and glucose in livers perfused for one minute with ^{14}C-bicarbonate. With ^{14}C-aspartate, treatment with glucagon for 8.5 minutes decreased the radioactivity of malate, citrate and oxalacetate and increased that of P-pyruvate and glucose.

These results suggest that an action of glucagon is exerted at the level of P-pyruvate carboxykinase. They also show an early hormonal stimulation of the utilization of glutamate and α-oxoglutarate leading to increased aspartate production.

Introduction

Glucagon has been reported to affect a wide variety of metabolic processes in the isolated liver (Table 1), and it is probable that cyclic AMP mediates most, if not all, of these effects. However, it is uncertain whether all of these responses can be elicited by concentrations of glucagon within the range occurring normally in portal blood in vivo. This paper presents data showing that glycogenolysis and gluconeogenesis have lower thresholds for glucagon activation than ketogenesis and ureogenesis in the perfused rat liver raising the possibility that the latter processes may not be influenced by glucagon under physiological conditions.

The paper also presents data relating to the action of glucagon on gluconeogenesis. Changes in the metabolism of ^{14}C-lactate and other labelled substrates in the perfused liver are examined in order to elucidate the mechanism of glucagon action. The results indicate that stimulation of gluconeogenesis from lactate is not due to inhibition of lactate oxidation, but is due, in part at least, to activation of P-pyruvate carboxykinase. Glucagon also appears to activate α-ketoglutarate utilization and to stimulate the transamination of oxalacetate to aspartate.

Materials and Methods

Livers from 90 to 130g Sprague Dawley rats fed ad libitum were perfused by the technique of Mortimore (1963) as modified by Exton and Park (1967). The perfusion medium was oxygenated Krebs-Henseleit bicarbonate buffer containing 3% bovine serum albumin (Pentex Inc., Kankakee, Ill., U.S.A.) and 18 - 22% bovine erythrocytes prepared according to Mallette et al. (1969a).

Most methods of medium and tissue analysis have been described by: Mortimore (1963), Mallette et al. (1969a), Exton and Park (1969), Exton and Park (in press), Ui and Park (in press). Radioactivity in lipid fractions, ketone bodies and protein was measured by methods based on those described by Regen and Terrell (1968). Ketone bodies and tissue metabolites were assayed enzymatically (Williamson et al. 1962). Radioactivity of tissue metabolites was determined in fractions of perchlorate extracts separated by column and thin-layer chromatography. Neutralized extracts were initially applied to Dowex-50 columns and fractions were eluted with water or N NH_4OH. The water eluates were next applied to Dowex-2 columns and eluted with HCOOH. The amino acids present in the NH_4OH eluates were separated by 2-dimensional, thin-layer chromatography on cellulose using phenol-NH_4OH (42-8) and butanol-acetone-diethylamine-H_2O (10-10-2-5) as solvent systems. The

Reported Effects of Glucagon on Isolated Liver

Process	Change	Involvement of Cyclic AMP
Glycogenolysis	Increase	Yes
Phosphorylase	Increase	Yes
Gluconeogenesis	Increase	Yes
Glycogen Synthesis	Decrease	Yes
Glycogen Synthetase	Decrease	Yes
Ureogenesis	Increase	Yes
Protein Synthesis	?Decrease	?
Protein Breakdown	?Increase	?
Ketogenesis	Increase	Yes
Lipolysis	Increase	Yes
Amino Acid Uptake	Increase	Yes
Tyrosine Aminotransferase	Increase	Yes
P-Pyruvate Carboxykinase	Increase	Yes
Lysosome Activation	Increase	?
K^+ Release	Increase	Yes
Ca^{++} Release	Increase	Yes
Mitochondrial Pyruvate Uptake	Increase	?
Krebs Cycle	Increase	?
Transamination	Increase	?
Lipoprotein Release	Decrease	Yes

Table 1.

organic acids in the HCOOH eluates were separated by thin-layer chromatography on silica gel using H_2O saturated ether - HCOOH (5-1) as solvent.

Unneutralized tissue extracts were also reacted with 2.4-dinitrophenyl hydrazine in HCl and the hydrazones of pyruvate, α-ketoglutarate and oxalacetate were extracted into ethyl acetate. The hydrazones were re-extracted with NH_4OH, concentrated by lyophilization and separated on cellulose chromatoplates using butanol-ethanol-NH_4OH (7-2). P-pyruvate in the unneutralized extracts was reacted with pyruvate kinase and the pyruvate formed was isolated as the hydrazone. Radioactive spots on chromatoplates were identified by comparison with authentic compounds, scraped off into vials and counted in toluene scintillation fluid.

Results

Sensitivity of Glycogenolysis, Gluconeogenesis, Ureogenesis and Ketogenesis to Activation by Glucagon in the Perfused Rat Liver

Table 2 shows the changes in glucose output, ureogenesis and ketone body production exerted by increasing concentrations of glucagon in livers perfused without added substrate. It is evident that glucagon activated glycogenolysis at a concentration (5×10^{-12} M) much lower than that producing significant stimulation of ketogenesis (1×10^{-10} M) or ureogenesis (2×10^{-10} M). Maximum stimulation of glycogenolysis was observed at 2×10^{-10} M glucagon and resulted in about a four-fold increase in glucose output whereas full activation of ketogenesis or ureogenesis required 2×10^{-9} M hormone and resulted in only two-fold changes in these parameters. To determine the sensitivity of gluconeogenesis, livers were perfused with 20 mM ^{14}C-lactate and the synthesis of ^{14}C-glucose was estimated. It is seen (Table 3) that this process was significantly stimulated by 2×10^{-11} M glucagon which is higher than that needed to increase glycogen breakdown (5×10^{-12} M). In the presence of lactate, 2×10^{-9} M glucagon was needed to stimulate ketogenesis (Table 3). This is 20-fold greater than the concentration required to increase ketogenesis in the absence of lactate (Table 2).

These results indicate that glycogenolysis is about four-fold more sensitive to glucagon than gluconeogenesis in the isolated liver and about 20-fold more sensitive than ketogenesis or ureogenesis raising doubts whether glucagon activates ketogenesis or ureogenesis under physiological conditions in vivo. The finding that gluconeogenesis from lactate is activated by concentrations of glucagon 100-fold lower than those stimulating ketogenesis in the presence of lactate provides further evidence that the changes responsible for the activation of the two processes are different.

Metabolic Changes Caused by Glucagon in Livers Perfused with 20 mM ^{14}C-Lactate

In an attempt to elucidate the mechanism of glucagon activation of gluconeogenesis, the effects of the hormone on lactate metabolism in the isolated liver were examined (Tables 4, 5). Since many of the metabolic changes were too small to be measured accurately in single passage (flow through) perfusions with physiological levels of

Glucagon Concentration	Glucose Production	Urea Production	Ketone body Production
(M)	(μmoles/100g body wt of rat/hr)		
0	189 ± 9	43 ± 2	10.5 ± 0.7
2×10^{-12}	206 ± 29	43 ± 3	10.8 ± 0.9
5×10^{-12}	236 ± 12*	43± 4	11.5 ± 0.8
1×10^{-11}	297 ± 41	44 ± 4	12.4 ± 1.2
2×10^{-11}	363 ± 29	41 ± 6	11.9 ± 0.9
5×10^{-11}	454 ± 24	48 ± 3	13.0 ± 1.1
1×10^{-10}	574 ± 41	54 ± 7	15.5 ± 1.5*
2×10^{-10}	720 ± 80	63 ± 4*	17.1 ± 2.1
2×10^{-9}	764 ± 80	80 ± 4	18.2 ± 2.9
2×10^{-8}	689 ± 103	90 ± 8	24.8 ± 3.4
2×10^{-7}	665 ± 25	72 ± 8	21.1 ± 3.7

Table 2. Dose Responses for Glucagon on Glycogenolysis, Ketogenesis and Ureogenesis in the Perfused Liver. + Significantly different from control $P < 0.05$

Glucagon Concentration	Glucose Concentration	Urea Production	Ketone body Production	^{14}C-Glucose Synthesis
(M)	(μmoles/100 g rat/hr)			(cpm$\times 10^{-3}$/100 g rat/hr)
0	317 ± 20	47 ±2	19.9 ± 0.7	122 ± 9
2×10^{-12}	377 ± 36	49 ± 5	-	136 ± 8
5×10^{-12}	439 ± 36*	46 ± 2	19.0 ± 1.3	131 ± 13
1×10^{-11}	429 ± 25	51 ± 5	20.2 ± 1.4	142 ± 14
2×10^{-11}	470 ± 25	56 ± 2	20.2 ± 1.5	166 ± 34*
5×10^{-11}	568 ± 25	53 ± 7	23.1 ± 1.7	192 ± 24
1×10^{-10}	662 ± 91	52 ± 4	21.0 ± 1.5	198 ± 18
2×10^{-10}	760 ± 43	59 ± 7*	20.9 ± 1.5	207 ± 9
2×10^{-9}	906 ± 29	82 ± 3	32.1 ± 1.4*	223 ± 10
2×10^{-8}	847 ± 121	95 ± 5	31.6 ± 1.8	240 ± 15
2×10^{-7}	750 ± 126	91 ± 7	30.5 ± 1.0	257 ± 36

Table 3. Dose Responses for Glucagon on Glucose Production, ^{14}C-Glucose Synthesis, Ureogenesis and Ketogenesis in Livers Perfused with 20 mM ^{14}C-Lactate. + Significantly different from control $p < 0.05$

METABOLITE	METABOLIC CHANGES		
	CONTROL	GLUCAGON	CHANGE DUE TO GLUCAGON
	μmoles/100g rat/hr		
Glucose	+227	+613	+386
Glycogen	-162	-508	-346
Lactate	-354	-453	- 99
Pyruvate	+ 19	+ 16	- 3
Oxygen	-564	-678	-114
Acetoacetate	+10.9	+13.8	+2.9
ß-Hydroxybutyrate	+ 8.5	+17.7	+9.2

Table 4. Metabolic Changes Produced by Glucagon in Livers Perfused with 20 mM Lactate

METABOLITE	RADIOACTIVITY		PERCENT CHANGE	
	CONTROL	GLUCAGON	RADIOACTIVITY	SPECIFIC ACTIVITY
	$cpm \times 10^{-3}$/100g rat		%	
Glucose	85.6	188.3	+119	-19
Glycogen	24.4	1.4	- 94	-84
CO_2	123.1	131.7	+ 7	
Ketone bodies	6.2	7.6	+ 23	-25
Total lipid	23.3	15.7	- 33	
Lipid glycerol	5.6	6.5	+ 16	
Lipid fatty acid	13.5	7.0	- 48	-50
Cholesterol	5.7	2.9	- 49	
Protein	7.2	4.6	- 36	
Lactate	-312	-399	- 28	

Table 5. Changes Produced by Glucagon in U-^{14}C-Lactate Metabolism

lactate, a perfusion system with recirculating medium was employed. The initial concentration of lactate was 20 mM and declined to about 10 mM during perfusion. Although these concentrations are far above the normal levels of lactate in vivo, there is no evidence that the effects of glucagon differ qualitatively from those occurring with levels of lactate below 2 mM or mixtures of 1 mM lactate and 0.1 mM pyruvate discussed later in this report. However, since the glucagon concentration employed in these experiments was very high (2×10^{-9} M), the physiological significance of some of the changes may be questioned.

Table 4 shows that glucagon produced the expected increase in glucose output, glycogenolysis, ketogenesis and ureogenesis. It also increased O_2 uptake and lactate utilization. The ^{14}C-glucose synthesis from ^{14}C-lactate was increased approximately two-fold whereas labeling of glycogen was markedly reduced (Table 5) consistent with activation of gluconeogenesis and glycogenolysis and perhaps inhibition of glycogen synthesis. The products of acetyl-CoA metabolism namely lipid fatty acid, ketone bodies and cholesterol were labeled to a relatively small extent relative to hexose and the total incorporation of ^{14}C into these products was reduced by about 30% by glucagon. Ketone radioactivity was increased slightly and the labeling of fatty acid and cholesterol was decreased by about 50%.

Protein radioactivity was diminished by about 40%. In contrast to results with U-^{14}C-alanine (Mallette et al. (1969a), $^{14}CO_2$ production from U-^{14}C-lactate was not significantly increased by glucagon. As will be discussed, the data indicate that activation of gluconeogenesis was not due to the small inhibition of lactate oxidation that apparently occurred. They also suggest that activation of the Krebs cycle was partly responsible for the stimulation of respiration and that the extra acetyl-CoA utilized in the cycle was derived from endogenous sources.

Effects of Glucagon on the Metabolism of 1 mM ^{14}C-Lactate Plus 0.1 mM ^{14}C-Pyruvate

Earlier studies have localized a major site of action of glucagon on gluconeogenesis to the sequence of reactions between pyruvate and P-pyruvate (Exton and Park 1969). In an attempt to identify the major reaction affected by glucagon, we examined the early changes induced by 5 x 10^{-9} M glucagon in the levels and radioactivities of gluconeogenic intermediates in livers perfused by the flow through (non-recirculating) technique with 1 mM U-^{14}C-lactate plus 0.1 mM 2-^{14}C-pyruvate. Livers were initially perfused for one hour with recirculating medium containing the labeled substrates with and without glucagon. Glucagon increased glucose labeling more than three-fold and increased the tissue concentrations and radioactivities of P-pyruvate and aspartate (Tables 6-8). The tissue level of malate was also increased slightly but its radioactivity was diminished. The specific activities of malate and P-pyruvate were decreased but that of aspartate was increased. The radioactivities and levels of glutamate, α-ketoglutarate and oxalacetate were reduced, whereas the radioactivities of succinate and fumarate were increased (Tables 7, 8).

It thus appeared that activation of gluconeogenesis was associated with rapid increases in the formation of P-pyruvate from pyruvate and endogenous sources. The increased formation of P-pyruvate did not result secondarily from increased synthesis of malate and oxalacetate from ^{14}C-pyruvate since the radioactivities of these compounds declined (Table 7).

Since it was possible that an activation of P-pyruvate carboxykinase might have masked a simultaneous stimulation of pyruvate uptake or carboxylation, the experiments were repeated in the presence of tryptophan, an inhibitor of P-pyruvate carboxykinase (Ray et al. 1966). Tryptophan markedly decreased the incorporation of ^{14}C from ^{14}C-lactate and ^{14}C-pyruvate into glucose and P-pyruvate and abolished the effects of glucagon (Fig. 1). It also caused the accumulation of malate, aspartate and citrate in control livers (Fig. 1). Glucagon caused marked increases in malate and aspartate in the presence of tryptophan but decreased the radioactivity of malate and did not affect that of aspartate (Fig. 2). The accumulation of these

^{14}C-SUBSTRATE	^{14}C INCORPORATED INTO GLUCOSE		GLUCAGON EFFECT
	CONTROL	GLUCAGON	
	cpm x 10^{-3}/2.5 min		% of control
^{14}C-U-L-Lactate	19.9 ± 2.5	64.2 ± 8.8	322
^{14}C-1-Pyruvate	15.5 ± 1.2	34.4 ± 3.8	223
^{14}C-Bicarbonate	5.5 ± 0.9	16.5 ± 3.1	302
^{14}C-U-Glutamate	2.5 ± 0.5	18.1 ± 1.8	736

Table 6. Effects of Glucagon on ^{14}C-Glucose Synthesis from Various ^{14}C-Substrates for 2.5 Minutes

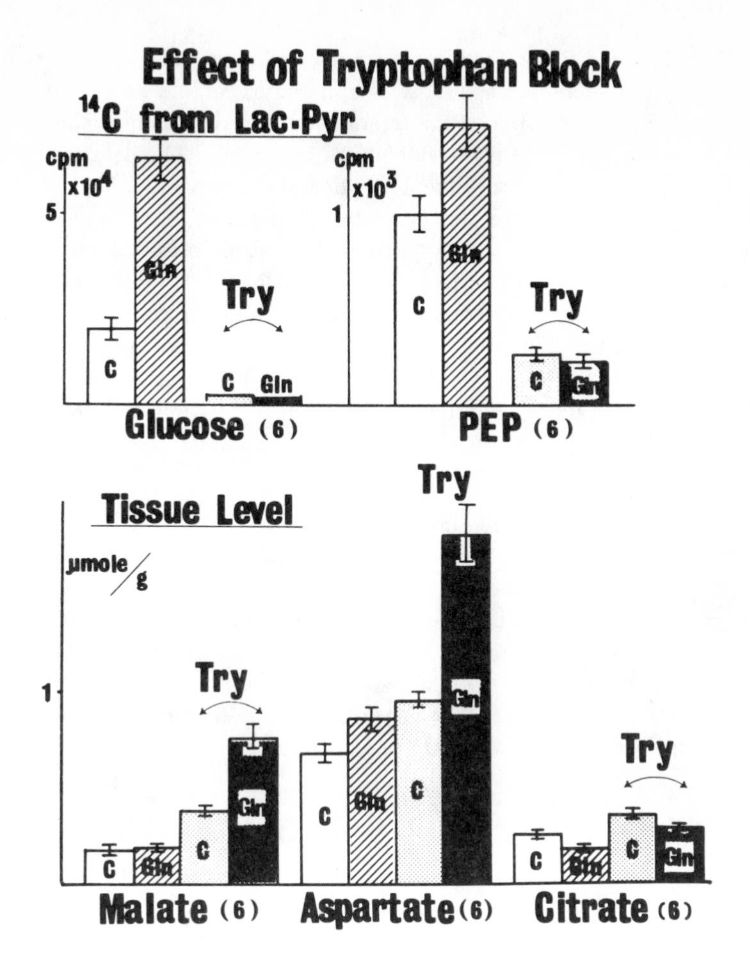

Fig. 1

intermediates therefore appeared to be due to their increased formation from unlabeled endogenous sources and not to increased carboxylation of pyruvate. The changes in the radioactivities or specific activities of glutamate and citric acid cycle intermediates in control livers indicated additional effects of glucagon on the Krebs cycle. To investigate these, experiments were carried out with ^{14}C-glutamate and ^{14}C-aspartate as described below.

Effects of Glucagon in Livers Perfused with ^{14}C-Glutamate

Livers were perfused by the flow-through technique with 1 mM lactate, 0.1 mM pyruvate and tracer amounts of U-^{14}C-glutamate. Measured during the first 2.5 minutes of treatment, glucagon increased ^{14}C-glucose synthesis more than sevenfold (Table 6) and produced changes in tissue metabolite levels identical to those observed with ^{14}C-lactate plus ^{14}C-pyruvate. It also caused marked changes in the ^{14}C-content of many intermediates (Table 8). The radioactivities of aspartate, succinate, fumarate and P-pyruvate were increased whereas those of α-ketoglutarate and oxalacetate were decreased and those of glutamate, malate and citrate were unchanged. The specific activities of the intermediates were largely unchanged except for those of aspartate and P-pyruvate which were increased. The data with this labeled substrate are thus consistent with the activation of P-pyruvate carboxykinase suggested by the studies with lactate plus pyruvate. In addition, they indicate possible activation of the utilization of α-ketoglutarate. To explore this possibility further, $^{14}CO_2$ production was examined in livers perfused with 1-^{14}C-glutamate in the presence of fluorocitrate. Figure 3 shows that glucagon caused a marked stimulation of $^{14}CO_2$ release which was abolished by arsenate. Since the response was observed in the presence of fluorocitrate, an inhibitor of citrate synthase, it was apparently due to activation of α-ketoglutarate utilization and not to increased isocitrate decarboxylation. Its abolition by arsenate an inhibitor of succinate thiokinase, confirms this view.

The activation of $^{14}CO_2$ formation from 1-^{14}C-glutamate by glucagon in the absence of fluorocitrate was less at six minutes than at two minutes and was not significant at 12 minutes. The loss of the glucagon response with prolonged exposure of the liver to isotopic glutamate was probably due to the protein catabolic action of the hormone since there were marked reductions in the specific activity of glutamate and α-ketoglutarate under these conditions (see below). Since the tissue levels of these metabolites were nevertheless decreased, it appeared that the increased formation of unlabeled glutamate was less than its increased utilization.

INCORPORATION OF ^{14}C-PYR IN TRY-TREATED LIVER

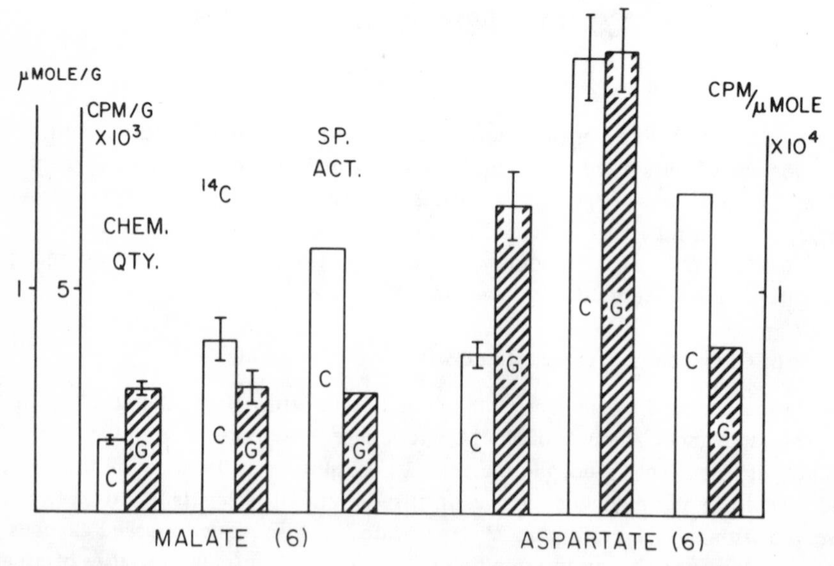

Fig. 2

Table 9 shows the changes in tissue levels, radioactivities and specific activities of gluconeogenic intermediates in livers perfused for 8.5 minutes with 1mM lactate plus 0.1 mM pyruvate and tracer amounts of ^{14}C-glutamate. It can be seen that the tissue metabolite concentrations were identical with those seen at 2.5 minutes except for glutamate which was further decreased. Compared with the changes observed at 2.5 minutes, the decreases in radioactivities and specific activities of glutamate and α-ketoglutarate produced by glucagon were pronounced. In contrast to the results at 2.5 minutes, the radioactivities and specific activities of malate and citrate were reduced by glucagon and the radioactivities of succinate and fumarate were unchanged. The increases in radioactivities of aspartate and P-pyruvate were less pronounced with glucagon at 8.5 minutes than at 2.5 minutes, and, at 8.5 minutes, increases in the specific activities of these metabolites were no longer evident. The changes at 8.5 minutes therefore provide clearer evidence of activation of aspartate and P-pyruvate formation with glucagon.

Effects of Glucagon in Livers Perfused with ^{14}C-Aspartate

To examine the effects of glucagon on P-pyruvate synthesis from a substrate which did not need to enter the mitochondria, livers were perfused for 2.5 minutes with 1 mM lactate plus 0.1 mM pyruvate and U-^{14}C-aspartate. It is seen from Table 10 that glucagon increased ^{14}C-glucose formation about four-fold. It also increased the radioactivity of P-pyruvate and decreased that of malate and citrate. The specific activity of aspartate was higher than that of malate and citrate and was not changed by glucagon. The results thus suggest that glucagon increased the conversion of cytosolic aspartate to P-pyruvate.

INTERMEDIATE	RADIOACTIVITY		SPECIFIC ACTIVITY	
	CONTROL	GLUCAGON	CONTROL	GLUCAGON
	cpm x 10^{-3}/g tissue		cpm/nmole	
Pyruvate	21.5	12.7	21.3	16.0
Oxalacetate	0.7	0.5		
Malate	2.7	2.3	20.0	12.3
Aspartate	2.8	8.5	6.6	12.5
Citrate	3.7	3.0	12.4	11.2
P-pyruvate	1.0	1.5	10.8	7.3
Glucose	19.9	64.2		
Glycogen	0.4	0.9		
Glutamate	21.4	14.3	9.4	8.9
a-Ketoglutarate	1.9	0.7	7.1	6.7
Succinate	3.4	4.2		
Fumarate	1.0	1.4		
Alanine	1.4	2.8	19.5	23.0

Table 7. Effects of Glucagon on Radioactivity of Intermediates in Livers Perfused with 1 mM ^{14}C-U-Lactate Plus 0.1 mM ^{14}C-U-Pyruvate for 2.5 Minutes.

INTERMEDIATE	CONCENTRATION		RADIOACTIVITY		SPECIFIC ACTIVITY	
	CONTROL	GLUCAGON	CONTROL	GLUCAGON	CONTROL	GLUCAGON
	μmole/g tissue		cpm x 10^{-3}/g tissue		cpm/nmole	
Glutamate	2.35	1.80*	107.5	90.7	45.7	50.5
α-Ketoglutarate	0.31	0.12*	25.9	6.7	75.6	61.0
Succinate			5.4	12.2*		
Fumarate			0.65	1.46*		
Malate	0.15	0.16	10.0	9.8	66.5	61.2
Oxalacetate			0.32	0.23		
Aspartate	0.53	0.72*	10.2	32.1*	18.5	41.6*
Citrate	0.34	0.29*	16.6	17.3	48.7	60.3
P-pyruvate	0.10	0.17*	10.0	25.1*	10.0	16.6*

Table 8. Effects of Glucagon on Levels and Radioactivity of Intermediates in Livers Perfused with U-^{14}C-Glutamate for 2.5 Minutes. + Significant effect

In all the preceding experiments, the specific activity of P-pyruvate in control livers was considerably lower than that of the initial substrate or of the 4 C-dicarboxylic acids. This was probably due to the formation of unlabeled P-pyruvate via glycolysis since in livers from fasted rats in which glycogen is depleted and the rate of glycolysis is low, the specific activity of P-pyruvate relative to malate or aspartate was much higher (data not shown).

Short-Term Effects of Glucagon in Livers Perfused with $H^{14}CO_3$

Further evidence that glucagon stimulated P-pyruvate carboxykinase was obtained in experiments in which livers were perfused for one hour with recirculating medium containing no added substrate and then for 11 minutes with non-recirculating medium containing 1 mM lactate plus 0.1 mM pyruvate with or without glucagon. The ^{14}C-bicarbonate was infused for one minute prior to liver freezing. Table 11 shows that glucagon produced changes in tissue metabolite levels similar to those in Table 8, namely increases in aspartate and P-pyruvate and decreases in α-ketoglutarate and citrate. It also reveals that the isotope was incorporated mainly into aspartate, citrate and malate which are the initial products of oxalacetate metabolism. The radioactivity of oxalacetate was very low and that of P-pyruvate was higher than that of succinate, fumarate or α-ketoglutarate. Glucagon increased the incorporation of ^{14}C into glucose three-fold (Table 6). The radioactivities of P-pyruvate and aspartate were increased by about 50% whereas those of malate, citrate, oxalacetate, fumarate and α-oxoglutarate were decreased.

The results are consistent with the activation of P-pyruvate carboxykinase postulated above. The indicate the existence of a very small pool of oxalacetate in the mitochondria which turns over extremely rapidly. They disprove the view that increased pyruvate carboxylation is solely responsible for the stimulation of gluconeogenesis by glucagon although they do not exclude the possibility of stimultaneous activation of P-pyruvate carboxykinase and of pyruvate carboxylase or pyruvate entry into the mitochondria.

Discussion

Differential Sensitivity of Metabolic Processes to Stimulation by Glucagon

The dose response curves for glucagon activation of glycogenolysis, gluconeogenesis, ketogenesis and ureogenesis shown in Tables 2 and 3 indicate the existence of a further basis for specificity of hormone action at the cellular level, namely differential sensitivity of cellular processes to activation by cyclic AMP. Other bases of specificity of action of hormones working through cyclic AMP are: 1) differential sensitivities of tissue adenyl cyclases to stimulation by hormones, and 2) differences in metabolic pathways of target tissues (Butcher et al. 1968).

It is of interest to relate the results of Tables 2 and 3 to the probable level of glucagon in portal venous plasma. However, the immunoassays currently available do not distinguish between pancreatic glucagon and glucagon-like material derived from the gut (Unger et al. 1968). This is a serious drawback since pancreatectomy produces little or no fall in the level of immunoreactive glucagon in jejunal or peripheral venous plasma (Buchanan et al. 1968, Buchanan et al. 1969) indicating that gut "glucagon" is the major component present. Ohneda et al. 1969 report levels of immunoreactive glucagon in pancreaticoduodenal venous blood from dogs ranging between 0.3×10^{-9} and 1.6×10^{-9} M, whereas Buchanan et al. (1969) give values between 0.1×10^{-9} and 0.3×10^{-9} M perhaps because of a lower cross-reactivity of their antisera with gut "glucagon." Since the true glucagon levels are

INTERMEDIATE	CONCENTRATION		RADIOACTIVITY		SPECIFIC ACTIVITY	
	CONTROL	GLUCAGON	CONTROL	GLUCAGON	CONTROL	GLUCAGON
	µmole/g		cpm x 10^{-3}/g		cpm/nmole	
Glutamate	3.42	1.96*	290.4	88.5*	114.3	53.5*
a-Ketoglutarate	.0.42	0.16*	44.9	5.9*	107.1	36.0*
Succinate			7.8	7.4		
Fumarate			1.1	1.1		
Malate	0.15	0.17	13.9	8.1*	97.4	50.5*
Oxalacetate			0.3	0.1*		
Aspartate	0.64	0.89*	28.7	38.5*	45.1	43.2
Citrate	0.44	0.38*	28.0	12.1*	64.4	31.7*
P-pyruvate	0.13	0.25*	0.8	1.8*	6.2	7.5

Table 9. Effects of Glucagon on Levels and Radioactivity of Intermediates in Livers Perfused with U-^{14}C-Glutamate for 8.5 Minutes. + Significant effect

INTERMEDIATE	RADIOACTIVITY		SPECIFIC ACTIVITY	
	CONTROL	GLUCAGON	CONTROL	GLUCAGON
	cpm x 10^{-3}/g tissue		cpm/nmole	
Aspartate	34.8	41.8	62.9	65.7
Malate	6.0	5.0	49.1	40.7
Oxalacetate	0.2	0.2		
Citrate	7.3	5.4	21.4	19.1
P-pyruvate	1.2	2.3	10.4	11.8
Glucose	21.9	88.1		
Glycogen	0.2	0.8		

Table 10. Effects of Glucagon on Radioactivity of Intermediates in Livers Perfused with ^{14}C-Aspartate for 2.5 Minutes.

probably lower than these estimates and since the glucagon concentration in portal blood would be expected to be less than one-tenth of that in pancreaticoduodenal blood, the present findings suggest that hepatic glycogenolysis and gluconeogenesis may be controlled by glucagon in vivo but perhaps not ketogenesis and ureogenesis. However, it should be emphasized that these studies have been carried out with semisynthetic perfusion medium and that the addition of substrates such as amino acids and fatty acids and of hormones such as corticosteroids and epinephrine may change the sensitivity of these processes to glucagon. In addition, blood glucagon levels in rats may be higher than in dogs.

The results of Table 3 show a 100-fold difference between the concentrations of glucagon required to stimulate ketogenesis and gluconeogenesis. This renders it improbable that the two processes are obligatorily linked.

Effects of Glucagon on Lactate Metabolism in the Perfused Liver

The results of Table 5 indicate that in livers from fed rats, lactate is metabolized by way of oxalacetate to products such as glucose, glycogen and lipid glycerol to a greater extent (about 40%) than by way of acetyl-CoA to products such as ketone bodies, lipid fatty acids and cholesterol (about 10%). Glucagon increases the flow of ^{14}C into the products of gluconeogenesis by a factor of about 1.7. This increase accounts for about 90% of the increased utilization of lactate. This must have involved activation of P-pyruvate formation since no comparable decrease of flow of lactate carbon into other pathways (CO_2, lipid, protein) could be detected. It should be noted in this regard that about 93% of the lactate carbon utilized was recovered in the products measured in the control and about 88% in the glucagon perfusions.

It is not possible to conclude definitely from the data of Table 5 whether glucagon inhibited or did not change pyruvate oxidation to acetyl-CoA. This is because CO_2

may be labeled from ^{14}C-pyruvate entering the cycle as oxalacetate or acetyl-CoA. It would seem, however, that the increased flow of isotope from pyruvate to P-pyruvate should have resulted in an increase in the specific activity of oxalacetate and hence increased $^{14}CO_2$ formation if the oxalacetate involved in gluconeogenesis mixes with that involved in the Krebs cycle, and there was no change in the rate of the cycle. However, since $^{14}CO_2$ formation was not changed by glucagon, it would appear that either the cycle was slowed or the entry of label via acetyl-CoA was inhibited. The following considerations render it improbable that the Krebs cycle was inhibited: 1) glucagon increased O_2 consumption to an extent greater than that attributable to the stimulation of ketogenesis and ureogenesis,+ 2) the increased ketogenesis and oxidation of amino acids outside the Krebs cycle would have supplied only a very small fraction of the extra ATP required for increased gluconeogenesis and ureogenesis, 3) the results of Regen and Terrell (1968) and the data obtained with ^{14}C-glutamate in the present study (Tables 8, 9, Fig. 3) suggest increased activity of the cycle with glucagon. In summary, it seems probable that the citric acid cycle was increased by glucagon, but that some extra substrate used was derived from endogenous protein resulting in increased synthesis of urea and unlabeled glucose.

Rapid Effects of Glucagon in Livers Perfused with Physiological Levels of Lactate and Pyruvate

The changes in tissue intermediates induced by glucagon treatment for 2.5 minutes in livers perfused with 1 mM lactate plus 0.1 mM pyruvate resemble those reported (by Exton and Park 1969) in livers perfused for much longer times and with higher concentrations of lactate. They are consistent with (1) the activation of glutamate and α-ketoglutarate utilization, (2) the stimulation of aspartate formation from oxalacetate, and (3) the activation of P-pyruvate formation from oxalacetate.

INTERMEDIATE	CONCENTRATION		RADIOACTIVITY	
	CONTROL	GLUCAGON	CONTROL	GLUCAGON
	nmoles/g tissue		cpm x 10^{-3}/g tissue	
Malate	172	144	17.0	11.4
Aspartate	727	904	38.9	54.9
Citrate	390	276	32.2	18.0
α-Ketoglutarate	378	119	1.5	0.6
Succinate			0.6	0.9
Fumarate			1.1	0.6
Oxalacetate			0.2	0.1
P-pyruvate	173	277	2.8	4.7

Table 11. Effects of Glucagon on Levels and Radioactivity of Intermediates in Livers Perfused with $H^{14}CO_3$ for one minute.

+ Unpublished calculations, this laboratory.

Additional data supporting the proposed activation of glutamate and α-ketoglutarate utilization are the following: 1) The increased release of $^{14}CO_2$ from 1-^{14}C-glutamate which is blocked by arsenate but is unaffected by fluorocitrate, 2) the fall in radioactivity of glutamate and α-ketoglutarate and the increases in ^{14}C-content of succinate and fumarate in livers perfused with U-^{14}C-glutamate, 3) the increased levels and decreased specific activities of malate and aspartate in livers perfused with tryptophan and ^{14}C-lactate plus ^{14}C-pyruvate.

The activation of 4C-dicarboxylic acid formation from glutamate and α-ketoglutarate does not lead to increased levels of malate and citrate apparently because of the concurrent activation of P-pyruvate and aspartate formation from oxalacetate.

The increased rate of glutamate-oxalacetate aminotransferase and activation of P-pyruvate carboxykinase are shown by the increased radioactivities and specific activities of aspartate and P-pyruvate and the decreased radioactivity and specific activity of malate in livers perfused with ^{14}C-lactate plus ^{14}C-pyruvate, $H^{14}CO_3$ or ^{14}C-glutamate. With the latter substrate, the activation of aspartate and P-pyruvate formation was more evident at 8.5 minutes, i.e., when the reduction in glutamate specific activity diminished the effect of glucagon on labeled 4C-dicarboxylic acid formation from this substrate. Although no reliable data were obtained for oxalacetate levels in these studies, the effect of glucagon to consistently lower the radioactivity of this intermediate points to the predominance of its action on oxalacetate conversion to P-pyruvate over its action on steps leading to the formation of oxalacetate.

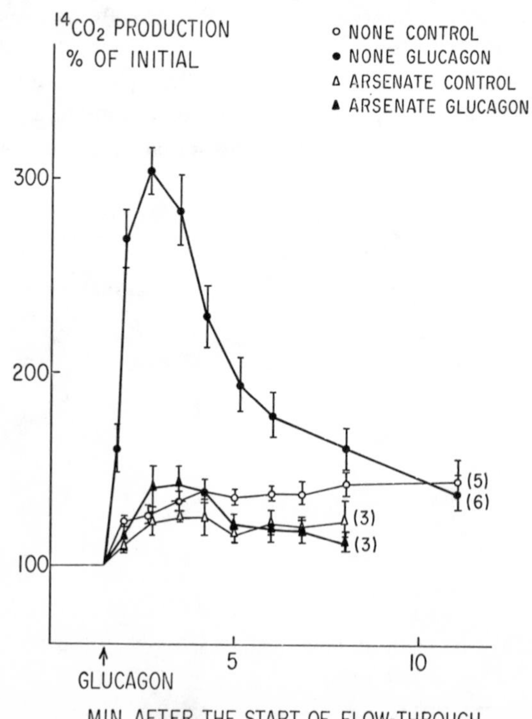

Fig. 3

Consideration of Possible Mechanisms of Action of Glucagon on Gluconeogenesis

There is much evidence indicating that glucagon does not act on gluconeogenesis by stimulating lipolysis. This has been discussed at length by Exton et al. (1969) and will not be recounted in this report.

More plausible proposals for the gluconeogenic action of glucagon include the following: 1) stimulation of pyruvate uptake by mitochondria (Adam and Haynes 1969), 2) Activation of pyruvate carboxylase (Williamson 1966), 3) Stimulation of 4C-dicarboxylic acid efflux from mitochondria, 4) Activation of P-pyruvate carboxykinase (Exton and Park 1969, Exton and Park 1966). None of these processes has been shown to be influenced by glucagon or cyclic AMP in vitro. However, the action of the hormone or the nucleotide may be indirect as in the case of glycogen phosphorylase.

Adam and Haynes (1969) have obtained evidence of increased pyruvate uptake by liver mitochondria isolated from glucagon-treated rats. This was associated with enhanced CO_2 fixation and pyruvate 1-^{14}C decarboxylation and was less apparent in mitochondria suspended in media of low osmolarity. It was suggested to be due primarily to stimulation of pyruvate transfer into mitochondria.

The present findings neither support nor exclude this hypothesis. In our studies, pyruvate decarboxylation appeared largely unaffected by glucagon in livers perfused with 20 mM ^{14}C-lactate. It is possible that activation of mitochondrial pyruvate uptake by glucagon resulted in increased pyruvate carboxylation but not increased pyruvate oxidation due to saturation or inhibition of the latter pathway and that oxalacetate and malate did not increase because of concurrent activation of aspartate and P-pyruvate formation. Furthermore, the studies with tryptophan did not reveal any stimulation of pyruvate uptake but conditions may have not been favorable. Similarly, an effect of glucagon on pyruvate carboxylase is also not supported by the studies with tryptophan.

No studies have been reported concerning possible effects of glucagon or cyclic AMP on the transfer of substances into or out of mitochondria. This process may be a major site of control for gluconeogenesis (and glycolysis). Recent studies have indicated differences in the rate of efflux of malate, aspartate and oxalacetate from mitochondria and have suggested that the transport of malate and oxalacetate into mitochondria may be energy-linked (Lardy et al. 1965, Walter 1966, Haynes 1965, Haslam and Krebs 1968, Haslam and Griffiths 1968). There is also evidence of competition between various anions for accumulation in mitochondria (Harris and Manger 1968, 1969) and of a possible role for mitochondrial cation content (e.g., K^+ or Ca^{++}) in the regulation of anion uptake (Haslam and Krebs 1968, Haslam and Griffiths 1968, Harris and Manger 1969, Harris 1968). It is therefore quite possible that cyclic AMP may effect this portion of the gluconeogenic pathway either directly or indirectly. Glucagon and cyclic AMP promote the release of K^+ from liver cells and increase the efflux of ^{45}Ca from livers previously perfused with this isotope (Friedmann and Park 1968). It is not known whether these changes result in or from alterations in mitochondrial K^+ or Ca^{++} content.

Although glucagon may have effects on the efflux of aspartate or malate from mitochondria, it appears that such changes are not fully responsible for the increased synthesis of P-pyruvate and glucose. This is because the hormone stimulates the incorporation of ^{14}C into P-pyruvate and glucose in livers perfused with ^{14}C-aspartate. Since it seems improbable that ^{14}C-P-pyruvate would be synthesized mainly from ^{14}C-oxalacetate produced in the mitochondria and not from ^{14}C-oxalacetate produced from ^{14}C-aspartate in the cytosol, it is concluded that glucagon probably affects P-pyruvate carboxykinase.

It is of interest to examine the possible relationships between the apparent effects of glucagon on α-ketoglutarate utilization, aspartate formation and P-pyruvate carboxykinase activation. In the first place, the action of glucagon on succinate formation from α-ketoglutarate appears to be independent of its effect on gluconeogenesis. This is indicated by the fact that concentrations of tryptophan sufficient to inhibit gluconeogenesis failed to block the increased labeling of succinate and fumarate and reduced labeling and concentration of α-ketoglutarate caused by glucagon in livers perfused with ^{14}C-glutamate. However, it is uncertain whether or not inhibition of α-ketoglutarate conversion to succinate would result in loss of the gluconeogenic action of glucagon.

Possible links between α-ketoglutarate oxidation and P-pyruvate synthesis are: 1) through increased generation of GTP by succinate thiokinase, 2) through changes in the levels of possible activators or inhibitors of P-pyruvate carboxykinase.

Since GTP is probably the physiological substrate of P-pyruvate carboxykinase, the continued synthesis of PEP from oxalacetate in the cytosol requires the production of GTP. This is probably largely formed from ATP and GDP by nucleoside diphosphate kinase since its formation by succinate thiokinase would require an extremely high, sustained utilization of α-ketoglutarate. In any case, some means of transferring GDP and GTP across the mitochondrial membrane is necessary. This transport problem does not apply to other species, such as the guinea pig or rabbit, in which P-pyruvate carboxykinase is mitochondrial and can be directly coupled to intramitochondrial GTP synthesis. Another possibility which cannot be ruled out is that the small amount of P-pyruvate carboxykinase in rat liver mitochondria, if capable of great activation (Nordlie and Lardy 1963), could be responsible for the action of glucagon on gluconeogenesis.

The second possible link between α-ketoglutarate oxidation and P-pyruvate synthesis remains equally speculative at the present time in view of the absence of known effectors of the carboxykinase.

There are a variety of possible explanations to explain the increased activity of glutamate-oxalacetate transamination following glucagon administration. Since the levels of intermediates measured in these studies presumably largely reflect changes of concentrations within the cytosol compartment rather than in the relatively very small mitochondrial compartment, it is not possible to know in which way mitochondrial transamination is driven toward aspartate formation by changes in the concentrations of other components of the transaminase reaction. It should be noted, however, that selective permeability changes between mitochondria and cytosol to these and other intermediates provide many possible explanations. For example, facilitated entry of glutamate and/or pyruvate (leading to a rise in oxalacetate) could drive transamination while facilitated egress of mitochondrial aspartate coupled with its faster utilization in P-pyruvate formation could pull the reaction. It is possible to construct various hypotheses based on alterations in permeability which could fit with the observed changes in intermediates. The principal experiments suggesting a role for permeability (1) are the studies with mitochondria from the livers of glucagon-treated rats which suggest increased permeability to pyruvate (Adam and Haynes 1969) (2) demonstrations of transport effects induced by glucagon and cyclic AMP at the level of the plasma membrane in rat liver to alanine, α-amino isobutyric acid and lysine (Mallette et al. 1969b) and (3) permeability effects of cyclic AMP in the toad bladder and, presumably, the mammalian kidney (Orloff and Handler 1967).

In this connection, it should be recalled that amino acids, and alanine in particular, provide the major source of carbon for gluconeogenesis in vivo (Mallette et al. 1969a). The stimulation of transport of alanine and other amino acids at the plasma membrane may prove, therefore, to be one of the most important physiological actions of glucagon and the catecholamines in gluconeogenesis.

In conclusion, it is apparent that glucagon has a wide variety of actions on the liver, not all of which may occur in vivo. Some of these, (e.g., the apparent activation of P-pyruvate carboxykinase) can be clearly related to the gluconeogenic effect of the hormone, but the significance of some of the others (e.g., increased aspartate formation and α-ketoglutarate utilization) is not understood and may be irrelevant to gluconeogenesis. Work is in progress to further elucidate the mechanism of action of glucagon on gluconeogenesis. Direct or indirect effects of cyclic AMP on more than one gluconeogenic reaction may be involved. Failure to observe in vitro effects of the nucleotide on purified enzymes does not necessarily exclude them from involvement in this action of glucagon.

Acknowledgment

Supported by Project Program Grant AM07462 from the National Institutes of Health, U. S. Public Health Service.

References

Adam, P. J., R. C. Haynes, Jr.: J. biol. Chem. 244 (1969) 6444

Buchanan, K. D., J. E. Vance, A. Morgan, R. H. Williams: Amer. J. Physiol. 215 (1968) 1293

Buchanan, K. D., J. E. Vance, K. Dinstl, R. H. Williams: Diabetes. 18 (1969) 11

Butcher, R. W., G. A. Robison, J. G. Hardman, E. W. Sutherland: Advances in Enzyme Regulation. Ed. by G. Weber, Vol. 6. Pergamon, New York 1968, p. 357

Exton, J. H., C. R. Park: Pharmacol. Rev. 18 (1966) 181
Exton, J. H., C. R. Park: J. biol. Chem. 242 (1967) 2622
Exton, J. H., J. G. Corbin, C. R. Park: J. biol. Chem. 244 (1969) 4095
Exton, J. H., C. R. Park: J. biol. Chem. 244 (1969) 1424
Exton, J. H., C. R. Park: J. biol. Chem. In press
Friedmann, N., C. R. Park: Proc. Natl. Acad. Sci. U.S.A. 61 (1968) 504
Harris, E. J.: Biochem. J. 109 (1968) 247
Harris, E. J., J. R. Manger: Biochem. J. 109 (1968) 239
Harris, E. J., J. R. Manger: Biochem. J. 113 (1969) 617
Haslam, J. M., D. E. Griffiths: Biochem. J. 109 (1968) 929
Haslam, J. M., H. A. Krebs: Biochem. J. 107 (1968) 659
Haynes, R. C., Jr.: J. biol. Chem. 240 (1965) 4103
Lardy, H. A., V. Paekau, P. Walter: Proc. Nat. Acad. Sci. (U.S.A.) 53 (1965) 1410
Mallette, L. E., J. H. Exton, C. R. Park: J. biol. Chem. 244 (1969a) 5713
Mallette, L. E., J. H. Exton, C. R. Park: J. biol. Chem. 244 (1969b) 5724
Mortimore, G. E.: Amer. J. Physiol. 204 (1963) 699
Nordlie, R. C., H. A. Lardy: J. biol. Chem. 238 (1963) 2259
Ohneda, A., E. Aguilar-Parada, A. M. Eisentraut, R. H. Unger: Diabetes. 18 (1969) 1
Orloff, J., J. Handler: Amer. J. Med. 42 (1967) 757
Ray, P. D., D. O. Foster, H. A. Lardy: J. biol. Chem. 241 (1966) 3904
Regen, D. M., E. B. Terrell: Biochem. biophys. Acta. 170 (1968) 95
Ui, M., C. R. Park: J. biol. Chem. In press
Unger, R. H., A. Ohneda, I. Valverde, A. M. Eisentraut, J. H. Exton: J. clin. Invest. 47 (1968) 48
Walter, P., V. Paetkau, H. A. Lardy: J. biol. Chem. 241 (1966) 2523
Williamson, D. H., J. Mellanby, H. A. Krebs: Biochem. J. 82 (1962) 90
Williamson, J. R.: Biochem. J. 101 (1966) 11c

Dissociation of Gluconeogenic and Ketogenic Action of Glucagon in the Perfused Rat Liver

J. Fröhlich and O. Wieland
Institut für Klinische Chemie und Forschergruppe Diabetes,
Krankenhaus München-Schwabing, Munich, FRG

Summary

The mode of action of glucagon in accelerating the biosynthesis of glucose in liver has been the subject of intensive work. There is now good evidence that fatty acid oxidation is intimately involved in the regulation of gluconeogenesis and that more fatty acid is made available from endogenous lipid in the presence of glucagon. The question whether glucagon exerts an effect on gluconeogenesis in addition to that produced by fatty acids (Exton et al. 1969, Ross et al. 1967) is still a matter of controversy. We have tried to get more insight into this problem by liver perfusion experiments studying the effect of glucagon on glucose formation from lactate, either in the presence of saturating concentrations of fatty acids or during inhibited fatty acid oxidation.

Maximal stimulation of gluconeogenesis by fatty acids (1.11 μmoles glucose formed per g liver wet weight per minute) was obtained in the presence of 1 mM oleate or 1.5 mM octanoate. This rate was further increased to 1.86 μmoles/g per minute when glucagon was given in addition to either oleate or octanoate. Ketogenesis from oleate was increased from 0.53 to 0.72 μmoles/g per minute by glucagon whereas ketone formation from octanoate was the same with or without glucagon (0.95 μmoles/g per minute).

In a second series of experiments endogenous lipolysis was inhibited by glycodiazine, an oral antidiabetic. No detectable amounts of ketone bodies were formed in the presence of 10 mM glycodiazine. At the same time, the rate of gluconeogenesis from lactate was decreased from normally 0.64 to 0.36 μmoles/per minute. Glucagon restored glucose formation to 0.61 μmoles/g per minute but did not release the inhibitory effect of glycodiazine on ketogenesis. In agreement with earlier findings from this laboratory (Menahan and Wieland 1969), the addition of oleate resulted in a rate of gluconeogenesis almost identical to that observed in the experiments without glycodiazine (1.10 μmoles/g per minute). As expected stimulation of gluconeogenesis by oleate was accompanied by the formation of considerable amounts of ketone bodies (0.96 μmoles/g per minute). Addition of glucagon together with oleate caused a further increase in gluconeogenesis (1.55 μmoles/g per minute) without affecting the rate of ketogenesis.

The reported results lend further support to the view that there exists another mechanism — not mediated by fatty acid oxidation — by which glucagon stimulates gluconeogenesis. A maximal effect of glucagon, however, is dependent upon the concomitant oxidation of fatty acids.

Introduction

During the last several years, considerable information has been gathered regarding the action of glucagon on different metabolic pathways in vivo. Among the well established effects, stimulation of glucose output, urea formation and ketogenesis could be observed in the perfused rat liver preparation (Schimassek and Mitzkat 1963, Sokal 1966, Miller 1965, Cahill 1965). Further investigations on the same system have demonstrated that the increased glucose output is a result of the stimulation of both glycogenolysis and gluconeogenesis (Foa 1964, Mortimore 1963). Ureogenesis is thought to be stimulated through activation of proteolysis, and ketogenesis by enhancement of lipolysis by glucagon (Miller 1965, Bewsher and Ashmore 1966, Struck et al. 1965). In contrast to these well known metabolic effects, little is known about the mechanisms by which the hormone displays its regulatory control. Special attempts have been made to elucidate the control of gluconeogenesis by glucagon. From their findings on fatty acid stimulation of gluconeogenesis Struck et al. (1965) suggested as one possible mechanism that glucagon action on gluconeogenesis may be mediated by an activation of hepatic lipolysis. On the other hand, recent findings lend support to a more direct action on the gluconeogenic pathway (Ross et al. 1967a, Exton and Park 1969). In order to get more insight into this problem, the following studies have been performed. Stimulated by previous findings (Ross et al. 1967b), the effect of glucagon on glucose formation from lactate in the presence of saturating concentrations of fatty acids has been investigated. Furthermore, glucagon action was studied under conditions when hepatic lipolysis was completely inhibited by glycodiazine. The results presented in this report are in support of the view that glucagon can stimulate gluconeogenesis in part by a mechanism which is not mediated by fatty acid oxidation.

Methods

The sources of materials and animals, the technique of perfusion and methods of analysis have been described by Teufel et al. (1967), Menahan and Wieland (1969a). Livers were taken from male Sprague-Dawley rats weighing between 140 and 170 g and starved for 24 hours. In all experiments, 10 mM lactate was added after 38 minutes of preperfusion and was followed by an infusion which approximately maintained this level. Synthetic glucagon was the kind gift of Dr. Wünsch (Institut für Eiweiß- und Lederforschung, München). Glucagon, oleic acid and octanoate were prepared according to Teufel et al. (1967), Menahan and Wieland (1969a), Löffler et al. (1965) and added at the following amounts: glucagon, 200 μg as single dose; oleic acid and octanoate, 100 and 150 μmoles respectively, as single doses with subsequent infusion of 100 and 300 μmoles per hour, respectively. Carnitine, 20 mg, was added together with oleic acid. Glycodiazine was a gift from Bayer AG, Leverkusen, and dissolved according to Menahan and Wieland (1969b). It was added as single dose (10 mM), 18 minutes prior to the addition of lactate in order to allow for complete equilibration of the drug. All data were calculated from the interval between 40 and 85 minutes of perfusion. The results were expressed as means $\pm$ standard deviation. Statistical significance was established by the student t-test.

Results and Discussion

An additive effect of different agents in metabolism is commonly regarded as indicating different mechanisms of action. This assumption can only be made under the condition that the agents are added in concentrations which yield maximal effects. Thus, in the present study, glucagon concentrations were employed which by far exceeded those previously shown to cause maximal stimulation of gluconeogenesis in the perfused rat liver (Ross et al. 1967a). Similarly the concentration of oleate 1 mM by 100% exceeded the concentration regarded to display maximal stimulation of gluconeogenesis by others (Williamson et al. 1969a). The comparison of ketogenesis and gluconeogenesis in the presence of either oleate or octanoate (Fig. 1) shows that while ketogenesis from octanoate was about twice that obtained with oleate the increase in gluconeogenesis was the same with either fatty acid. This would indicate that the rate at which oleate is oxidized is sufficiently high in order to produce maximal stimulation of glucose formation. In the following experiment the effect of glucagon on gluconeogenesis from lactate in the presence of oleate was studied (Fig. 2). Glucagon was added either at the beginning or 32 minutes after oleate addition. As may be seen, glucagon increased gluconeogenesis markedly beyond the already elevated rate produced by oleate, irrespectively of the time the hormone was added. As shown in Fig. 3, the additive effect on gluconeogenesis by glucagon could also be observed in the presence of octanoate instead of oleate. In these studies fatty-acid-dependent stimulation of gluconeogenesis was preceded by a characteristic lag phase.

The rates of gluconeogenesis and ketogenesis obtained in these studies are summarized in Table 1. Basal glucose output agrees quite well with earlier data reported by Struck et al. (1965) and Williamson et al. (1968) but is somewhat lower than that published by Ross et al. (1967b) and Menahan et al. (1968). Glucagon produced about 100% stimulation accompanied by a slight decrease of the C_3/glucose ratio. The stimulatory effect of oleate or octanoate was smaller and amounted to about 75%. These findings agree with previously reported results (Menahan 1968). The overall rate of gluconeogenesis in the presence of both glucagon and fatty acid arrives at the same value with either oleate or octanoate and does exceed the figure which is obtained by summing up the respective data for fatty acid stimulation and glucagon stimulation.

Table 1 also shows the corresponding rates of ketogenesis. Glucagon stimulated ketone body formation to a small degree and had no detectable effect on the ß-hydroxybutyrate/acetoacetate ratio. As expected, oleate markedly increased ketogenesis as well as the ß-hydroxybutyrate/acetoacetate ratio. As already observed by Krebs (1968) glucagon, in addition to oleate caused some further stimulation of ketogenesis. This may be due to additional supply of fatty acids from endogenous lipid resulting from stimulation of lipolysis or inhibition of fatty acid re-esterification (Bewsher and Ashmore 1966, Menahan and Wieland 1969b, Claycomb et al. 1969). The additional fatty acid oxidation produced by glucagon discussed before does not explain the excess stimulation of gluconeogenesis due to the hormone. This view is

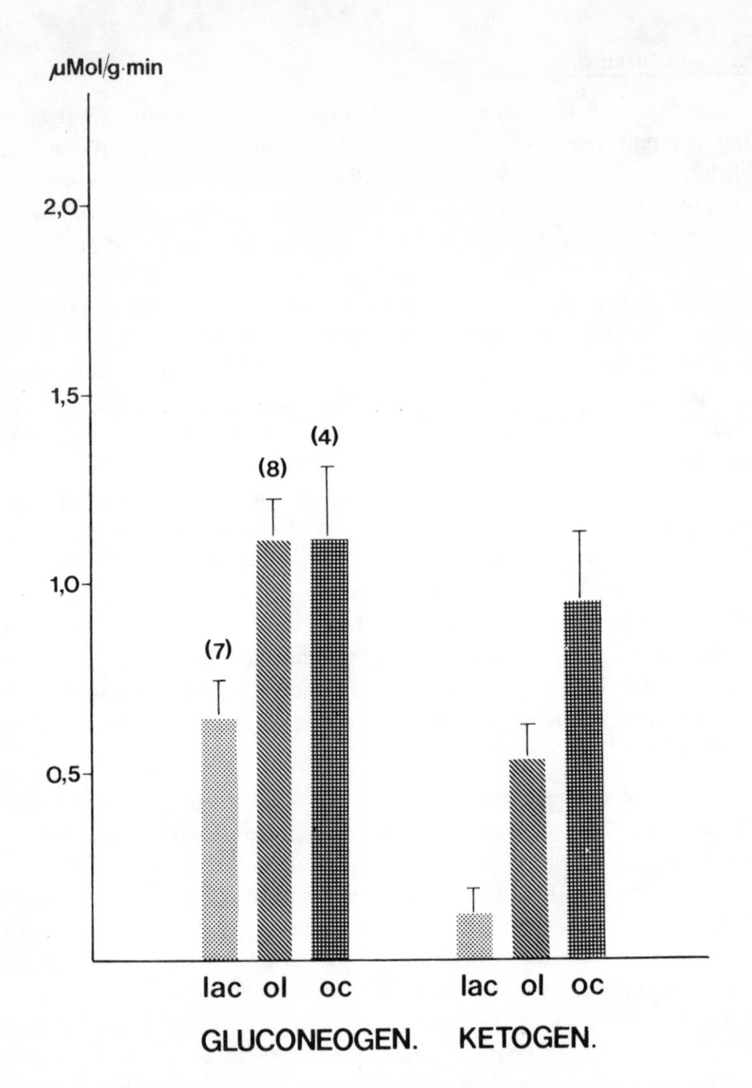

Fig. 1. Effect of Oleate and Octanoate on Gluconeogenesis and Ketogenesis with Lactate as Substrate. Rats were starved for 24 hours. Lactate and fatty acids were added at 38 minutes of perfusion. Oleic acid and octanoate were given as single dose (100 and 150 μmoles, respectively) followed by a constant infusion (100 and 300 μmoles/60 min.). The columns represent the means ± standard deviation, with the number of perfusions in parentheses; Lac (control), ol (oleate), oc (octanoate).

also supported by data from Exton et al. (1969) who found an additive stimulation of gluconeogenesis by glucagon without a further increase in ketogenesis. In agreement with earlier observations (Exton et al. 1969, Krebs et al. 1969), in this report the rate of ketogenesis in the presence of octanoate was considerably higher as compared

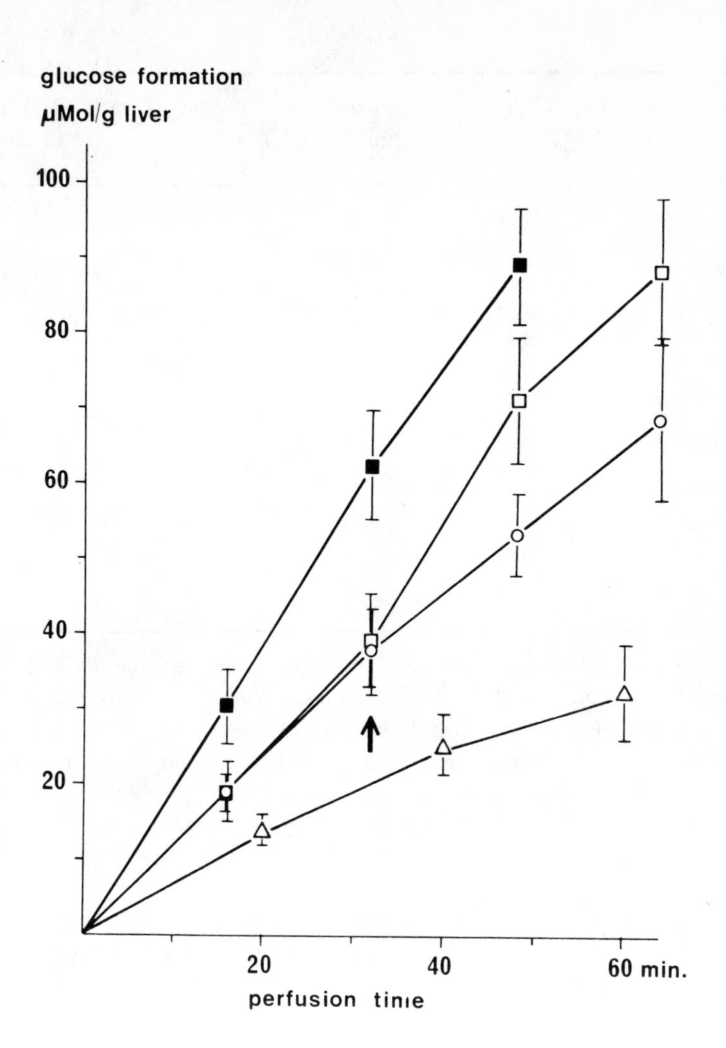

Fig. 2. Effect of Oleate and Oleate plus Glucagon on Gluconeogenesis from Lactate. Rats were starved for 24 hours. Lactate and oleic acid were added after 38 minutes of preperfusion, 100 µmoles of the fatty acid as single dose with subsequent infusion of 100 µmoles/60 min. Glucagon was added as single dose (200 µg) either after 38 minutes or after 70 minutes[+] of perfusion. The values represent the means ± standard deviation, with the number of perfusions in parentheses. + (at arrow) △ control (7); o oleate (8); □ oleate + glucagon (arrow) (6); ■ oleate + glucagon (4)

Experimental Conditions (n)	Glucose formed	ratio C_3/ glucose	Ketones formed	ratio ß-HOB/ AcAc in medium
Control (7)	0.64	2.5	0.12	0.7
Glucagon (7)	1.21	2.3	0.19	0.7
Oleate (8)	1.12	2.8	0.53	1.1
Oleate + glucagon (4)	1.86	2.4	0.72	1.0
Octanoate (4)	1.11	2.6	0.95	1.1
Octanoate + glucagon (4)	1.81	2.5	0.94	1.1

Table 1. Rates of Glucose and Ketone Formation in the Presence of Fatty Acids and Glucagon. Livers from rats starved for 24 hours were perfused with lactate as substrate. Additions were made after 38 minutes of perfusion as follows: glucagon, 200 μg; oleate, 100 μmoles as single dose and 100 μmoles/60 minutes as infusion; octanoate, 150 μmoles as single dose and 300 μmoles/60 minutes as infusion. The rates are expressed as μmoles/min./g liver fresh weight ± standard deviation.

with oleate. Glucagon did not further increase either ketogenesis or ß-hydroxybutyrate/acetoacetate ratio in the presence of octanoate. Hence, the gluconeogenic action of glucagon could be dissociated from its ketogenic effect. In support of previous findings (Regen and Terrell 1968, Exton and Park 1968), it would therefore appear that glucagon is capable of stimulating gluconeogenesis independently from its lipolytic and ketogenic action.

Special attempts in order to clarify this matter have been made by applying inhibitor of hepatic fatty acid oxidation (Menahan and Wieland 1969b, Williamson et al. 1969a, Burges et al. 1968, Ruderman et al. 1968). Only a few of these have so far been investigated especially with regard to glucagon action (Menahan and Wieland 1969b, Williamson et al. 1969b). In extending recent work by Menahan and Wieland (1969b), we investigated the influence of glycodiazine, a potent and selective inhibitor of lipolysis, (Hasselblatt 1969) upon the rate of gluconeogenesis in the presence of fatty acid and/or glucagon.

As illustrated in Fig. 4, glycodiazine completely suppressed ketone body formation and this also occurred in the presence of glucagon. Nevertheless, stimulation of gluconeogenesis by glucagon was preserved (Fig. 5).

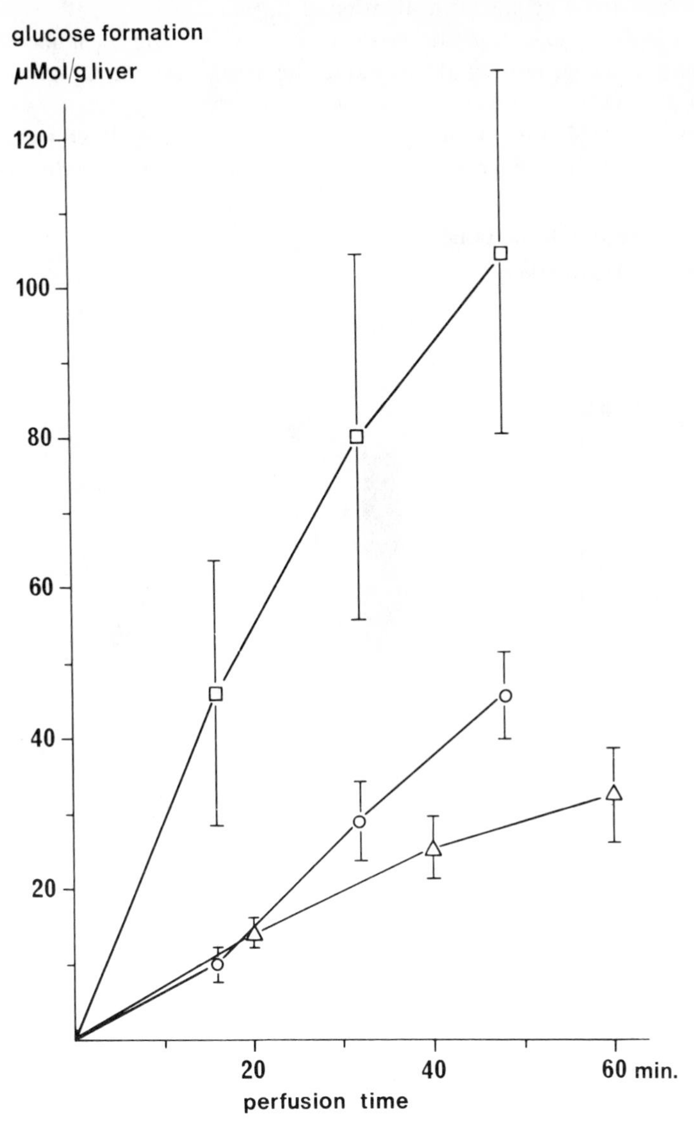

Fig. 3. Effect of Octanoate and Octanoate plus Glucagon on Gluconeogenesis from Lactate. Experimental conditions as given in the legend to Fig. 2, except that octanoate was used instead of oleate (150 μmoles as single dose with subsequent infusion of 300 μmoles/60 min.). △ control (7); ○ octanoate (4); □ octanoate + glucagon (4)

Table 2 summarizes the rates of glucose and ketone body formation obtained from the studies with glycodiazine. As may be seen, gluconeogenesis from lactate was suppressed by 40% as compared to the control without glycodiazine. C_3- uptake was not diminished correspondingly resulting in an increase in the C_3/glucose ratio. This may be explained by greater utilization of C_3-units by other pathways than gluconeogenesis probably oxidation via the citric acid cycle. Glucagon abolished the inhibition of gluconeogenesis by glycodiazine and normalized the C_3/glucose ratio. This agrees with earlier observations from our laboratory with pyruvate as substrate (Menahan and Wieland 1969b). In the experiments with oleate, there was no difference in the rates of gluconeogenesis and C_3 utilization whether glycodiazine was

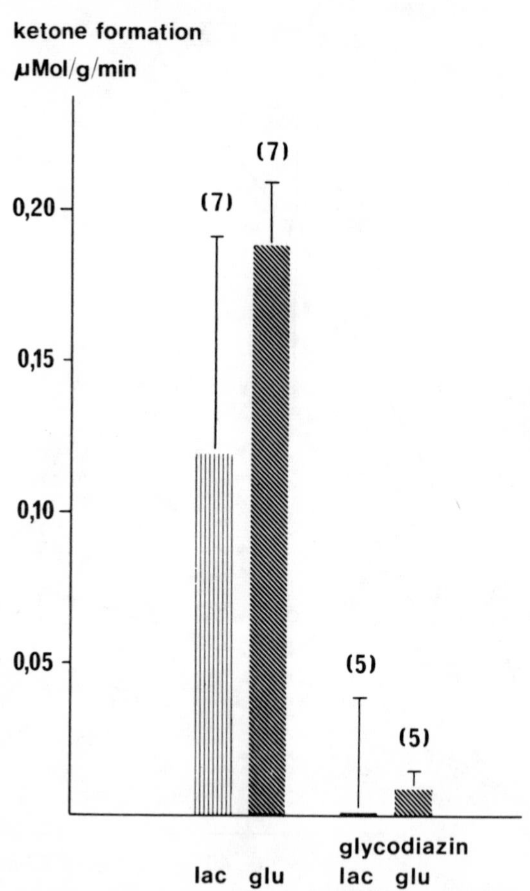

Fig. 4. Effect of Glycodiazine on Ketone Formation in the Absence and Presence of Glucagon. Rats were starved for 24 hours. Lactate and glucagon (200 µg) were added after 38 minutes of preperfusion. Glycodiazine was given as single dose (1 mmole) after 20 minutes of preperfusion. The columns represent the means ± standard deviation, with the number of perfusions in parentheses; Lac (control), glu (glucagon).

Experimental Conditions (n)	Glucose formed	ratio C_3/ glucose	Ketones formed	ratio ß-HOB/ AcAc in medium
Control (5)	0.37	3.4	-0.01	0.2
Glucagon (5)	0.61	2.4	0.01	0.3
Oleate (4)	1.10	3.1	0.96	0.9
Oleate + glucagon (5)	1.55	2.4	1.00	1.1

Table 2. Effect of Glycodiazine on Rates of Glucose and Ketone Formation in the Presence of Oleate and Glucagon. For experimental conditions, see legend to Table 1. Glycodiazine (1 mmole) was added as single dose after 20 minutes of preperfusion. Results are given as μmoles/min./g liver ± standard deviation.

present or absent. Thus, the suppression of basal gluconeogenesis by glycodiazine is overcome by oleate which also displays its undiminished stimulatory effect on glucose formation. Again, glucagon given together with oleate produced an additional stimulation which, however, was somewhat smaller as compared to the studies without glycodiazine. The data in Table 2 once again show complete inhibition of endogenous ketone body formation by glycodiazine (see also Fig. 4). As expected, the ß-hydroxybutyrate/acetoacetate ratio decreased to a very low value. Perfusion with oleate resulted in an increase of ketogenesis which was the same in the presence and absence of glucagon. Simultaneously, the ß-hydroxybutyrate/acetoacetate ratio arrived at the same level as observed without glycodiazine. From these studies with glycodiazine, there is additional evidence that glucagon can exert its stimulatory action on gluconeogenesis also under conditions where fatty acid oxidation is suppressed. Yet, the effect of the hormone and also the rate of basal gluconeogenesis are lower than under conditions of unrestricted fatty acid oxidation. As an explanation, mitochondrial reducing equivalents normally derived from fatty acid oxidation may limit intramitochondrial reduction of oxaloacetate to malate (Struck et al. 1965, Mehlman 1967) thus interfering with gluconeogenesis at the malate shuttle step. The low ß-hydroxybutyrate/acetoacetate ratios observed in the presence of glycodiazine would seem to favor this view (Klingenberg and Häfen 1963). The lowering in mitochondrial NADH (Bremer 1966, Wieland et al. 1969, Garland and Randle 1964) and/or acetyl-CoA could also, by feedback release on the pyruvate dehydrogenase reaction, lead to an increased rate of pyruvate oxidation thus removing pyruvate from the gluconeogenic pathway. Besides, a lowering in acetyl-CoA would possibly also result in decreased activity of pyruvate carboxylase (Keech and Utter 1963). Exton and Park (1967), in discussing the energy needs of gluconeogenesis, concluded that glucose formation from lactate is dependent on ATP supply from fatty acid oxidation. Thus, lack of ATP may be a further limiting factor when the supply of fatty acids is hindered by the presence of glycodiazine. Glucagon possibly stimulates gluconeogenesis by direct activation of a step between pyruvate and phosphoenol-pyruvate as suggested by Ross et al. (1967a) and Exton and Park (1969). Our studies indicate that the action of glucagon is only fully expressed when fatty acid oxidation occurs at undiminished rates. This raises the question where the auxiliary factors for

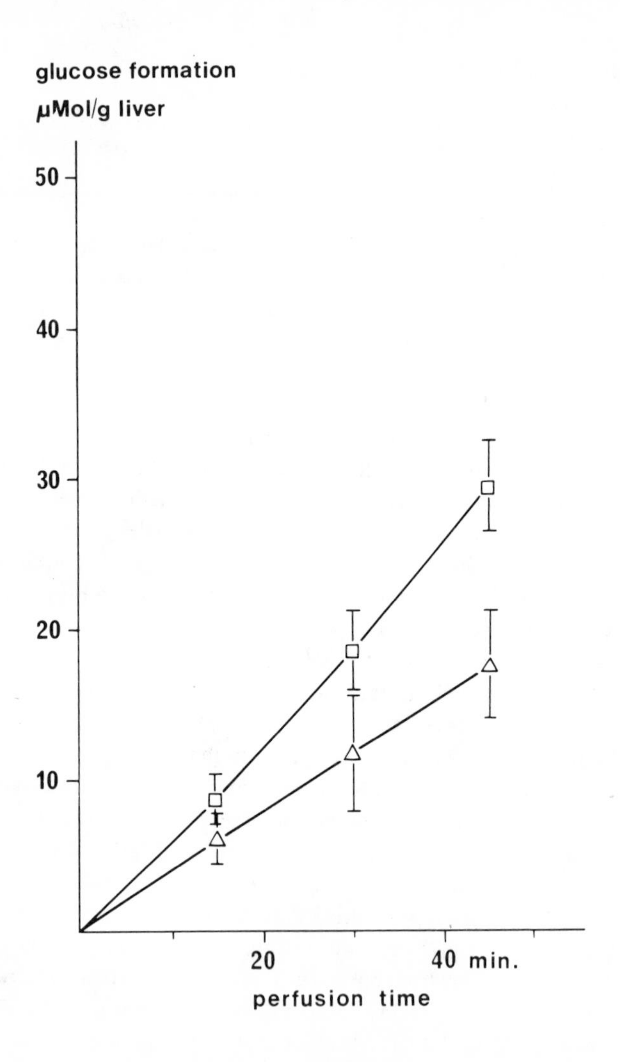

Fig. 5. Effect of Glucagon on Gluconeogenesis from Lactate in the Presence of Glycodiazine. Experimental conditions as given in the legend to Fig. 4. The values represent the means ± standard deviation with the number of perfusions in parentheses. △ glycodiazine + lactate (5); □ glycodiazine + lactate + glucagon (5).

gluconeogenesis are derived from under conditions of inhibited fatty acid oxidation. Stimulation of proteolysis and the urea cycle by glucagon may be a mechanism by which NADH could be supplied to a certain extent. By the same route substrate for oxidation within the tricarboxylic acid cycle would be made available resulting in generation of ATP. All these mechanisms, however, do not seem to permit unlimited stimulation of gluconeogenesis by glucagon.

In contrast to glucagon, oleate completely restored gluconeogenesis to a rate comparable with that obtained at perfusion without glycodiazine. It would, therefore, appear that by fatty acid oxidation, all of the factors essential in gluconeogenesis are now provided in sufficient amounts. The additive effect of glucagon on fatty acid stimulated gluconeogenesis was also seen in the presence of glycodiazine. As expected, ketogenesis remained unchanged in these experiments. This finding again demonstrates that glucagon can stimulate gluconeogenesis by a mechanism which is not necessarily linked to the simultaneous enhancement of ß-oxidation. In summary, our results seem to indicate that stimulation of gluconeogenesis by glucagon may be observed in the absence of any detectable lipolytic action of the hormone. Nevertheless, the degree of this effect on the rate of gluconeogenesis appears to depend on the rate of simultaneous fatty acid oxidation in so far as the stimulation does increase along with elevated ß-oxidation.

Acknowledgment

The authors wish to thank Mrs. R. Milfull for her excellent and conscientious technical assistance. This work was supported by the Deutsche Forschungsgemeinschaft, Bad Godesberg, FRG.

References

Bewsher, P. D., J. Ashmore: Biochem. biophys. Res. Commun. 24 (1966) 431

Bremer, J.: Biochim. biophys. Acta 116 (1966) 1

Burges, R. A., W. D. Butt, A. Baggaley: Biochem. J. 109 (1968) 38 P

Cahill, G. F., Jr.: In: Advances in Enzyme Regulation. Ed. by G. Weber, Vol. 3. Pergamon, New York 1965, p. 145

Claycomb, W. C., J. K. Bynagle, G. S. Kilsheimer: Biochem. biophys. Res. Commun. 36 (1969) 414

Exton, J. H., C. R. Park: J. biol. Chem. 242 (1967) 2622

Exton, J. H., C. R. Park: J. biol. Chem. 243 (1968) 4189

Exton, J. H., J. G. Corbin, C. R. Park: J. biol. Chem. 244 (1969) 4095

Exton, J. H., C. R. Park: J. biol. Chem. 244 (1969) 1424

Foa, P. P.: In: The Hormones. Ed. by G. Pincus, K. V. Thiman, and E. B. Astward, Academic, New York 1964, p. 531

Garland, P. B., P. J. Randle: Biochem. J. 93 (1964) 678

Hasselblatt, A.: Naunyn-Schmiedebergs Arch. exp. Path. Pharmakol. 262 (1969) 152

Keech, D. B., M. F. Utter: J. biol. Chem. 238 (1963) 2609
Klingenberg, M., H. v. Häfen: Biochem. Z. 337 (1963) 120
Krebs, H. A.: In: Stoffwechsel der isoliert perfundierten Leber. 3. Konferenz der Gesellschaft für Biologische Chemie Ed. by W. Staib and R. Scholz, Springer, Berlin 1968, 129
Krebs, H. A., P. G. Wallace, R. Hems, R. A. Freedland: Biochem. J. 112 (1969) 595
Löffler, G., F. Matschinsky, O. Wieland: Biochem. Z. 342 (1965) 76
Mehlman, M. A., P. Walter, H. A. Lardy: J. biol. Chem. 242 (1967) 4594
Menahan, L. A., B. D. Ross, O. Wieland: In: Stoffwechsel der isoliert perfundierten Leber. 3. Konferenz der Gesellschaft für Biologische Chemie Ed. by W. Staib and R. Scholz, Springer, Berlin 1968, 142
Menahan, L. A., O. Wieland: Europ. J. Biochem. 9 (1969a) 55
Menahan, L. A., O. Wieland: Europ. J. Biochem. 9 (1969b) 182
Miller, L. L.: Fed. Proc. 24 (1965) 737
Mortimore, G. E.: Am. J. Physiol. 204 (1963) 699
Regen, D. M., E. B. Terrell: Biochim. biophys. Acta 170 (1968) 95
Ross, B. D., R. Hems, H. A. Krebs: Biochem. J. 102 (1967a) 942
Ross, B. D., R. Hems, R. A. Freedland, H. A. Krebs: Biochem J. 105 (1967b) 869
Ruderman, N., E. Shafrir, R. Bressler: Life Sciences 7 (1968) 1083
Schimassek, H., H. J. Mitzkat: Biochem. Z. 337 (1963) 510
Sokal, J. E.: Endocrinology 78 (1966) 538
Struck, E., J. Ashmore, O. Wieland: Biochem. Z. 343 (1965) 107
Teufel, H., L. A. Menahan, J. C. Shipp, S. Böning, O. Wieland: Europ. J. Biochem. 2 (1967) 182
Wieland, O., B. v. Jagow-Westermann, B. Stukowski: Hoppe-Seylers Z. physiol. Chem. 350 (1969) 329
Williamson, J. R., E. T. Browning, R. Scholz, R. A. Kreisberg, J. B. Fritz: Diabetes 17 (1968) 194
Williamson, J. R., E. T. Browning, R. Scholz: J. biol. Chem. 244 (1969a) 4607
Williamson, J. R., E. T. Browning, R. G. Thurman, R. Scholz: J. biol. Chem. 244 (1969b) 5055

Discussion to Exton, Ui and Park and Fröhlich and Wieland

Seubert: I have a question for Dr. Exton concerning the regulatory point at the site of PEP-carboxykinase. Have you done enzymological studies on this enzyme which could explain your proposed mechanism?

Exton: We, in addition to a lot of people, for example Dr. H. A. Lardy and Dr. M. F. Utter, have looked for effects of cyclic AMP on PEP carboxykinase in vitro. So far, no one has found any evidence of activation of this enzyme. But one has to be careful, because in the only well-characterized system by which cyclic AMP acts on an enzyme, namely phosphorylase, the action is very indirect. Cyclic AMP does not act directly on phosphorylase but through a complicated sequence of additional enzymes. In the case of muscle, the kinase is the initial enzyme affected. Its activation leads in turn to activation of phosphorylase-b-kinase, which then acts on phosphorylase b. So when you are looking at a highly purified enzyme preparation you may miss evidence of activation. Of course, if the PEP carboxykinase system were analogous to phosphorylase, it would be necessary for there to be two forms of the enzyme with different activity and possibly differing in phosphorylation or, maybe, adenylylation.

The only experiment in our studies which really suggested that the enzyme affected was PEP-carboxykinase was the last experiment with labeled aspartate. The reason for this is the following: A consistent feature of the experiments with ^{14}C-pyruvate, lactate, glutamate and bicarbonate was an increase in the specific activity of aspartate. Because of this, the apparent increase in the radioactivity incorporated into PEP might have been merely due to an increase in the specific activity of oxaloacetate in the cytosol. Thus we cannot conclude from these studies whether the rate of PEP synthesis was actually increased.

Several explanations may be advanced for why the specific activity of aspartate was increased with glucagon. One feature that you may have observed in the tables and figures was that the specific activity of aspartate in the control livers was always much lower than that of malate or citrate. A possible reason for this is that the bulk of aspartate in the cell lies outside the mitochondria in the cytosol, whereas malate and citrate do not lie outside the mitochondria to the same extent as aspartate. Oxaloacetate may be mainly intra-mitochondrial. For the studies with ^{14}C-lactate, pyruvate, glutamate and HCO_3, we may be measuring mainly intra-mitochondrial radioactivity in malate and aspartate. If glucagon caused an acceleration of the formation and efflux of labeled aspartate from the mitochondria this would enter the large pool of unlabeled aspartate in the cytosol resulting in a large increase in its specific activity. This would lead to an increase in the specific activity of oxaloacetate in the cytosol and hence an increase in the labeling of phosphoenolpyruvate. It is only in the ^{14}C aspartate experiments that aspartate in the cytosol is labeled directly (i.e.) not via mitochondrial reactions. In these experiments there was no change in the specific activity of aspartate. The increase in PEP labeling is therefore better evidence of PEP carboxykinase activation. Additional support for an effect on PEP synthesis comes from the changes in the chemical levels of the intermediates.

J. R. Williamson: My own data (Williamson et al.: Proceedings of Detroit Symposium on the Action of Hormones from Molecules to Population Control, July 1969, P. P. Foa (ed.), in press; Williamson et al.: J. Biol. Chem. 244 (1969) 5055) is best interpreted as showing that glucagon produces an activation of pyruvate carboxylase. In any case, the primary control site cannot be at PEP-carboxykinase. Without a stimulus to increase flux through pyruvate carboxylase, flux would be limited by the rate of formation of oxalacetate, since pyruvate carboxylase is an irreversible enzyme. I agree that it would be an elegant solution to the problem of the mode of action of glucagon if pyruvate carboxylase existed as phosphorylated and non-phosphorylated forms having different kinetic properties. Citric acid cycle activity is increased after glucagon administration as shown by the increased rate of oxygen consumption, but control cannot be primarily at α-ketoglutarate dehydrogenase as this would rapidly deplete the earlier steps of substrate. The overall rate of the citric acid cycle can only be increased by increased activity of citrate synthase, which itself is controlled by the oxalacetate concentration. This is controlled by the intramitochondrial malate concentration and by the state of reduction of the mitochondrial pyridine nucleotides.

It is a very difficult problem to deduce sites of interactions in the citric acid cycle from isotope studies in the whole liver since each intermediate is compartmented and is subject to different exchange reactions at varying rates according to its distribution. In relation to the aspartate distribution, we come to opposite conclusions, since on the basis of the calculations given in my paper we consider that it is mostly in the mitochondria along with glutamate, while α-ketoglutarate and oxalacetate are mainly extra-mitochondrial. However, calculations are no proof of compartmentation. They require validification by other methods. We are presently tackling this problem by the use of reconstituted systems.

Krebs: Dr. Exton, would the following consideration explain the postulated second effect of glucagon. The first effect — the acceleration of the carboxylation of pyruvate — would cause an increased concentration of oxaloacetate in the mitochondria. This in turn might accelerate the degradation of glutamate, because mitochondrial glutamate transaminates very readily (e. g. H. A. Krebs and D. Bellamy: Biochem. J. 75 (1960) 523). Would this not explain the increased label in aspartate? When liver is perfused with a variety of substrates glucagon stimulates gluconeogenesis only when lactate and alanine are precursors (B. D. Ross, R. Hems and H. A. Krebs: Biochem. J. 102 (1967) 942). This is difficult to reconcile with the suggestion that there is a primary effect of glucagon at the stage between oxoglutarate and succinate.

Exton: Of course it is correct that flux through pyruvate carboxylase must be increased. The point is how does this come about? If the actual rate of carboxylation is determined by the rate of entry of pyruvate into the mitochondria as hypothesized by Haynes (Adams, P. J., R. C. Haynes, Jr.: J. Biol. Chem. 244 (1969) 6444) control may be not on the carboxylase itself, but largely on the entry of pyruvate into the mitochondria. The only data which appear to be against an effect on pyruvate carboxylation are the tryptophane studies. However, these do not rule it out. For instance, if the action on pyruvate carboxylase or pyruvate entry is not a primary action of glucagon, but is secondary to other changes resulting from the activation of gluconeogenesis in general, then a glucagon effect on pyruvate carboxylase or entry would have been knocked out by using tryptophane. Thus, in the final analysis, we have not obtained evidence for such an effect nor can we exclude it.

Now, in answer to the question about activation of the citric acid cycle: What we observed was an increase of the radioactivity of succinate in livers perfused with ^{14}C-glutamate, an increase in α-ketoglutarate decarboxylation and an increase in glutamate utilization. These data indicate increased conversion of α-ketoglutarate to succinate but do not indicate that the cycle as a whole was increased. If glutamate was converted to α-ketoglutarate, and a large percentage of this was converted via the cycle to oxaloacetate which then left the mitochondrion as aspartate, then only a portion of the cycle would have been accelerated. Perhaps the rate of conversion of citrate to α-ketoglutarate was unchanged, or possibly slowed.

Söling: Under two circumstances, namely with uniformly labeled lactate and with uniformly labeled glutamate, you have observed the same pattern of changes under the influence of glucagon: The specific activity of malate decreased, that of aspartate increased while the specific activity of α-oxoglutarate remained constant. I can explain these findings only by assuming malate and aspartate to be compartmentalized in a different way, since both share the oxaloacetate pool.

Krebs: May I try to offer an explanation of the findings? Aspartate could be derived from oxaloacetate formed by carboxylation of pyruvate. Malate could come from glutamate by transamination with oxaloacetate. Thus aspartate and malate would come from different pools.

Söling: Yes, I agree, but then you would not expect the specific radioactivity of α-oxoglutarate to remain constant when the specific activity of malate decreases. May I ask a second question? I would like to know whether you have the same pattern of specific radioactivities of malate and aspartate if you use uniformly labeled lactate or pyruvate?

Exton: We have never used lactate or pyruvate alone, the substrate was always a 10:1 mixture of lactate and pyruvate. So we do not know the answer.

Seubert: I would just like to ask whether this scheme of transport mechanisms through the mitochondrial membranes is still right, especially with regard to Dr. Müllhofer's data. Dr. Müllhofer showed that aspartate had a lower specific activity than PEP or glucose. I think we should first clarify this point.

Exton: There is no question that the interpretation of this type of work would occupy a week's time. One has always to be aware of the limitations of the conclusions that one can draw.

Inhibition of Hepatic Gluconeogenesis by Pent-4-enoic acid: Role of Redox State, Acetyl-CoA and ATP

N. B. Ruderman, C. J. Toews, C. Lowy, and E. Shafrir
Joslin Research Laboratory, Department of Medicine, Harvard Medical School, Peter Bent Brigham Hospital and The Diabetes Foundation Inc., Boston, Massachusetts, U.S.A.

Summary

Pent-4-enoic acid inhibited gluconeogenesis from alanine, alanine oxidation to CO_2 and ketone body production from linoleate by 60-80% in the isolated perfused rat liver. When octanoate, a short-chain fatty acid, was used in place of linoleate, neither gluconeogenesis nor ketone body production was impaired. Assays of tissue metabolites indicated that gluconeogenesis was inhibited first at glyceraldehyde-P-dehydrogenase and secondarily at pyruvate carboxylase. The former block was evident after only two minutes of perfusion with pentenoic acid and appeared to be related to a lack of NADH as evidenced by the pronounced decrease in the tissue lactate/pyruvate ratio. The latter block correlated well with a gradual decrease in acetyl-CoA an obligatory activator of pyruvate carboxylase. After 45 minutes of perfusion, ATP levels were decreased, but only by 20%. When ethanol (8 mM) was added to the perfusate of livers that had been exposed to pentenoic acid for 15 minutes or more, the lactate/pyruvate ratio returned towards normal, but acetyl-CoA levels remained depressed. In association with these changes, the inhibition at glyceraldehyde-P-dehydrogenase was reversed, but gluconeogenesis was still suppressed because of the inhibition at pyruvate carboxylase. When ethanol was added after ten minutes of exposure to pentenoic acid, gluconeogenesis was substantially restored. Acetyl-CoA levels fell, but only to 1/3 of control values and the block at pyruvate carboxylase was less marked. Ketone body production remained depressed. Correlation of rates of gluconeogenesis with tissue levels of acetyl-CoA in these studies revealed a sigmoid relationship with an apparent activation constant for acetyl-CoA of 1.8×10^{-5} M. These findings suggest that pentenoic acid suppresses glucone-genesis by inhibiting long-chain fatty acid oxidation, thereby making NADH and ace CoA limiting for key steps in the gluconeogenic pathway. The data also suggest that pyruvate carboxylase may be activated by acetyl-CoA in intact tissue in much the same fashion as it is in cell-free preparations.

+ Unusual Abbreviations:

Enzyme Code Numbers: Glyceraldehyde-3-phosphate dehydrogenase (EC 1.2.1.12 Pyruvate carboxylase (EC 6.4.1.1), 3-phosphoglyceric acid kinase (EC 2.7.2.3), Glucose-6-phosphate dehydrogenase (EC 1.1.1.49).

Introduction

Pent-4-enoic acid is one of a series of intermediate-chain length fatty acids which cause hypoglycemia in animals and man (Hassal et al. 1954, De Renzo et al. 1958, Chen et al. 1957, Bressler et al. 1969). In general, these compounds possess a terminal vinyl group separated by at least two carbon atoms from the carboxyl group. Chemical analogs of pent-4-enoic acid such as pentanoic (valeric) acid, which lacks the double bond and pent-2-enoic acid whose double bond is only one carbon atom removed from the carboxyl group have not been found to cause hypoglycemia (Chen et al. 1957) (Table 1).

Interest in these substances was initially generated by a syndrome referred to as "Jamaican Vomiting Sickness" (Hassal et al. 1954, De Renzo 1958, Chen 1957, Bressler et al. 1969, von Holt 1966, Entman and Bressler 1967). For many years, it has been known that residents of the island of Jamaica, particularly poorly nourished children, who ingested the unripe fruit of the ackee tree, developed an illness characterized by vomiting, followed by convulsions, coma and sometimes death. Profound hypoglycemia invariably accompanied these symptoms.

The substance in the ackee fruit which caused these effects was identified as methylenecyclopropyl alanine, more commonly referred to as hypoglycin (Hassal et al. 1954, von Holt et al. 1966). The studies of von Holt and more recently those of Bressler and his colleagues indicated that hypoglycin and its chemical analog pent-4-enoic acid inhibit long-chain fatty acid oxidation in liver and muscle (von Holt 1966, Entman and Bressler 1967, Corredor et al. 1968) and that this effect is evident prior to the onset of hypoglycemia (Entman and Bressler 1967, Corredor et al. 1968). The latter workers (Corredor et al. 1968) also suggested, on the basis of in vivo studies, that the hypoglycemic effect of these agents may in large part be due to an inhibition of hepatic gluconeogenesis.

The aim of the present study was to determine the effect of pentenoic acid on hepatic glucose production in an in vitro system, and, using pentenoic acid as a tool, to explore the relationship between fatty acid oxidation and the liver's capacity to carry on gluconeogenesis. The results of these studies will be presented in greater detail in subsequent publications (Ruderman et al. 1970, Toews et al. 1970).

Materials and Methods

Perfusion Technique

Male Sprague Dawley rats (200-250 g), starved overnight, were used in all studies. Details of the perfusion apparatus, the preparation of the perfusate and of the operative procedure have been described previously (Ruderman and Herrera 1968). Livers were perfused with 100 ml of cell free, Krebs-Ringer bicarbonate buffer (Krebs and Henseleit 1932), containing 4 g bovine serum albumin, 10 mM alanine or another gluconeogenic substrate and 1.2 mM sodium linoleate. An additional 200 μmole/h of the fatty acid was infused into the perfusing medium during the experiment. Sodium pent-4-enoate and ethanol were added to the perfusate as described in the results.

Determinations

Glycogen (Good et al. 1933) was determined on portions of the liver excised at the beginning and end of each experiment. Perfusate glucose was determined with an autoanalyzer (Technicon Co., Chauncey, N.Y., U.S.A.) or by the glucose oxidase method (Huggett and Nixon 1957). ß-Hydroxybutyrate and acetoacetate were measured by a fluorometric adaptation of the method of Williamson et al. (1962). The sum of ß-hydroxybutyrate and acetoacetate release into the perfusate was regarded as total ketone body production. The latter was used as an index of fatty acid oxidation.

Liver metabolite concentrations were determined fluorometrically on tissue frozen *in situ*, or excised and frozen. The details of these procedures have been described elsewhere (Ruderman et al. 1970, Toews, 1970).

When alanine-U-^{14}C was added with the unlabelled alanine, perfusate glucose and glucose present in the glycogen hydrolysates were isolated as the phenylosazone derivatives. The osazones were plated in steel planchettes and radioactivity determined in a gas flow counter (Karnofsky et al. 1955). The techniques used for collecting and counting $^{14}CO_2$ are described in a separate communication (Ruderman et al. 1970).

Reagents

Linoleic acid, octanoic acid and pent-4-enoic were obtained from K & K Laboratories, Plainview, N.Y., U.S.A. They were all prepared as the sodium salt (pH 9-11) with NaOH. Alanine-U-^{14}C and aspartate-U-^{14}C were obtained from New England Nuclear Corporation, Boston, Massachusetts, U.S.A. Enzymes for metabolite determinations were obtained from Boehringer-Mannheim, New York, U.S.A. except for glucose-6-phosphate dehydrogenase (Type VI or X) which was obtained from Sigma, St. Louis, Missouri, U.S.A. Substrates and cofactors for metabolite assays were purchased from either Sigma or Boehringer-Mannheim. Bovine serum albumin (Cohn fraction V) was obtained from Armour Pharmaceuticals, Kankakee, Illinois, U.S.A. and absolute ethanol from Commercial Solvents Corp., Terre Haute, Indiana, U.S.A.

Results and Discussion

Gluconeogenesis from Alanine and Ketone Body Production

Pentenoic acid rapidly suppressed both gluconeogenesis and ketone body production in livers perfused with a linoleate containing medium (Fig. 1, Table 2). After 15 minutes of perfusion, net glucose release was about 60% of that of control livers. Thereafter, livers perfused with pentenoic acid produced essentially no glucose, whereas their control counterparts continued to produce glucose at their initial rate. In association with the decrease in hepatic ketone body production, there was a sharp fall in the ß-hydroxybutyrate/acetoacetate ratio of the perfusate, suggesting that intra cellular NADH levels were probably diminished.

Effect of Pentenoic Acid in the Presence of Octanoate

Corredor et al. (1969) have related the inhibition of long-chain, fatty-acid oxidation by pentenoic acid to the fact that pentenoic acid and acrylic acid, a slowly metabolized product of pentenoate oxidation, bind free carnitine and extramitochondrial CoA, and as a consequence, retard the entrance of long-chain, fatty-acids into the mitochondria. In line with these observations, they found no inhibition of hexanoate oxidation in myocardial homogenates obtained from pentenoic acid treated mice (Corredor et al. 1968). Brendel et al. (1969) and Williamson et al. (1969) have observed lowered levels of both free carnitine and coenzyme A in liver tissue that had been exposed to pentenoic acid.

In the present study, when octanoate, an eight-carbon fatty acid, was substituted for linoleate, pentenoic acid no longer suppressed gluconeogenesis from alanine. In addition, it had no significant effect on either ketone body production or the ß-hydroxybutyrate/acetoacetate ratio. These data strongly suggest that the effect of pentenoic

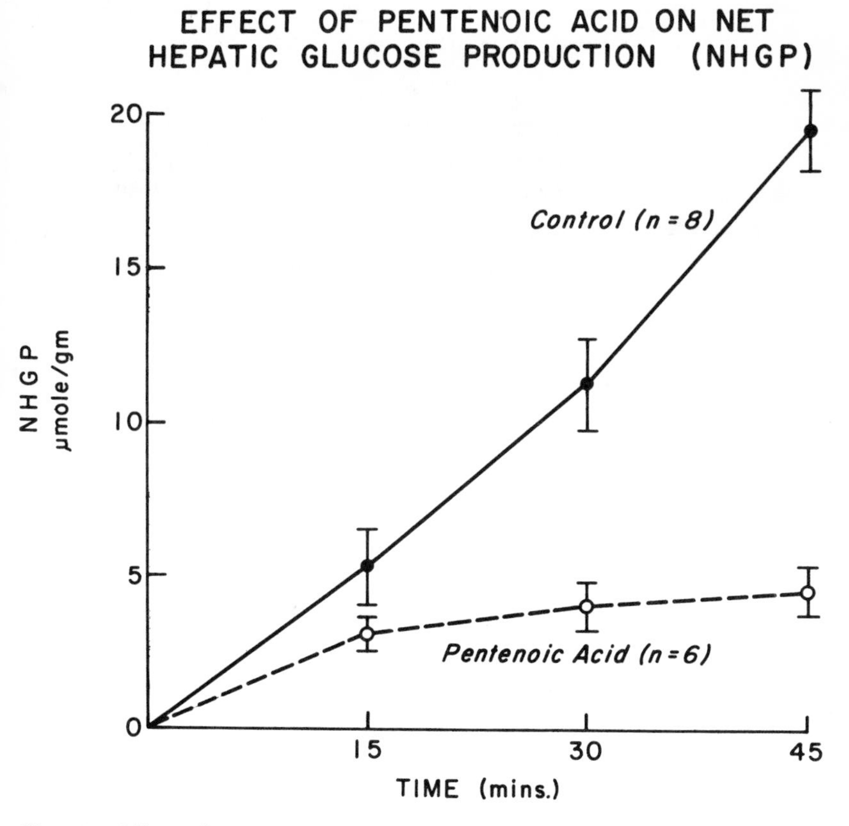

Fig. 1. Effect of Pent-4-enoic Acid on Net Hepatic Glucose Release. Livers were perfused in the presence of alanine and linoleic acid as described in Table 2. Pentenoic acid (0.5 mM) was added to the initial medium. Results are expressed as mean ± s.e.m.

Hypoglycemic	Non-hypoglycemic
$CH_2 = CH\text{-}CH_2\text{-}CH_2\text{-}COOH$ Pent-4-enoic Acid	$CH_3\text{-}CH_2\text{-}CH_2\text{-}CH_2\text{-}COOH$ Pentanoic (valeric) Acid
CH_2 NH_2 $CH_2 = C\text{-}CH\text{-}CH_2\text{-}CH$ COOH Methylenecyclopropyl Alanine (Hypoglycin)	$CH_3\text{-}CH_2\text{-}CH = CH\text{-}COOH$ Pent-2-enoic Acid

Table 1. Pent-4-enoic Acid and Related Compounds.

	Glucose + glycogen production	Ketone body production	ß-Hydroxybutyrate / Acetoacetate
	moles/g wet wt. per 45 min. ± s.e.		
Linoleate			
Control (6)	22 ± 2	58 ± 9	0.33 ± 0.03
Pentenoic Acid (6)	4.4 ± 1.8	7.5 ± 2.5	0.08 ± 0.04
Octanoate			
Control (6)	23 ± 6	79 ± 12	0.39 ± 0.13
Pentenoic Acid (5)	29 ± 5	59 ± 11	0.37 ± 0.18

Table 2. Effect of Pent-4-enoic Acid on Hepatic Gluconeogenesis and Ketone Body Production. Livers were perfused for 45 minutes with 100 ml of a cell free medium composed of KRB buffer, 4% bovine serum albumin, 10 mM alanine and 1.2 mM linoleate or octanoate. An additional 150 μmoles of fatty acid were added to the perfusate during the experiment. Pentenoic acid (0.5-1.0 mM) was present in the initial medium. Number of livers in parentheses.

acid on gluconeogenesis is related to its ability to inhibit fatty acid oxidation.

Effect of Pentenoic Acid on Alanine Oxidation

In accord with the observations of others (Corredor et al. 1969, Senior and Sherratt 1968a), pentenoic acid significantly depressed alanine (pyruvate) oxidation (Table 3). The $^{14}CO_2$ production from alanine-U-^{14}C was reduced to 1/3 of the control value. Part of this difference could be accounted for by the higher rate of gluconeogenesis in the control livers, since for every five alanine carbons incorporated into glucose, one theoretically should have been lost as CO_2 when oxaloacetate was converted to phosphoenolpyruvate. When this correction was taken into account, the pentenoic acid treated livers still produced considerably less $^{14}CO_2$ than the control group (9.6 versus 18 μmole/g per 45 minutes). It seems unlikely that inhibition of pyruvate oxidation played a major role in the suppression of gluconeogenesis. Senior and Sheratt (1968b) have demonstrated that agents such as pentanoic (valeric) acid, which can inhibit pyruvate but not fatty acid oxidation in rat liver mitochondria, have no effect on gluconeogenesis. Furthermore, in this study, we found that alanine oxidation could account for at most about 12% of the oxygen consumed by control livers (data not shown in this report). Hence, in contrast to fatty acid oxidation, which accounted for at least 75% of O_2 consumption, it was a relatively unimportant source of acetyl-CoA and NADH, the rate limiting factors for gluconeogenesis in our experiments (*vide infra*). Conceivably, gluconeogenesis might have been less inhibited if pyruvate oxidation was not suppressed; however, this point requires further study.

Tissue Metabolites

In other studies (Ruderman et al. 1970, Ruderman et al. 1968) we demonstrated that pentenoic acid depresses gluconeogenesis from aspartate by nearly 75%, but has not effect on glucose production from glycerol or fructose. This suggested that the site at which gluconeogenesis was inhibited lay between oxaloacetate and glyceraldehyde-3-phosphate. The observation that the ß-hydroxybutyrate/acetoacetate ratio was markedly reduced in the presence of pentenoic acid suggested that the block might be at the NADH dependent step catalyzed by glyceraldehyde-phosphate dehydrogenase. To evaluate this point further and to determine whether gluconeogenesis might be inhibited at other sites, tissue metabolite levels were determined (Table 4, Fig. 2).

In this series of livers, pentenoic acid depressed gluconeogenesis from alanine by 80% and ketone body production by 95% during a 45-minute perfusion. The concentrations of 3-phosphoglycerate and all the metabolites below it were markedly increased whereas the levels of gluconeogenic intermediates between glyceraldehyde-phosphate and glucose 6-phosphate were either decreased or not significantly changed. These findings are consistent with inhibition at either glyceraldehyde phosphate dehydrogenase or 3-P-glyceric acid kinase. Since the concentration of 1-3-diphosphoglycerate could not be determined, it was not possible to further define this block on the basis of gluconeogenic metabolite levels. However, the relatively small decrease in the levels of ATP and of the ATP/ADP ratio, in comparison with the pronounced fall in the ratio of lactate to pyruvate, suggests that the glyceraldehyde phosphate dehydrogenase block was more likely to have been the dominant one.

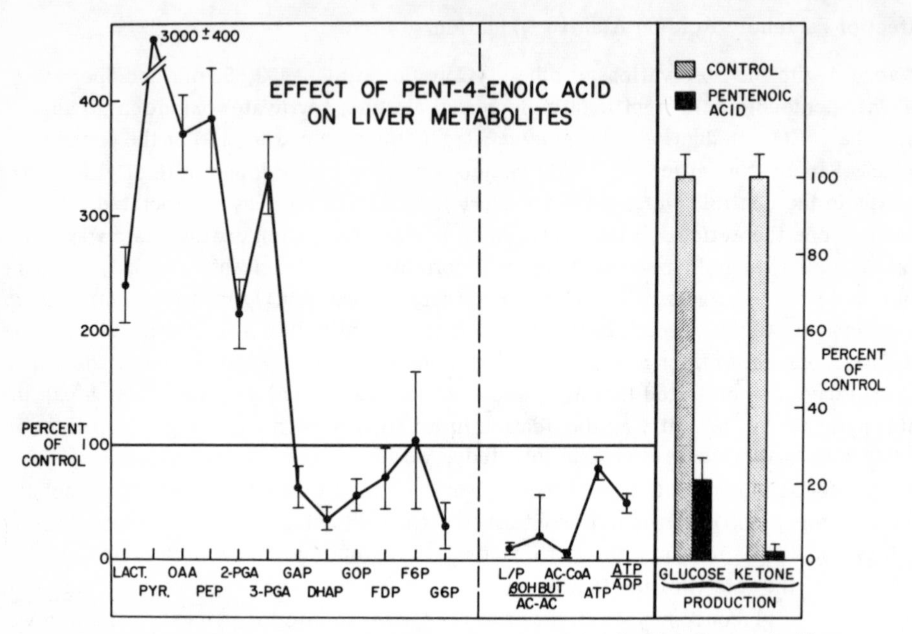

Fig. 2. Effect of Pent-4-enoic Acid on Rat Liver Metabolite Concentrations. Frozen tissue biopsies were obtained after 45 minutes perfusions in the presence of alanine, linoleate and, where indicated, pentenoic acid (0.1 mM). Experimental conditions are described in Table 2 and the Methods section. The data are expressed as percent ± s.e.m. of the corresponding control values.

Pyruvate increased to a much greater extent than other metabolites, suggesting a second site of inhibition at pyruvate carboxylase, possibly due to the marked decrease in acetyl-CoA, an obligatory activator of pyruvate carboxylase (Utter and Keech 1963). This block was partially masked by the elevated level of oxaloacetate. The latter, however, in all likelihood was largely due to the more oxidized state of the cell, since the concentrations of malate and aspartate, with which oxaloacetate is in equilibrium, were reduced to less than 1/2 the control level (Table 4). A similar conclusion was drawn on the basis of studies with ethanol which will be described in Fig. 3.

Biopsies taken after shorter periods of perfusion with pentenoic acid (Toews et al. 1970) revealed that the block at glyceraldehyde-phosphate dehydrogenase and the changes in redox state were fully established after only two to five minutes of perfusion. The block at pyruvate carboxylase, as gauged from the rise in tissue pyruvate, was more gradual in onset, as was the fall in acetyl-CoA. After ten minutes of perfusion, the concentration of acetyl-CoA was still sufficient to maintain a near normal rate of pyruvate carboxylase activity in the perfused liver.

	Alanine carbon to CO_2	Alanine carbon to glucose + glycogen
	μ moles/g wet wt. per 45 min ± s.e.	
Control	32 ± 5 (5)	85 ± 8 (6)
Pentenoic Acid	11 ± 2.7 (5)	6 ± 2.5 (5)

Table 3. Effect of Pent-4-enoic Acid on Alanine Conversion to Glucose and CO_2. Livers were perfused for 45 minutes with 100 ml of perfusate containing 10 mM alanine and 1.2 mM linoleate. An amount of 8.3 μc of alanine-U-^{14}C and pent-enoic acid (1 mM) were present in the initial medium. An additional 150 μmoles of linoleate were added by infusion during the experiment. Number of livers in parentheses.

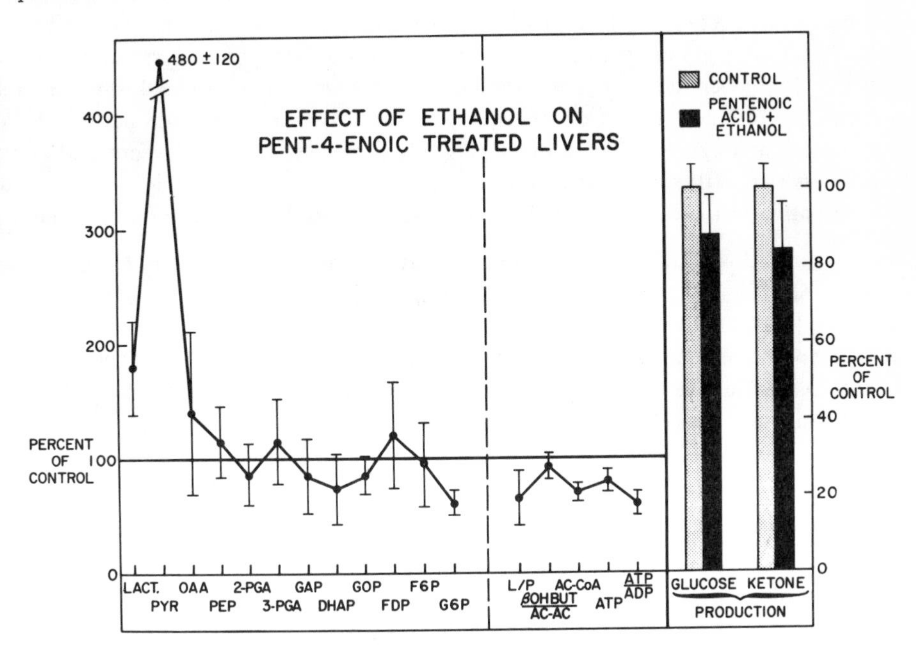

Fig. 3. Effect of Pent-4-enoic Acid + Ethanol on Rat Liver Metabolite Concentrations. Alanine, linoleate, ethanol (8 mM) and pent-4-enoic acid (1.0 mM) were present in the initial perfusing medium. Biopsies were taken after 45 minutes of perfusion. Experimental conditions are described in Table 2 and Methods section. The data are expressed as percent ± s.e.m. of the corresponding control values.

Effect of Ethanol Added Simultaneously With Pentenoic Acid

If the initial and major block in gluconeogenesis was due to a lack of NADH at glyceraldehyde-phosphate dehydrogenase, one would expect that the provision of an electron donor whose initial oxidation did not require carnitine or coenzyme A might release this block. Indeed, in earlier studies (Ruderman et al. 1968), we observed that ethanol substantially prevented the inhibition of gluconeogenesis by pentenoic acid in the perfused liver. To evaluate the mechanism for this effect and also to evaluate the relative importance of the inhibition at pyruvate carboxylase, a series of experiments was carried out in which ethanol was added to the perfusate after varying periods of perfusion with pentenoic acid. The results of one such study are presented in Fig. 3.

In these experiments, 8 mM ethanol and pentenoic acid were simultaneously added to the perfusate at the beginning of the perfusion. When ethanol was added in this way, it prevented the inhibition of gluconeogenesis and, surprisingly, also of ketone body production, the latter being maintained at 80% of the control rate. The crossover previously present at glyceraldehyde-P-dehydrogenase was not observed and the NADH/NAD ratio in the cytosol, as reflected by the lactate/pyruvate ratio, remained normal. The concentrations of lactate and pyruvate were both elevated whereas oxaloacetate had returned to control levels, suggesting that there was still a partial block at pyruvate carboxylase. Such a block might have been related to the fact that acetyl CoA levels were still somewhat below normal. The contribution of the block in pyruvate oxidation to the elevated levels of pyruvate in these livers was not ascertained.

The concentration of ATP and the ATP/ADP ratio were not significantly higher in the presence of ethanol + pentenoic acid than in the presence of pentenoic acid alone. Since the NADH/NAD ratio was maintained in the presence of ethanol, we once again have compelling evidence that the primary block in gluconeogenesis was the inhibition of glyceraldehyde 3-P-dehydrogenase due to a lack of NADH and not an inhibition at 3-phosphoglyceric acid kinase due to decreased levels of ATP.

Effect of Ethanol Added After 30 Minutes of Perfusion with Pent-4-enoic Acid

When ethanol was added to the perfusate after 30 minutes of perfusion with pentenoic acid, strikingly different findings were obtained (Fig. 4). Prior to the addition of ethanol, the crossover at glyceraldehyde-P-dehydrogenase had developed and the secondary block at pyruvate carboxylase was also present. Following ethanol addition, neither gluconeogenesis nor ketone body production were restored although the crossover at glyceraldehyde-P-dehydrogenase had disappeared and the lactate/pyruvate ratio had returned to normal. Acetyl-CoA levels did remain low, however, and the block at pyruvate carboxylase persisted. Hence, it appears that once the block at pyruvate carboxylase is fully established, as it seems to be after 30 minutes of perfusion with pentenoic acid, ethanol no longer has the ability to restore gluconeogenesi

Metabolite	Tissue concentration (nmole/g wet wt. ± s.e.)	
	Control	Pentenoic acid
Glucose-6-phosphate	48 ± 18 (8)	4.3 ± 3 (3)
Fructose-6-phosphate	10.5 ± 3 (7)	5.6 ± 4 (3)
Fructose diphosphate	3.1 ± 1 (7)	1.4 ± 0.5 (10)
Dihydroxyacetone phosphate	4.1 ± 1 (7)	2.6 ± 0.8 (9)
D-glyceraldehyde phosphate	2.3 ± 0.5 (6)	2.2 ± 0.8 (8)
3-phosphoglycerate	225 ± 34 (11)	750 ± 92 (17)
2-phosphoglycerate	47 ± 7 (11)	106 ± 16 (16)
Phosphoenolpyruvate	134 ± 19 (11)	516 ± 69 (14)
Pyruvate	24 ± 3 (12)	720 ± 107 (17)
Lactate	444 ± 84 (10)	1019 ± 187 (13)
Oxaloacetate	2 ± 0.9 (5)	7.8 ± 1 (5)
Malate	321 ± 27 (6)	124 ± 9 (6)
Aspartate	857 ± 69 (6)	396 ± 17 (6)
Acetyl-CoA	34 ± 5 (12)	4.3 ± 0.9 (10)
ATP	1034 ± 114 (12)	800 ± 82 (17)
ADP	650 ± 77 (10)	1000 ± 51 (17)
AMP	521 ± 116 (10)	654 ± 80 (17)
Glycerol phosphate	127 ± 14 (6)	51 ± 14 (4)

Table 4. Effect of Pent-4-enoic Acid on Rat Liver Metabolite Concentrations. Livers were perfused for 45 minutes in the presence of linoleate and 1.0 mM pentenoic acid as described in Table 3 and in Methods section. Number of livers in parentheses.

Effect of Ethanol Added After Ten Minutes of Perfusion with Pent-4-enoic Acid

Another series of livers was perfused with pentenoic acid for only ten minutes before adding ethanol (Table 5). Other studies suggested that we would find pyruvate carboxylase only partially inhibited at this time. In this case gluconeogenesis was partially, although not completely restored to normal; however, fatty acid oxidation, as assessed by ketone body production, remained almost totally depressed. The inhibition at glyeraldehyde-phosphate dehydrogenase was reversed, while some inhibition seemed to remain at pyruvate carboxylase as indicated by the elevated pyruvate and decreased acetyl-CoA concentrations.

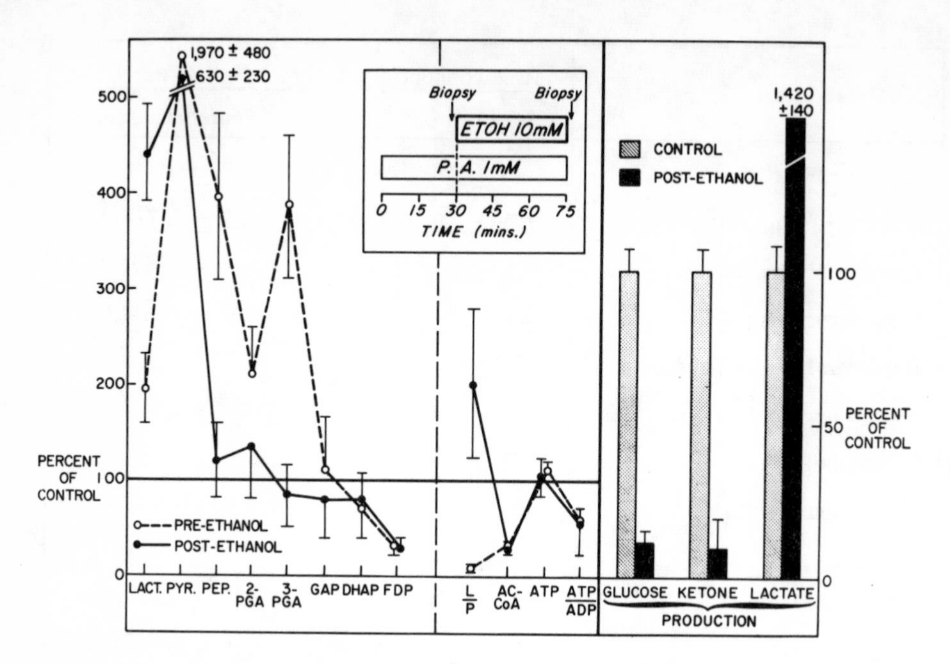

Fig. 4. Effect of Ethanol Added 30 Minutes After Pent-4-enoic Acid. Livers were perfused with a medium containing alanine and linoleate (Table 2) and pentenoic acid (1.0 mM) for 30 minutes. At this time 8 mM ethanol was added and the perfusion allowed to continue for an additional 45 minutes. Tissue was biopsied for metabolite determinations, immediately prior to the addition of ethanol and at the end of the experiment. The data are expressed as percent ± s.e. of the corresponding control values.

	Tissue Concentration (nmoles/g wet wt.)			Lactate	Glucose	Ketone Body
	Pyruvate	3PGA	Ac-CoA	Pyruvate	Prod.	Prod.
					(μ moles/g per 45 mi	
10 min. (6)	265 ± 47	910 ± 266	14.5 ± 4	4.7 ± 0.8		
55 Min. (6)	48 ± 5	245 ± 40	12.5 ± 3.6	14.4 ± 4.6	15 ± 5	1.1 ± 0.9
Control (11)	24 ± 3	225 ± 34	34 ± 5	26 ± 7	20 ± 1.2	49 ± 6

Table 5. Effect of Ethanol (8 mM) Added Ten Minutes After Pent-4-enoic Acid (1.0mM
Experimental model is the same as that described in Fig. 4.

When ethanol was added after five minutes of perfusion with pentenoic acid, both the rate of gluconeogenesis and the tissue level of acetyl-CoA were completely restored to normal (data not shown in this report). Ketone body production remained inhibited, however, but only by 50%.

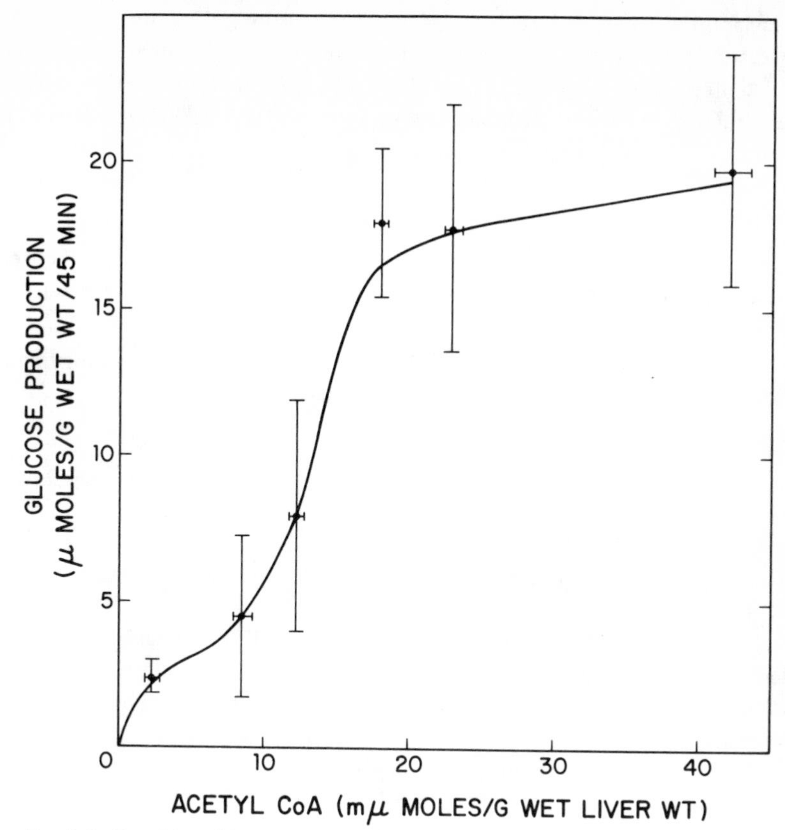

Fig. 5. Relationship of Liver Acetyl-CoA Concentration and the Rate of Gluconeogenesis in the Presence of 10 mM Alanine. The data were obtained from experiments in which 8 mM ethanol was added to the perfusate 5-30 minutes after 1.0 mM pentenoic acid. The vertical and horizontal bars represnt ± s.e.m.

Acetyl-CoA and Rate of Gluconeogenesis

The role of acetyl-CoA in regulating gluconeogenic flux in an intact tissue cannot be assessed unless it is the rate limiting factor for gluconeogenesis. This seemed to be the case in the experiments where ethanol was added after a period of perfusion with pentenoic acid, since there was no longer a block at glyceraldehyde phosphate dehydrogenase and the concentration of the substrates for pyruvate carboxylase, ATP, HCO_3^- and pyruvate did not appear to be limiting.

In Fig. 5, the rate of gluconeogenesis is plotted against the tissue level of acetyl-CoA at the end of these perfusions. The rate of gluconeogenesis appears as a sigmoidal function of acetyl-CoA concentration with the maximal effect of acetyl-CoA being evident at about 20 nmoles/g wet wt. Although the concentration of acetyl-CoA at the site of pyruvate carboxylase activity was not known, it is noteworthy that the shape of the plot and the apparent K_a of 1.8×10^{-5} M, when corrected for cell water concentration, w not very dissimilar from the findings with a purified rat liver pyruvate carobxylase des cribed by Seufert et al. in press. These findings suggest that acetyl-CoA may regulat pyruvate carboxylase activity in intact tissue, in much the same way it does in a cell free system.

Acknowledgment

We are very grateful to Mrs. Zenta Skulte and Mrs. Ludmilla Klavins for their expert technical assistance. This work was supported by U.S. Public Health Service Grants AM-09587 and TI-AM 5077.

Figure 1 was reproduced by permission of the American Journal of Physiology and Figures 2-5 by permission of the Journal of Biological Chemistry.

References

Brendel, K., C. F. Corredor, R. Bressler: Biochem. biophys. Res. Commun. 34 (1969) 340

Bressler, R., C. Corredor, K. Brendel: Pharmacol. Rev., 21 (1969) 105

Chen, K. K., R. C. Anderson, M. D. McCowne, R. N. Harris: J. Pharmacol. exp. Ther. 121 (1957) 272

Corredor, C., K. Brendel, R. Bressler: Proc. Nat. Acad. Sci. (Wash.), 58 (1968) 2299

Corredor, C., Brendel, K., R. Bressler: J. biol. Chem. 244 (1969b) 1212

De Renzo, E. C., K. W. McKerns, H. H. Bird, W. P. Cekleniak, B. Coulomb, E. Kaleifa: Biochem. Pharmacol. 1 (1958) 236

Entman, M., R. Bressler: Molec. Pharmacol. 3 (1967) 333

Good, C. A., H. Kramer, M. Somogyi: J. biol. Chem. 100 (1933) 485

Hassal, C. H., K. Reyle, P. Feng: Nature, 173 (1954) 356

Huggett, A. St. G., D. A. Nixon: Lancet 2 (1957) 368

Karnofsky, M. L., J. M. Foster, L. T. Gidez, D. D. Hagerman, C. U. Robinson, A. K. Solomon, C. E. Villee: Anal. Chem. 27 (1955) 852

Krebs, H. A., K. Henseleit: Z. Physiol. Chem. 210 (1932) 33

Ruderman, N. B., M. G. Herrera: Am. J. Physiol. 214 (1968) 1346

Ruderman, N. B., E. Shafrir, R. Bressler: Life Sci. 7 (1968) 783

Ruderman, N. B., C. J. Toews, C. Lowy, I. Vreeland, E. Shafrir: Amer. J. Physiol. 219 (1970) 51

Senior, A. E., H. S. A. Sherratt: Biochem. J. 110 (1968a) 499

Senior, A. E., H. S. A. Sherratt: Biochem. J. 110 (1968b) 521

Seufert, D., E. Herlemann, W. Seubert: Hoppe-Seylers Z. physiol. Chem. in press

Toews, C. J., C. Lowy, N. B. Ruderman: J. biol. Chem., J. Biol. Chem. 245 (1970) 818

Utter, M. F., D. B. Keech: J. biol. Chem. 238 (1963) 2603

Von Holt, C. M., M. Von Holt, H. Bohm: Biochem. biophys. Acta. 125 (1966) 11

Williamson, D. H., J. Mellanby, H. A. Krebs: Biochem. J. 82 (1962) 90

Williamson, J. R., M. H. Fukami, M. J. Peterson, S. G. Rostand, R. Scholz: Biochem. biophys. Res. Commun. 36 (1969) 407

Discussion to Ruderman, Toews, Lowy and Shafrir

J. R. Williamson: I find it interesting that you observe a crossover at GAPDH, although, as you say, the rate-controlling interaction is at pyruvate carboxylase, which does not appear as a crossover. It is incorrect to consider that a fall of the NADH/NAD can directly cause an inhibition of gluconeogenesis. Some feedback or additional effects must be occurring to diminish the rate of PEP formation. Otherwise, intermediates up to 1,3 PGA would pile-up and eventually eliminate the crossover at GAPDH as GAP, DAP and FDP levels also increased above controls. The essence of the problem is that the changes of acetyl-CoA which you observe are not always adequate to account for an inhibition of pyruvate carboxylase. The answer may come from two possibilities: 1) the increased PEP concentration in conjunction with a fall of ATP could increase pyruvate kinase activity, and 2) a lowered mitochondrial NADH/NAD ratio would increase the ratio of acetoacetyl-CoA/ß-hydroxybutyryl-CoA, which as Dr. Utter has told us, would result in an inhibition of pyruvate carboxylase even if the acetyl-CoA level did not change.

Ruderman: We observed an inhibition of gluconeogenesis after five minutes of perfusion with pentenoic acid. At this time the redox changes were well established, but acetyl-CoA had not yet fallen to levels at which pyruvate carboxylase would have been inhibited. Theoretically, the mechanism proposed by Dr. Utter could in part account for this rapid inhibition of gluconeogenesis: however, the finding that aspartate conversion to glucose was inhibited suggests that the gluconeogenic pathway was inhibited at some step beyond pyruvate carboxylase. In addition, tissue levels of oxaloacetate were three to four times higher than those found in control livers. If there was no block above oxaloacetate, then we should have had a three- to four-fold increase in gluconeogenesis.

J. R. Williamson: The problem is, why if there is an inhibition at pyruvate carboxylase should the total tissue oxalacetate levels rise? In our experiments (Williamson et al.: J. Biol. Chem., August 1970), we perfused livers with pyruvate, and obtained results somewhat similar to yours upon addition of 4-pentenoic acid. We also found an increase of total tissue oxalacetate, but when calculations of intra- and extra-mitochondrial oxalacetate concentration were made, it turned out that the increase of oxalacetate was entirely extra-mitochondrial while the intra-mitochondrial concentration fell. Thus, a crossover between pyruvate and oxalacetate was obtained on the basis of oxalacetate compartmentation. With regard to flux through PEP carboxykinase, possibly the rise of cytoplasmic oxalacetate was compensated by a fall of GTP. The inhibition of gluconeogenesis from aspartate with 4-pentenoic acid could be caused by a fall of α-ketoglutarate, or increased pyruvate kinase activity.

D. H. Williamson: When you added ethanol you obtained a restoration of ketogenesis. Do you think these ketone bodies come from alanine or from free fatty acids?

Ruderman: I think they came from restoration of fatty acid oxidation; however, I have no conclusive evidence. We carried out some studies with uniformly labeled alanine. In these the generation of $C^{14}O_2$ was almost totally inhibited in the presence of ethanol and pentenoic acid, even when ketogenesis was restored. This would mean that the ketone bodies would have had to come from a substrate other than alanine.

Exton: Is there any possibility that they came from the ethanol?

Ruderman: This is a possibility. It is also conceivable that ethanol was the source of acetyl-CoA in those experiments in which it maintained gluconeogenesis, but did not restore ketone body production. We have not studied the disposition of labeled ethanol in these circumstances.

Söling: I would like to ask Dr. Williamson: Do you really believe that under conditions of decreased fatty acid oxidation in the presence of pentenoic acid the level of acetoacetyl-CoA can reach a point where it is able to inhibit pyruvate carboxylase? I cannot believe that this could play a role even when the redox systems are more oxidized.

J. R. Williamson: I have not measured the concentration of acetoacetyl-CoA under the conditions of 4-pentenoic acid inhibition. But one can imagine that the levels of these ketone-CoA compounds are fairly independent of the total ketone body concentration and the rate of fatty acid oxidation.

Seubert: Is it known at which site pentenoic acid does inhibit fatty acid oxidation?

Ruderman: There appear to be at least two mechanisms by which pentenoic acid inhibits fatty acid oxidation. Pentenoic acid is initially converted to pentenoyl CoA which is in turn oxidized to acrylyl-CoA and acetyl CoA. Acrylyl-CoA, or some 3-carbon derivative of it, is metabolized very slowly and as a result, acrylyl-CoA and pentenoyl CoA accumulate within the cell and there is marked decrease in free CoA. Simultaneously there is an increase in pentenoyl-carnitine and acrylyl carnitine, with a resultant decrease in free carnitine. As a consequence of the deficits in free CoA and carnitine, the transport of long chain fatty acids into the mitochondria and hence their subsequent oxidation, are inhibited. If one incubates for a sufficiently long time with pentenoic acid, the oxidation of short chain fatty acids and α-ketoglutarate are also inhibited, presumably because of a lack of intra-mitochondrial CoA. However, this is a late phenomenon, which does not seem to be relevant to our studies.

Krebs: If this were correct carnitine might be expected to abolish the effect of pentenoic acid.

Ruderman: This is a rather complicated problem. In vivo, Corredor et al. (Proc. nat. Acad. Sci. 58 (1968) 2299) were able to prevent hypoglycemia in mice by administering carnitine together with pentenoic acid. On the other hand the same investigators were unable to restore gluconeogenesis in pigeon liver homogenates, that had been preincubated with pentenoic acid by adding carnitine. When they added CoA together with carnitine, however, both gluconeogenesis and long chain fatty acid oxidation were restored. The reason for the discrepancy between the in vivo and in vitro findings has not been established.

Hanson: We could show a very marked effect of pentenoic acid on lipogenesis in adipose tissue as well as pyruvate carboxylation in isolated adipose tissue mitochondria. Perhaps this compound acts on lipogenesis as well as on gluconeogenesis.

Utter: How did you get your curve which showed the dependency of the rate of gluconeogenesis on the acetyl-CoA concentration?

Ruderman: We took all values of acetyl-CoA in the same range, i. e. between 0-5 μmol/g, 5-10, 10-15 etc., calculated the mean and standard error and plotted this against the corresponding rates of gluconeogenesis. Each value on the curve represents at least four to five experiments.

Utter: But in these cases ethanol was always present?

Ruderman: Yes, if there indeed was a limiting step due to the redox change, this would not be a factor.

Wieland: Horecker's group (Nakashima et al.: Proc. nat. Acad. Sci. 64 (1969) 947) has shown a stimulatory effect of free CoA on fructose-1, 6-diphosphatase. I would like to ask whether this could also be related to the decrease in gluconeogenesis under these conditions with a lack of free CoA?

Ruderman: I doubt it, since gluconeogenesis from glycerol and fructose were not inhibited.

Regulation of Gluconeogenesis in Rat and Guinea Pig Liver +

H. D. Söling, B. Willms, and J. Kleineke
Abteilung für Klinische Biochemie, Medizinische Universitätsklinik Göttingen, FRG

Summary

The stimulation of gluconeogenesis from L-lactate by isolated perfused livers of fed or starved rats by fatty acids is well established.

Similar experiments with isolated perfused guinea pig livers showed a significantly higher rate of net formation of glucose from 10 or 20 mM L-lactate compared with rat liver, but hexanoate as well as albumin-bound oleate inhibited the net formation of glucose and the uptake of L-lactate in guinea pig liver under conditions where both parameters were stimulated in rat liver. Glucagon as well as 3′, 5′-AMP or its dibutyryl-derivative failed to stimulate gluconeogenesis from L-lactate in guinea pig liver. Ketogenesis by rat liver from oleate was considerable, that from hexanoate moderately higher compared with guinea pig liver, but differences with respect to glucose formation remained even under conditions of similar rates of ketogenesis. Changes of the levels of acetyl-S-CoA and free CoA-SH during stimulated fatty acid oxidation were similar in both species. An analysis of activities of enzymes involved in gluconeogenesis and glycolysis in the fed and the 48-hour starved state revealed in guinea pig liver a significantly lower glucokinase + hexokinase/glucose-6-phosphatase ratio and a significantly higher phosphofructokinase/FDPase ratio.

The pyruvate kinase/pyruvate carboxylase ratio was 4.00 in the fed (rat: 9.50), 4.75 (rat: 3.22) in the 48-hour starved state. The pyruvate kinase/phosphoenolpyruvate carboxykinase ratio was 0.65 in the fed (rat: 11.90), 0.82 (rat: 4.55) in the 48-hour starved state. Partially purified pyruvate carboxylase from guinea pig and rat livers behaved similarly with respect to apparent K_M′s for pyruvate and ATP and to the apparent K_a for acetyl-S-CoA at different pH values, but exhibited different pH optima with Mn^{++} (rat: 8.05, guinea pig: 7.25) or Mg^{++} (rat: 8.50, guinea pig: 7.95). It seems unlikely that pyruvate carboxylation is a site of regulation of gluconeogenesis in guinea pig liver.

\+ Unusual Abbreviations: MICA, 5-methoxy-indol-2-carbonic acid

Enzyme Code Numbers: Fructose-1,6-diphosphate aldolase (EC 4.1.2.12), Fructose-1, 6-diphosphatase (EC 3.1.3.11), Glucokinase (EC 2.7.1.2), Glucose-6-phosphatase (EC 3.1.3.9), Hexokinase (EC 2.7.1.1), Lactate dehydrogenase (EC 1.1.1.27), Nucleoside diphosphate kinase (EC 2.7.4.6), Phosphoenolpyruvate carboxykinase (EC 4.1.1.32), Phosphofructokinase (EC 2.7.1.11), Pyruvate carboxylase (EC 6.4.1.1), Pyruvate kinase (EC 2.7.1.40), Triosephosphate dehydrogenase (EC 1.2.1.12).

Inhibition of gluconeogenesis from L-lactate by 5-methoxy-indol-2-carbonic acid and the release of this inhibition by fatty acids occurred similarly in rat and guinea pig livers.

Quinolinic acid, on the other hand, exhibited a much stronger inhibitory effect in rat liver than in guinea pig liver, while it was without any effect at all in the isolated perfused pigeon liver. Studies with ^{14}C-quinolinic acid showed that this agent does not enter the mitochondrial compartment at a measurable rate. The degree of inhibition of gluconeogenesis seems to be inversely related to the proportion of phosphoenolpyruvate carboxykinase located in the mitochondria.

Introduction

Looking for an experimental model to learn more about regulation of gluconeogenesis in man, we started to study this regulation in guinea pig liver. This seemed to be reasonable as in human liver several parameters related to gluconeogenesis and glycolysis are similar to those in guinea pig livers, but different from those in rat livers (Table 1).

In guinea pig livers (Nordlie and Lardy (1963) as well as in human livers (Shrago personal communication), phosphoeonolpyruvate carboxykinase activity can be found intramitochondrially as well as in the cytosol, whereas in rat livers it is found nearly exclusively in the cytosol (Nordlie and Lardy 1963). Pyruvate carboxylase activity is high in rat livers, but low in human and guinea pig livers (Böttger et al. 1969). Nucleoside diphosphate kinase activity is high in rat livers, but low in guinea pig livers (Ishihara and Kikuchi 1968) and -- as we found recently — in human livers. Glucokinase activity is high in rat livers, but low in guinea pig (Lawris et al. 1966) and human livers (Willms et al. in press).

Gluconeogenesis from L-lactate and Ketogenesis in Guinea Pig Livers

In vivo studies showed that the guinea pig, a herbivorous animal, can starve for more than 72 hours without developing severe hypoglycemia. The incorporation of (^{14}C) activity from (2-^{14}C) pyruvate into glucose and glycogen in vivo increased significantly after 48 hours of starvation (Table 2).

Isolated perfused livers from guinea pigs starved for 48 hours produced significantly more glucose from L-lactate compared with livers from rats starved for 48 hours (Fig. 1).

We have previously reported that hexanoate significantly stimulated the incorporation of radioactivity from (2-^{14}C) or (1-^{14}C) pyruvate into glucose by isolated perfused livers from fed rats (Söling et al. 1968). Under similar conditions, the intraportal fusion of 0.72 mmole/hr. of sodium hexanoate did not stimulate the incorporation of radioactivity into glucose, instead it inhibited it (Fig. 2). The formation of $^{14}CO_2$ from (1-^{14}C) pyruvate was not inhibited, whereas it was inhibited in rat livers. The changes in the concentrations of acetyl-S-CoA and free CoA-SH on the other hand, were similar to those observed in rat livers (Söling et al. 1968) under similar conditions (Fig. 3).

	Rat	Guinea Pig	Man
Glucokinase	high	low	low
Pep. Carboxykinase (Mitochondr.)	absent	present	present
Pep. Carboxykinase (Cytoplasmic)	present	present	present
Pyr. Carboxylase	high	low	low
Pyr. Carboxylase/ Pep. Carboxykinase Ratio	high	low	low
Nucleoside diphosphate Kinase	high	low	low

Table 1.

	Incorporation into glucose	Incorporation into liver glycogen
	(percentage of injected (^{14}C) radioactivity)	
Chow fed animals	3.96 ± 0.48	0.41 ± 0.24
48-hour-starved animals	12.69 ± 2.21	2.26 ± 0.69

Table 2. Incorporation of ^{14}C radioactivity from 2-^{14}C pyruvate into glucose and liver glycogen by chow fed and 48-hour-starved guinea pigs. One mmole of 2-^{14}C pyruvate was injected intraperitoneally. The animals were killed one hour later.

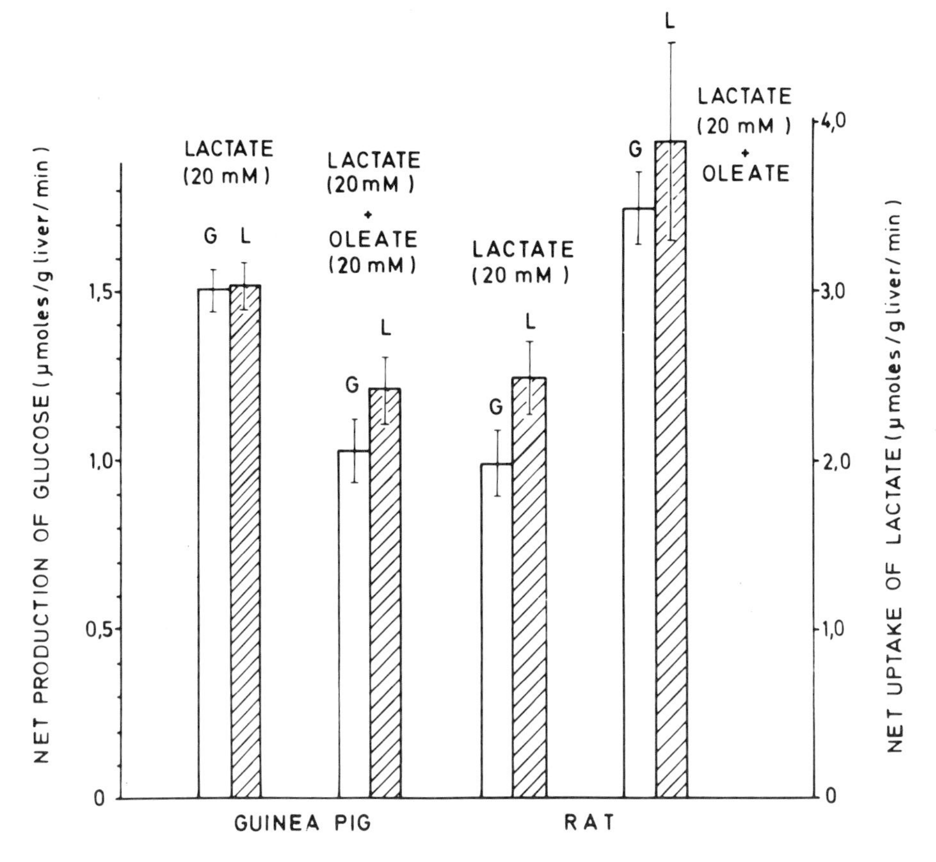

Fig. 1. Effect of albumin bound oleate (2 mM) on the net production of glucose (white columns) and the net uptake of L-lactate (hatched columns) by isolated perfused livers from rats starved for 48 hours and guinea pigs with L-lactate (20 mM) as the gluconeogenic precursor.

Albumin bound oleate added to the perfusion medium to give a final concentration of 2 mM significantly enhanced the net production of glucose and the net uptake of lactate by isolated perfused livers from rats starved for 48 hours when lactate was the precursor, but it inhibited net glucose production and lactate uptake in isolated perfused guinea pig livers (Fig. 1).

Similarly, glucagon, 3′, 5′-AMP, and dibutyryl-3′, 5′-AMP stimulated the net production of glucose from L-lactate in the isolated perfused rat liver, but failed to do so in guinea pig liver (Fig. 4). Both agents, on the other hand, exerted their glycogenolytic action in isolated perfused livers from fed guinea pigs in the same way as in rat liver, and enhanced ketogenesis and urea production (Table 3).

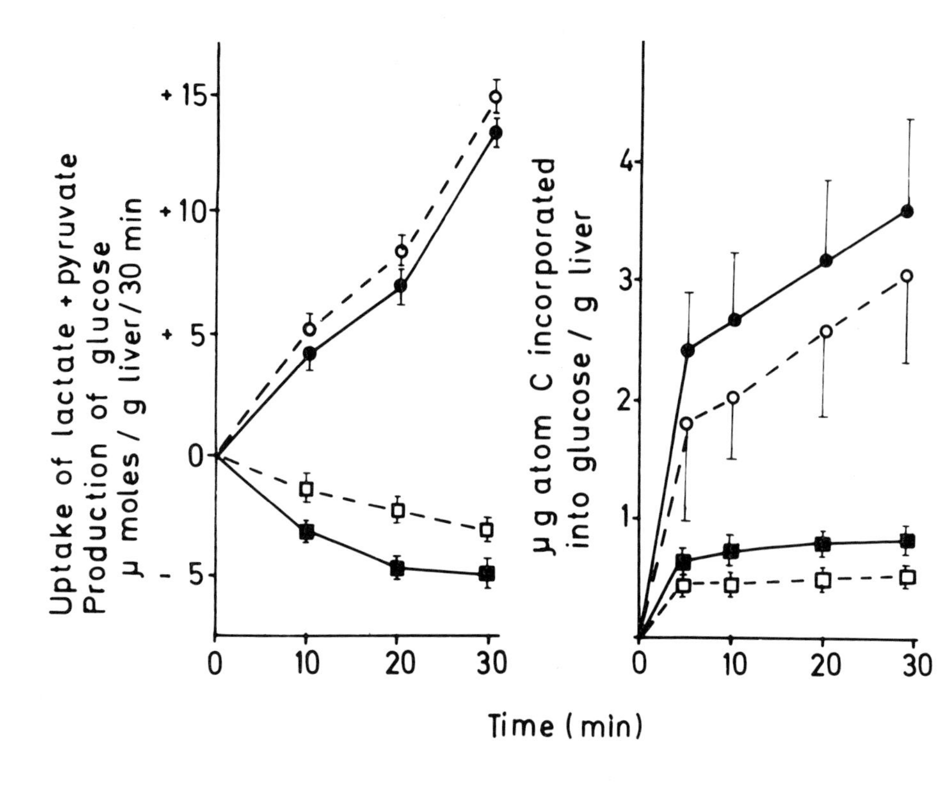

Fig. 2. Lack of an effect of an intraportal infusion of sodium hexanoate (0.72 mmol/hr) on the net uptake of lactate + pyruvate (lower part, left side) the net production of glucose (upper part, left side) and the incorporation of radioactivity from ($1-^{14}C$) pyruvate (lower part, right side) or ($2-^{14}C$) pyruvate (upper part, right side) into glucose by isolated perfused livers from chow fed guinea pigs.

The labeled pyruvate was added with cold lactate and pyruvate (lactate/pyruvate = 10/1 on a molar basis) to give a final concentration of 2 mM. Solid black symbols = control experiments, open symbols = experiments with hexanoate.

When L-carnitine (1.5 mM) was added to the perfusion medium together with oleate gluconeogenesis again was inhibited but not stimulated.

Isolated perfused rat livers produced considerably more ketone bodies from oleate than guinea pig livers, whereas ketogenesis from hexanoate was only slightly lower in guinea pig livers (Table 4). The lower rate of ketogenesis from oleate in guinea pig livers may result from a lower long chain fatty acid thiokinase activity as can be concluded from the lower rate of oleate uptake.

When, in experiments with isolated perfused guinea pig livers, the rate of hexanoate infusion was raised in a way that the rate of ketogenesis reached or did exceed those found in isolated perfused rat livers, gluconeogenesis from L-lactate was still inhibited and not stimulated.

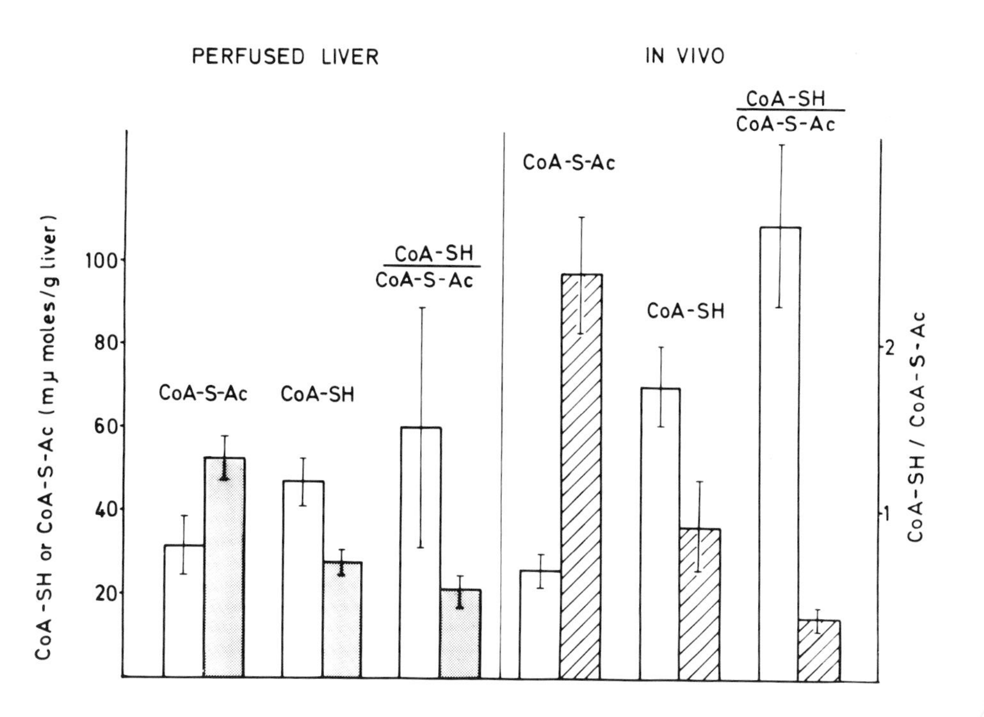

Fig. 3. Changes in the concentrations of acetyl-S-CoA and free CoA-SH and the CoA-SH/acetyl-S-CoA ratio in isolated perfused livers from chow fed guinea pigs during an intraportal infusion of 0.72 mmoles/hr of sodium hexanoate (left side) and in livers from fat fed guinea pigs in vivo (right side). The fat fed animals were starved for 24 hours and then received 4 ml of olive oil by tube feeding three times a day for two days.

White columns = controls, grey columns = perfusions with hexanoate, hatched columns = fat fed animals in vivo.

	Total ketone body production	Urea production	n
	(μ moles/g liver/min)		
20 mM L-lactate	0.17 ± 0.02	0.17 ± 0.04	15
20 mM L-lactate + glucagon	0.22 ± 0.05	0.27 ± 0.04	12
20 mM L-lactate + dibutyryl-3′, 5′-AMP	0.47 ± 0.09	0.33 ± 0.11	5

Table 3. Stimulation of ketogenesis and urea production by glucagon and dibutyryl-3′, 5′-AMP in isolated perfused livers from guinea pigs starved for 48 hours.

In the glucagon experiments, 350 μgs of glucagon were added at the beginning of the perfusion. Thereafter, 150 μgs of glucagon were added every 15 minutes.

Dibutyryl-3′, 5′-AMP was added to the medium at the beginning of the perfusion in a final concentration of $1 \cdot 10^{-4}$M.

Enzyme Activities in Rat and Guinea Pig Liver

From these experiments it follows that guinea pig liver gluconeogenesis from L-lactate seems not to be regulated at the pyruvate carboxylase step in the way it is supposed to be regulated in other livers. Therefore, kinetic studies with partially purified pyruvat carboxylase from rat and guinea pig livers were undertaken. The results have been re ported by Kleineke and Söling (1970). According to these studies, the enzymes from both sources behave similarly, especially with respect to the dependence of enzyme activities on the concentration of acetyl-S-CoA. Therefore the activities of several other enzymes involved in gluconeogenesis and glycolysis were studied in livers from chow fed and from rats and guinea pigs starved for 48 hours.

The glucose-6-phosphatase/hexokinase + glucokinase ratio was significantly higher in guinea pig livers (Fig. 5), but fructose-1, 6-diphosphatase activity and fructose-1, 6-diphosphatase/phosphofructokinase ratio were significantly higher in rat liver (Fig. 6). In accordance with Böttger et al. (1969), the total pyruvate carboxylase activity was found to be very low in guinea pig livers (Fig. 7). In contrast to rat livers, there was no increase in pyruvate carboxylase activity after 48 hours of starva-tion (Fig. 7). The same was found for guinea pig phosphoenolpyruvate carboxy-kinase (Fig. 7). The ratio of the activity of the rate-limiting enzyme (phosphoenol pyruvate carboxykinase in rat liver, pyruvate carboxylase in guinea pig liver) and py-ruvate kinase was much lower in rat livers in the fed state, but did not show a signif icant difference after 48 hours of starvation (Fig. 7). With no experimental conditio

Experimental Condition	Species	Net uptake of fatty acid	Total ketone body formation	Radioactivity in total ketone bodies	Radioactivity in CO_2	Net production of glucose
		(μmoles/g/min)		(% of total radioactivity/g liver)		(μmoles/g/min)
	Rat	6.29	1.15	3.84	0.80	1.70 (Controls: 0.97)
Oleate (2 mM)	Guinea pig	3.49	0.41	0.99	0.50	1.29 (Controls: 1.51)
Hexanoate (2 mM)	Rat	--	2.35	2.99	0.59	1.65
+ 0.54 mmole/h	Guinea pig	--	1.72	2.10	0.34	1.10 (Controls: 1.51)

Table 4. Uptake of (1-^{14}C) oleate, as well as ketogenesis and $^{14}CO_2$ -formation from (1-^{14}C) oleate and (1-^{14}C) hexanoate by isolated perfused livers from 48-hour-starved rats and guinea pigs with 20 mM L-lactate as a precursor for glucogeneogenesis.

Oleate was added to the medium in an albumin bound form as described by Ross et al. (1967). Hexanoate was added to the medium at the beginning of the experiments in a final concentration of 2 mM and was further infused intraportally at a rate of 0.54 mmole/hr.

Note that the rate of ketogenesis by guinea pig livers with hexanoate is higher than that of rat livers with oleate, whereas gluconeogenesis is inhibited in guinea pig livers, but stimulated in rat livers.

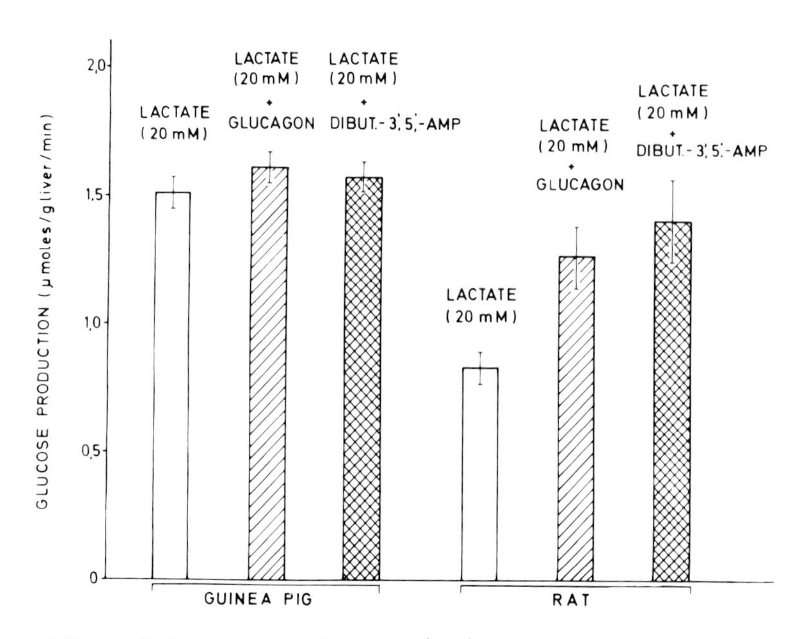

Fig. 4. Effects of glucagon and dibutyryl-3′, 5′-AMP on the net production of glucose by isolated perfused livers from rats and guinea pigs starved for 48 hours with L-lactate (20 mM) as the gluconeogenic precursor. The experimental conditions were the same as described in the legend for Table 3.

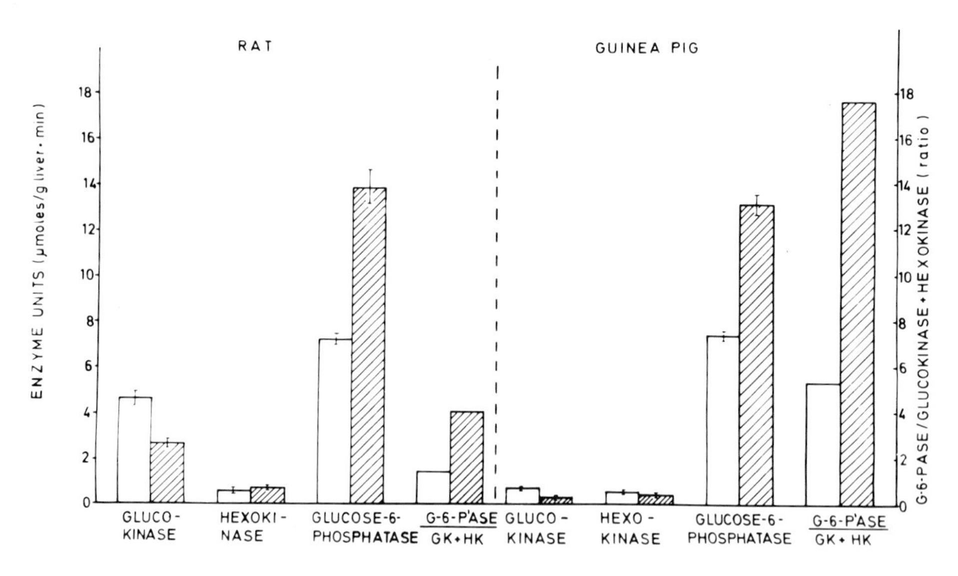

Fig. 5. V_{Max}-activities of hexokinase, glucokinase and glucose-6-phosphatase in livers from chow fed (white columns) and from 48-hour -starved rats and guinea pigs (hatched columns).

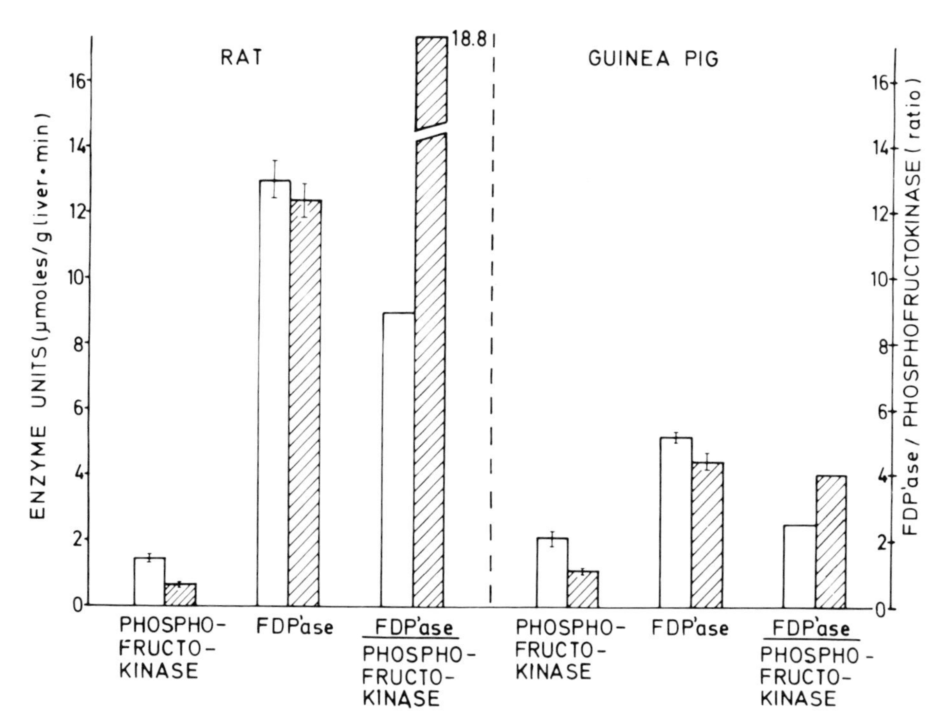

Fig. 6. V_{Max}-activities of phosphofructokinase and fructose-1, 6-diphosphatase in livers from chow fed (white columns) and from 48-hour-starved rats and guinea pigs (hatched columns).

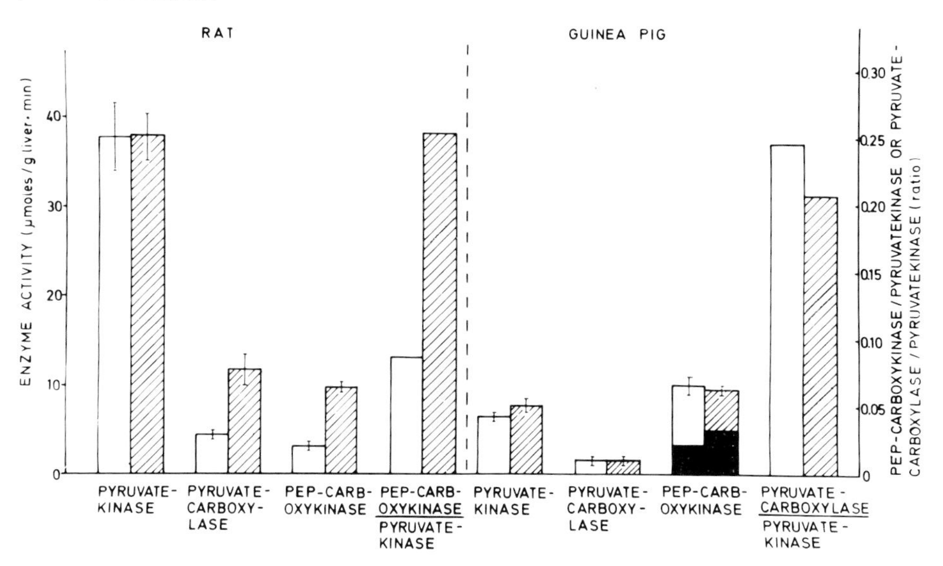

Fig. 7. V_{Max}-activities of pyruvate kinase, phosphoenolpyruvate carboxykinase, and pyruvate carboxylase in livers from chow fed (white columns) and from 48-hour-starved rats and guinea pigs (hatched columns).

the activity of pyruvate carboxylase in guinea pig livers did exceed 2 µ moles/g liver/min. As glucose production from L-lactate was higher than 1.5 µ moles/g liver/min., it turns out that pyruvate carboxylase activity must have been underestimated. Further studies have to be made to find a proper extraction procedure for this enzyme.+

But even under the assumption that pyruvate carboxylase activity would be twice the activity measured by Böttger et al. (1969) and by us, this still would mean that pyruvate carboxylation must proceed at a maximum rate already in absence of exogenous fatty acids to allow for the observed rate of gluconeogenesis from L-lactate.

Of the bifunctional enzymes measured, the activity of lactate dehydrogenase and triosephosphate dehydrogenase were significantly, that of aldolase slightly lower in guinea pig livers compared with rat livers (Fig. 8).

Redox State and Gluconeogenesis from Pyruvate

When guinea pig livers were perfused with a medium containing high amounts of pyruvate (20 mM), a slight stimulation of net production of glucose was found with high loads of hexanoate. Ethanol at a concentration of 10 mM had a similar effect. From the latter finding as well as from crossover plots (Fig. 9), we conclude that the effect of hexanoate on gluconeogenesis from pyruvate under these conditions results from a stimulation at the triosephosphate dehydrogenase step and not at the site of pyruvate carboxylase.

Inhibitors of Gluconeogenesis

As described for rat livers by Hanson et al. (1969), 5-methoxy-indol-2-carbonic acid inhibited gluconeogenesis from L-lactate by isolating perfused guinea pig livers nearly completely and this inhibition was relieved by fatty acids (Fig. 10).

Inhibition of gluconeogenesis by quinolinic acid (Veneziale et al. 1967), on the othe hand, revealed significant quantitative species differences (Fig. 11).

At low concentrations of quinolinic acid, the inhibitory effect of quinolinate was significantly stronger in rat than in guinea pig livers. Gluconeogenesis by isolated perfused pigeon livers could not be inhibited even by concentrations as high as 28 mM.

The inhibitory effect of quinolinic acid is positively related to the percentage of phos phoenolpyruvate carboxykinase activity located in the cytosol.

Therefore, isolated mitochondria from rat and guinea pig livers were incubated with (^{14}C) quinolinic acid. The uptake of quinolinic acid was measured after separation of the mitochondria from the incubation medium by the filter technique described by

+) Note added in proof: By changing the extraction procedure an activity of pyruvate carboxylase of 5.14 and 7.26 µ moles/g liver/min was found in livers of fed and 48 hours starved guinea pigs respectively. (Söling et al. Europ. J. Biochem., in press.)

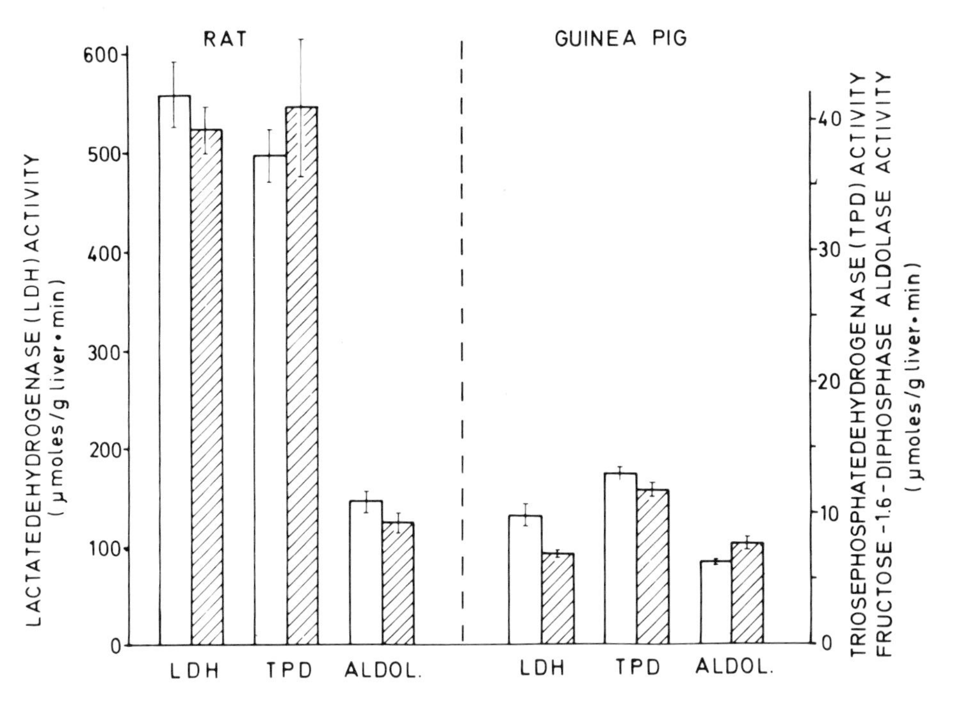

Fig. 8. V_{Max}-activities of lactate dehydrogenase, triosephosphate dehydrogenase and fructose-1, 6-diphosphate aldolase in livers from chow fed (white columns) and from 48-hour-starved rats and guinea pigs (hatched columns).

Garber and Ballard (1969). Corrections for extra-mitochondrial water space contamination were made by aid of labeled Dextran-70. With concentrations of quinolinic acid up to 14 mM, no measurable uptake of quinolinic acid by rat or guinea pig liver mitochondria could be found (Table 5).

Conclusions

1) In the isolated perfused guinea pig liver, gluconeogenesis from L-lactate is not regulated by changes in the concentration of acetyl-S-CoA. The rate of pyruvate carboxylation depends mainly on the concentration of pyruvate at the site of pyruvate carboxylase. As the redox state of the lactate/pyruvate system in guinea pig livers is already more negative compared with rat livers (Willms et al. 1970), the inhibition of gluconeogenesis form L-lactate during increased fatty acid oxidation most probably results from a decrease in the concentration of pyruvate due to a further reduction of the lactate/pyruvate system during increased fatty acid oxidation.

2) A slight stimulation of gluconeogenesis from pyruvate by high concentrations of fatty acids does not result from an allosteric stimulation of pyruvate carboxylase but from the supply of reduction equivalents to the $NAD^+/NADH$ system in the cytosol.

3) The fact that glucagon and 3′, 5′-AMP stimulate gluconeogenesis in rat livers but not in guinea pig livers supports the idea that these agents act in rat livers at the site of pyruvate carboxylase.

4) The experiments with quinolinic acid make it probable that in guinea pig livers the extra-mitochondrial as well as the intra-mitochondrial phosphoenolpyruvate carboxykinase are participating in the formation of phosphoenolpyruvate for gluconeogenesis.

5) Because of the similarities between human and guinea pig livers, it seems doubtful that the rate of fatty acid oxidation exerts a direct effect on regulation of gluconeogenesis in human livers.

Acknowledgment

Supported by the Deutsche Forschungsgemeinschaft Bad Godesberg, FRG, under the grant numbers So 43/6 and Cr 20/5.

The authors wish to thank Miss G. Janson and Mrs. H. Peters for their excellent technical assistance.

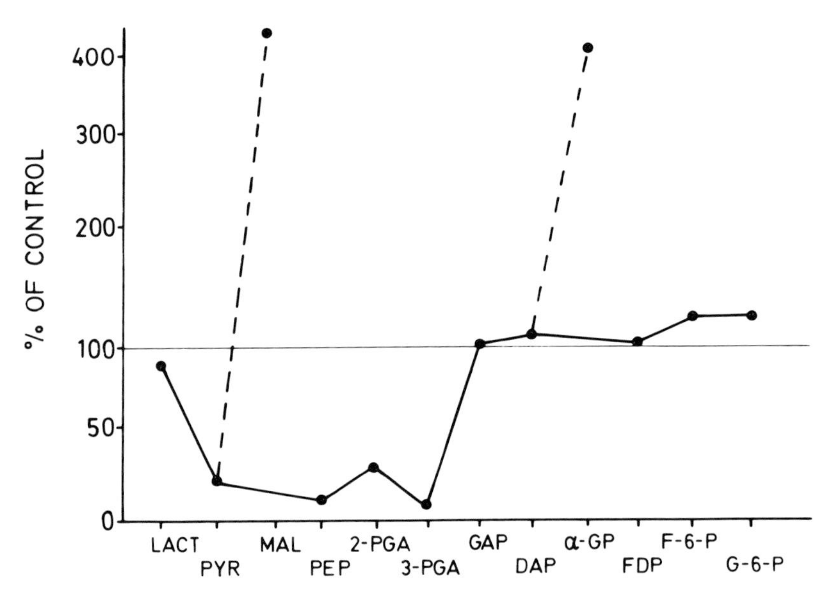

Fig. 9. Effects of sodium hexanoate on the metabolic pattern (crossover plot) in isolated perfused guinea pig livers when pyruvate (20 mM) was used as the gluconeogenic precursor. A primer dose of hexanoate (1 mM final concentration) was given at the beginning of the experiment followed by an intraportal infusion of 1.2 mmole hexanoate/hr. The livers were freeze-stopped after 60 minutes of perfusion. (Mean values from three experiments under each condition.)

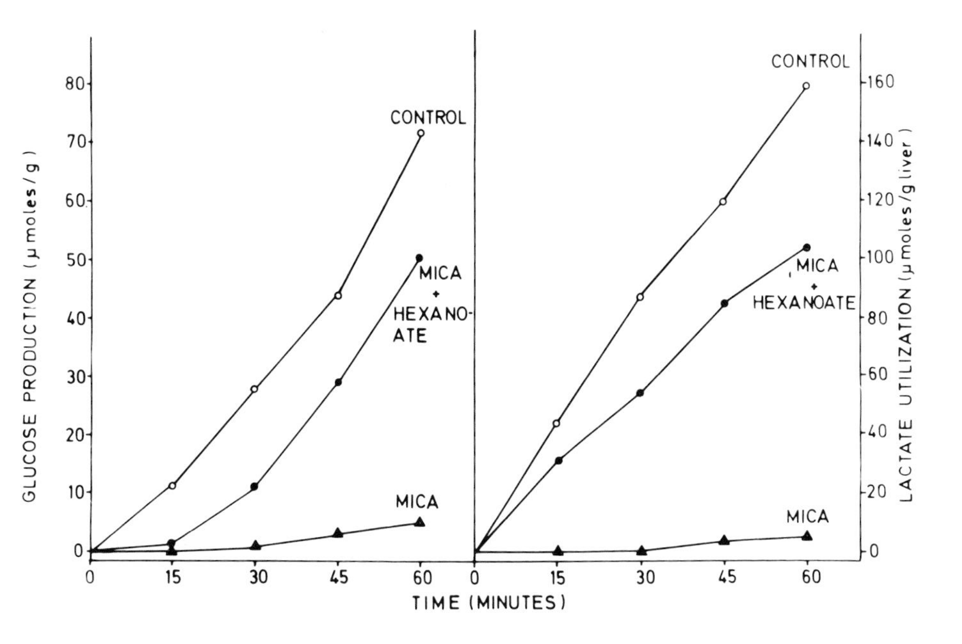

Fig. 10. Effects of 5-methoxy-indol-2-carbonic acid (MICA) (0.4 mM) on the net formation of glucose (left side) and the uptake of L-lactate (right side) by isolated perfused livers from guinea pigs starved for 48 hours. Lactate was added to the medium to give a final concentration of 20 mM. The intraportal infusion of 0.72 mmoles sodium hexanoate/hr releaves the inhibition.

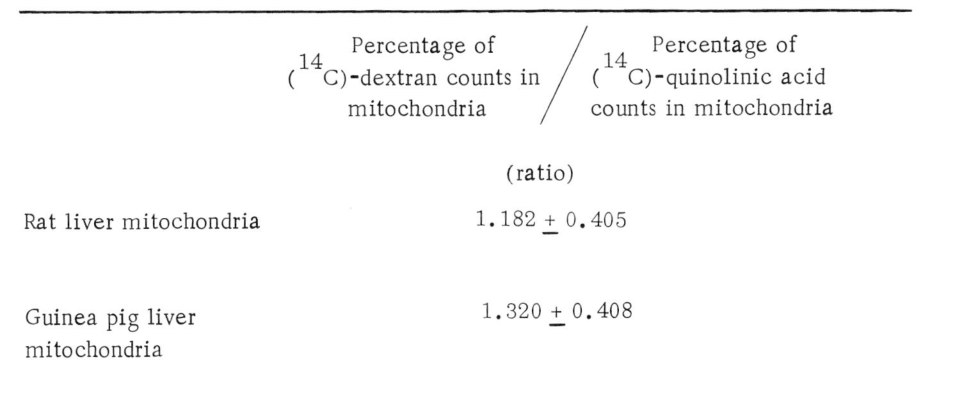

	Percentage of (^{14}C)-dextran counts in mitochondria / Percentage of (^{14}C)-quinolinic acid counts in mitochondria
	(ratio)
Rat liver mitochondria	1.182 ± 0.405
Guinea pig liver mitochondria	1.320 ± 0.408

Table 5. Mitochondria from rat or guinea pig livers were incubated with Dextran-70 and quinolinic acid. Additionally to one group of incubations ^{14}C-quinolinic acid; to another group ^{14}C-dextran was added. The mitochondria were separated from the medium, and extracted according to Garber and Ballard (1969).

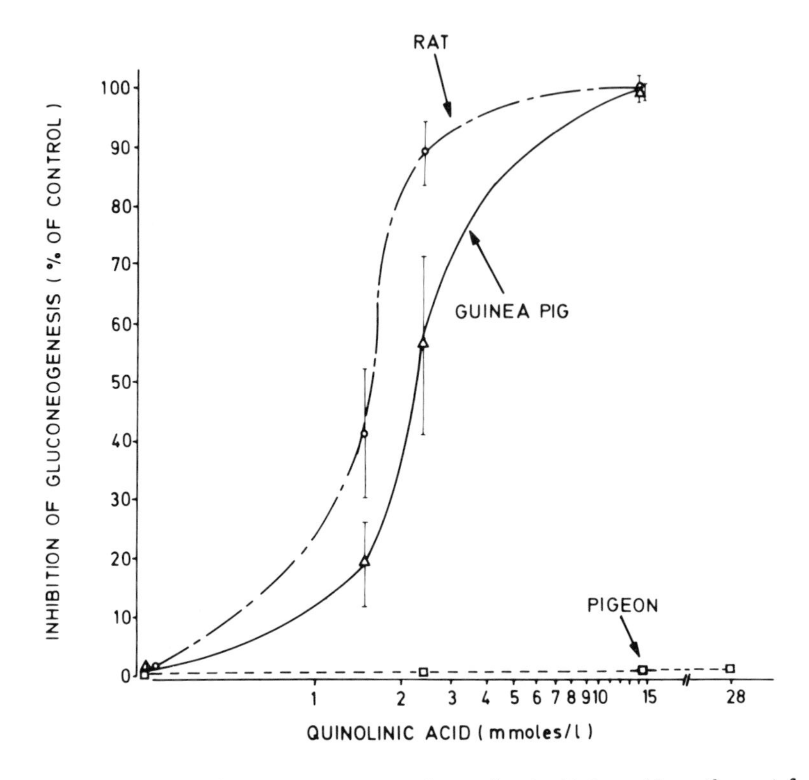

Fig. 11. Effects of various concentrations of quinolinic acid on the net formation of glucose from 10 mM L-alanin (guinea pig, rat) or 20 mM L-lactate (pigeon) by isolated perfused livers from guinea pigs, rats and pigeons. Rats and guinea pigs were starved for 48 hours; the pigeons for 72 hours.

References

Böttger, I., O. Wieland, D. Bridiczka, D. Pette: Europ. J. Biochem. 8 (1969) 113

Garber, A. J., F. J. Ballard: J. biol. Chem. 244 (1969) 4696

Hanson, R. L., P. D. Ray, P. Walter, H. A. Lardy: J. biol. Chem. 244 (1969) 435

Ishihara, H., G. Kikuchi: Biochem. biophys. Acta. 153 (1968) 733

Kleineke, J., H. D. Söling: in: Regulation of Gluconeogenesis, 9th Conference of the German Society for Biological Chemistry, Reinhausen, 1970, Thieme, Stuttgart, 1970, p. 22

Lauris, V., M. Cahill, G. F. Cahill: Diabetes 7 (1966) 475

Nordlie, R. C., H. A. Lardy: J. biol. Chem. 238 (1963) 2259

Ross, B., R. Hems, R. A. Freedland, H. A. Krebs: Biochem. J. 105 (1967) 869
Shrago, E.: personal communication
Söling, H. D., B. Willms, D. Friedrichs, J. Kleineke: Europ. J. Biochem. 4 (1968) 364
Veneziale, C. M., P. Walter, K. Kneer, H. A. Lardy: Biochem. 6 (1967) 2129
Willms, B., J. Kleineke, H. D. Söling: in: Regulation of Gluconeogenesis, 9th Conference of the German Society for Biological Chemistry, Reinhausen, 1970, Thieme, Stuttgart, 1970, p. 113
Willms, B., P. Ben-Ami, H. D. Söling: Hormones and Metabolism, 2 (1970) 135

Discussion to Söling, Willms and Kleineke

Utter: I find it difficult not to put together your observations with those reported by Dr. Willms at this symposium concerning the ratios of acetoacetate and ß-hydroxybutyrate. In guinea pig liver, unlike rat liver, the administration of hexanoate failed to increase the proportion of ß-hydroxybutyrate. If this situation is also reflected in the CoA derivatives of these acids, the failure of hexanoate to stimulate gluconeogenesis and perhaps pyruvate carboxylation in the guinea pig might be due to an unfavorable change in the ß-hydroxybutyryl-CoA to acetoacetyl-CoA ratio even though acetyl-CoA might actually increase.

Walter: I want to make a comment concerning the quinolinic acid data. Hagino and co-workers (J. biol. Chem. 243 (1968) 4980) have measured the uptake of quinolinic acid into the liver cell and found this to be very slow. And I think an alternative interpretation would be that the permeability to quinolinic acid of the rat liver cell is different compared with that of a guinea pig liver cell.

Krebs: Dr. Söling mentioned that he used the guinea pig in order to find something about the regulation of gluconeogenesis in man. For the very same reason we took the rat. There are of course numerous similarities between man and guinea pig, but there is one great difference which is relevant in this context: The guinea pig is herbivorous and the rat, like man, is omnivorous. Therefore, it is possible with rats to vary the diet a great deal, for example completely excluding carbohydrates and giving a diet consisting only of fat and protein. We have tried to do similar feeding experiments on guinea pigs, but they refused to eat a diet which did not contain relatively large quantities of carbohydrates. This means that in the guinea pig the need for gluconeogenesis is rather limited. Do perhaps some of the differences between the rat and the guinea pig arise from the different dietary habits of these animals, especially from the circumstance that the rat is much more adaptable than the guinea pig to the lack of carbohydrates in the diet?

Söling: I think the fact that the guinea pig can very quickly adapt to a state of starvation seems to demonstrate that at least the apparatus for gluconeogenesis functions very well. So at least the metabolic outfit of the guinea pig is such that it can produce enough glucose and even more glucose than the rat liver for some reasons.

The fat feeding in the experiments Dr. Willms talked about previously was done by tube feeding.

The Effect of 3', 5'-cyclic AMP on Glucose Synthesis in Isolated Rat Kidney Tubules

W. Guder and O. Wieland
Institut für Klinische Chemie und Forschergruppe Diabetes, Munich, FRG

Summary

In order to study the regulation of glucose synthesis in rat kidney a method has been developed which allows for sampling of numerous tissue aliquots from one pool.

After incubation of kidney cortex particles with crude collagenase the tubules are separated by decanting and low speed centrifugation procedures. Aliquots of the tubule suspension were incubated with different substrates (glutamate, oxoglutarate, fumarate, malate, aspartate, proline, succinate, lactate, pyruvate, glycerol, dihydroxyacetone and fructose) in Krebs Henseleit buffer to study the effects of cyclic nucleotides on glucose synthesis.

Cyclic 3', 5'-adenosinemonophosphate stimulated glucose formation from nearly all substrates which are thought to enter gluconeogenesis via oxaloacetate. Maximal stimulation was achieved at 10^{-4} M cAMP. This effect was not further stimulated by addition of aminophyllin. From the other nucleotides cyclic-GMP and IMP (Inosin) also seem to be effectual. The dibutyryl derivative of cyclic AMP stimulated glucose formation only in a concentration higher than 10^{-4}M.

The results are discussed in connection with the possible role of cyclic nucleotides in the hormonal regulation of kidney metabolism.

Introduction

During the past ten years gluconeogenesis has been studied extensively in slices of rat kidney cortex (Krebs 1969) and in experiments with the isolated perfused kidney (Nishiitsutsuji-Uwo et al. 1967). The highest rates of glucose formation from a variety of substrates were observed with the slice system. We have reported another method of kidney preparation which allows for sampling of a large number of aliquots from one single pool (Guder et al. 1969). Here tissue fragments with a diameter of less than 0.5 mm were prepared by forcing the cortex portion through a suitable nylon sieve. After two washings of the resulting cell brei these particles could be suspended and pipetted into the incubation vessels. Although the metabolic rates obtained with this preparation were comparable to those of incubated cortex slices, reproducibility of pipetting was limited due to the relative size of the particles thus preventing us to achieve a variability of less than 10% within one series. Consequently, in this system metabolic effects with changes smaller than 20% could hardly be detected.

We therefore looked for a method that would yield smaller tissue particles or even isolated cells. Further mechanical disruption of the tissue by different methods was associated with a rapid loss of gluconeogenic capacity. In the course of this work we also tried to separate kidney cells by enzymatic treatment. However, instead of yielding single cell preparations this procedure resulted in isolated tubule fragments which could not be disaggregated further without loss of metabolic activity. The following method led to kidney tubule fractions which exhibited high metabolic activity and could be easily and reproducibly dispensed by pipetting. In principle, a similar method has been employed before by Burg and Orloff (1962) in their physiological studies on tubular transport of paraaminohippuric acid.

Methods

In our studies male albino rats of 100-200 g body weight were used after starvation for 24 hours. Under light anesthesia the kidneys were perfused *in situ* with cold Krebs-Henseleit medium. The organs were then excised, cut in halves and the medulla was roughly dissected from the cortex tissue. The combined cortex chips were forced through a commercial tea sieve and the tissue particles were prepared from the resulting cell brei as described (Guder et al. 1969). These particles of about 0.5 mm diameter were suspended in five volumes of the medium and placed into a 100 ml polyethylene bottle. After addition of crude collagenase (Worthington CLS) at a final concentration of 1 mg (corresponding to 150-200 U/mg) per 100 mg of original tissue brei the bottle was gassed with a mixture of CO_2/O_2 (5:95), capped and incubated at 37^0C in a metabolic shaker. After 45 minutes, the incubation was stopped and the tissue suspension diluted with the medium to give a final volume of about 30 ml. Longer incubation periods led to a decrease in metabolic activity. The content of the bottle was transferred to a 50 ml glass tube and the unbroken particles were allowed to sediment for about 20 seconds. The supernatant containing most of the isolated tubules, cells and subcellular particles was decanted. From this supernatant, the tubules were now separated from broken cells and subcellular particles by low speed centrifugation at about 10 g for 30 seconds, washed twice on the centrifuge and suspended in about twice the volume of the original tissue weight. The final protein concentration of the suspension was between 5 and 10 mg per ml. Microscopically these suspensions contained isolated tubules of different length, contaminated with some debris and single cells (Fig. 1). An amount of 0.1 ml of this suspension was added to an incubation mixture containing 10 mM substrate and other additions in neutral solutions together with Krebs-Henseleit medium, in a final volume of 1 ml. For incubation, 15 ml polyethylene vials with screw caps (Packard Corp. Chicago, Ill.,U.S.A) were used. The incubation mixture was gassed with 95% O_2:5% CO_2 while standing on ice, then capped and incubated in a shaking water bath at 37^0C. The incubation was stopped by addition of 0.2 ml of 30% perchloric acid and the glucose determined enzymatically (Slein 1962) in the neutralized extracts. The procedure allows for preparation of about 30 samples from

1 g of fresh kidney and for large series of aliquots during one experiment.

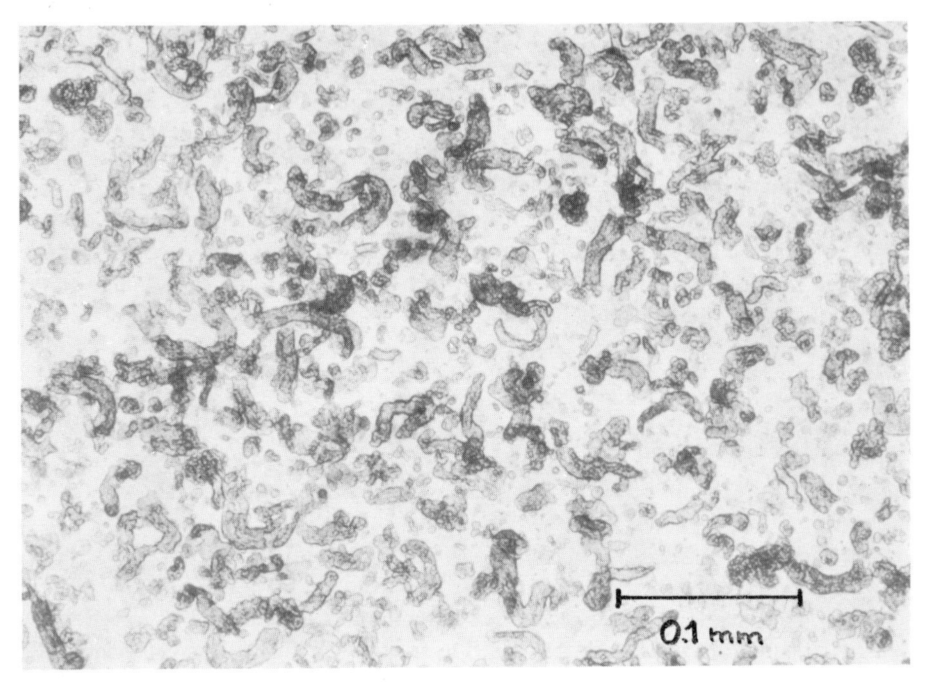

Fig. 1. Microscopic appearance of isolated kidney tubules. Photomicrograph taken from an unstained tubule suspension with a Leitz Photomicroscope. The authors wish to thank Dr. E. Siess for taking the picture.

Results

With this method glucose formation from various substrates was linear over a period of at least 60 minutes. In the experiment illustrated in Fig. 2, glucose formation from glutamate was determined with four different amounts of tissue from the same tubule preparation. As may be seen there is rather good proportionality between the rate of glucose formation and the amount of tubules added.

In order to study the reproducibility of the procedure the experiment shown in Table 1 was performed. Within a series of ten identical assays the variation coefficient was as low as 1.65%. In the presence of 10^{-4} M cyclic AMP which stimulated glucose formation the variation coefficient was somewhat higher (3.86%).

Table 2 summarizes the rates of glucose formation starting from various substrates, expressed as μmoles/g protein per hour. Protein was determined with a modified biuret method using dioxan which prevents turbidity caused by lipids (Weiss, unpublished data). Under the conditions applied, the protein represented 60-70% of the tissue dry weight. On this basis the rates measured (although strong variations

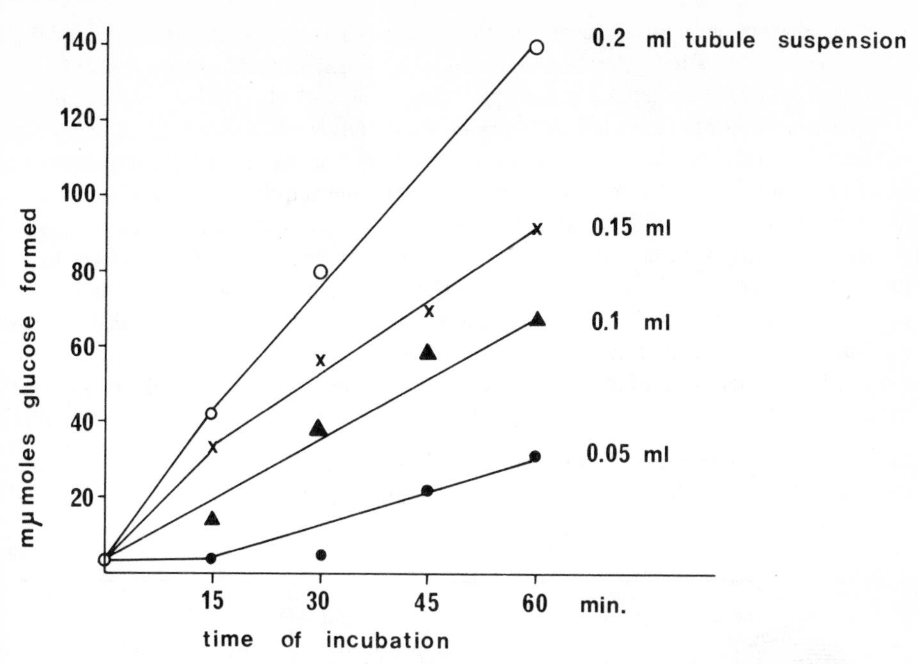

Fig. 2. Time dependency of glucose formation by rat kidney tubules.

Tubules prepared from kidney cortex from starved rats as described in the Methods section were incubated with 10 mM glutamate. One ml of tubule suspension contained 2.32 mg tissue protein.

	- 3′, 5′-AMP	+ 3′, 5′-AMP
glucose formed (mean of ten samples	236.7	329.14
S.D.	3.9	12.72
variation coefficient	1.6%	3.86%

Table 1. Reproducibility of glucose formation by isolated rate kidney tubules.

Ten aliquots from one tubule preparation were incubated with 10 mM α-oxoglutarate in the absence and presence of 10^{-4}M 3′, 5′-cyclic AMP. Each sample contained 0.66 mg of tissue protein. Glucose formation is expressed as mμ moles/assay in 60 minutes.

were found from preparation to preparation) were generally higher than those obtaine earlier with cortex slices (Krebs et al. 1963), with the perfused kidney (Nishiitsutsuj Uwo et al. 1967) or suspended kidney particles (Guder et al. 1969). For example with lactate, the rates reported by Krebs et al. (1963) were 200 μmoles/g dry weight per hour in cortex slices. The perfused kidney from starved rats as reported from the same laboratory (Nishiitsutsuji-Uwo et al. 1967) produced 82 μmoles/g per hour and the suspended particles in our former studies (Guder et al. 1969) 180 μmoles/ g tissue per hour. Similar to earlier observations with different kidney preparations, the rate of glucose synthesis is higher from pyruvate than from lactate. This is in contrast to the liver where the rate is almost the same with either substrate. Another substrate pair, dihydroxyacetone/glycerol shows the same tendency, namely that the oxidized precursor is transformed to glucose at a higher rate than its reduced counterpart.

In the case of glutamate and α-oxoglutarate no such difference could be observed. The highest rates of glucose formation were obtained with fructose, pyruvate, dihydroxyacetone and succinate as the precursor.

substrate added	glucose formed μmoles per g protein in 60 min.	
none	61.6 ± 5.8	(24)
glutamate	309.6 ± 97.3	(4)
proline	218.4 ± 38.0	(7)
α-oxoglutarate	293.6 ± 30.1	(4)
succinate	514.3 ± 49.7	(6)
fumarate	403.3 ± 81.6	(4)
malate	408.7 ± 27.0	(4)
pyruvate	632.6 ± 176.5	(7)
lactate	434.0 ± 43.0	(10)
glycerol	243.1 ± 49.7	(4)
dihydroxyacetone	523.9 ± 86.1	(4)
fructose	855.8 ± 76.3	(4)

Table 2. Glucose formation from various substrates in isolated rat kidney tubules from starved rats. Tubules (0.5-1.0 mg protein) were incubated with 10 mM substrate over a period of 60 minutes. The results are given as mean ± s.e.m. with the numbers of observations in parentheses.

Previous results of Pagliara and Goodman (1969) obtained with kidney slices prompted us to study the effect of cyclic AMP on glucose production in our system. Figure 3 represents an experiment where glucose formation from lactate in the presence of various concentration of 3′, 5′-cyclic AMP was investigated. The results are expressed as percent changes relative to the rate of glucose formation without cyclic AMP which is set 100%. Cyclic AMP stimulated glucose formation at concentrations between 10^{-5} to 10^{-4} M, the rate decreasing at higher concentrations. The significant inhibition of gluconeogenesis observed at 5 mM cyclic AMP may be explained by the studies of Weidemann, Hems and Krebs (1969a, b) in which 5′-AMP inhibited glucose formation in kidney slices and perfused kidney at concentrations above 1 mM.

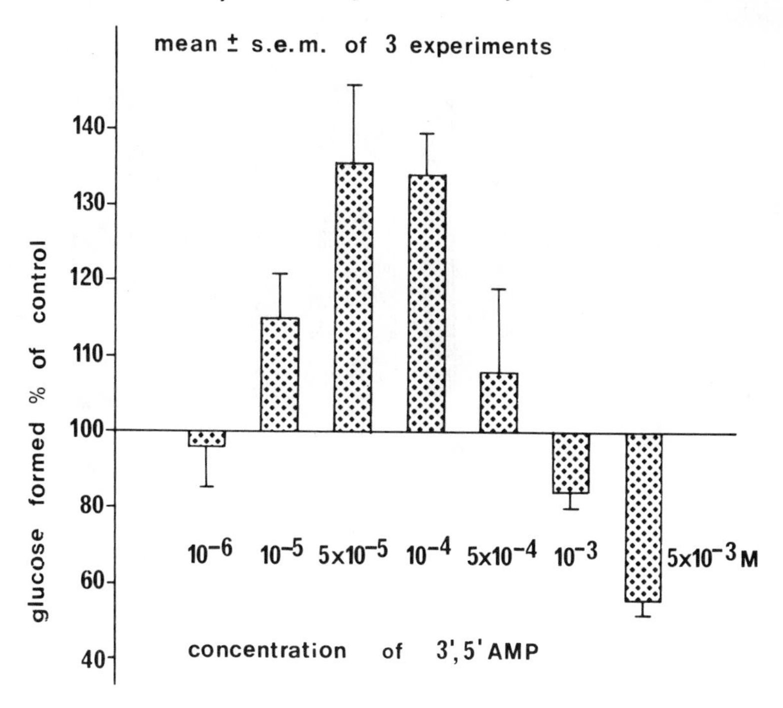

Fig. 3. Effect of various concentrations of 3′, 5′-cyclic AMP on glucose formation from 10 mM lactate.

Glucose formation in 60 minutes is given as percent of the control rate incubated without the nucleotide.

Thus, as a reaction product of cyclic nucleotide phosphodiesterase which is very active in kidney (Senft et al. 1968) inhibitory amounts of 5′-AMP could possibly be formed from the high concentrations of 3′, 5′-cyclic AMP used in this assay.

In the following experiments the action of 3′, 5′-cyclic AMP on the rate of glucose synthesis from various precursors was studied. Since the stimulatory action of 10^{-4}M cyclic AMP was linear over one hour we routinely chose that incubation period. The results are illustrated in Fig. 4. Corresponding to Fig. 3, the data were not corrected for the endogenous rate of glucose formation which was low compared with the rates in the presence of substrate. Although endogenous glucose formation was stimulated by about 50% by cyclic AMP this cannot account for the effects observed in the presence of substrate. In fact, the stimulation expressed in percent of control remains nearly the same after substraction of the endogenous glucose formation from the rate obtained in the presence of precursor.

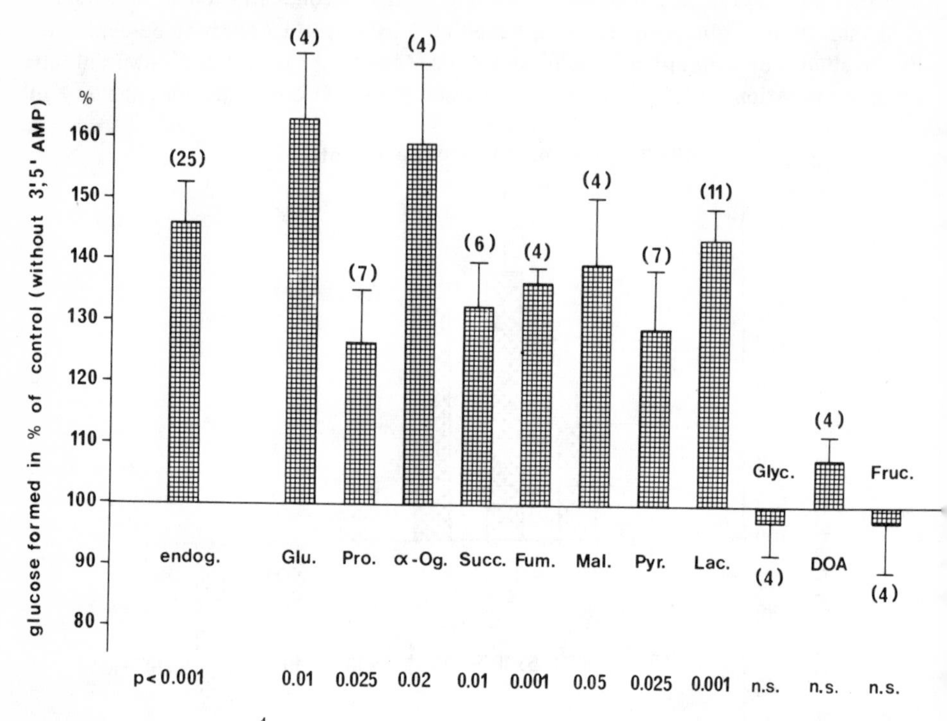

Fig. 4. Effect of 10^{-4} M 3′, 5′-cyclic AMP on glucose formation from various substrates in kidney tubules from starved rats.

Results are represented in percent of the basal rate as given in Table 2. Means ± s.e.m. are given with the number of pairs in parentheses.

As may be noted from Fig. 4, gluconeogenesis is regularly stimulated by the cyclic nucleotide if a substrate is present that has to enter the citric acid cycle before being converted to glucose. No effect of cyclic-AMP was seen in the presence of glycerol, dihydroxyacetone and fructose that are precursors that enter the gluconeogenic pathwa between the step of phosphoenolpyruvate and fructosediphosphate. Similar observations have been reported by Bowman (1969) from his studies with the perfused kidney. The ineffectiveness of the cyclic nucleotide on glucose formation from glycerol and fructose would imply an action at a site somewhere between the citric acid cycle and the triosephosphates. Phosphoenolpyruvate carboxykinase would be a good candidate, since this enzyme can be considered to represent the rate limiting step between malate and the triosephosphates. In a study by Alleyne and Scullard (1969), PEP-carboxykinase has been reported to be responsible for the stimulatory effect of metabolic acidosis on renal gluconeogenesis. Pyruvate carboxylase would seem of minor importance in this connection because glucose formation from intermediates of the tricarboxylic acid cycle was stimulated by cyclic AMP to the same extent as that of pyruvate and lactate.

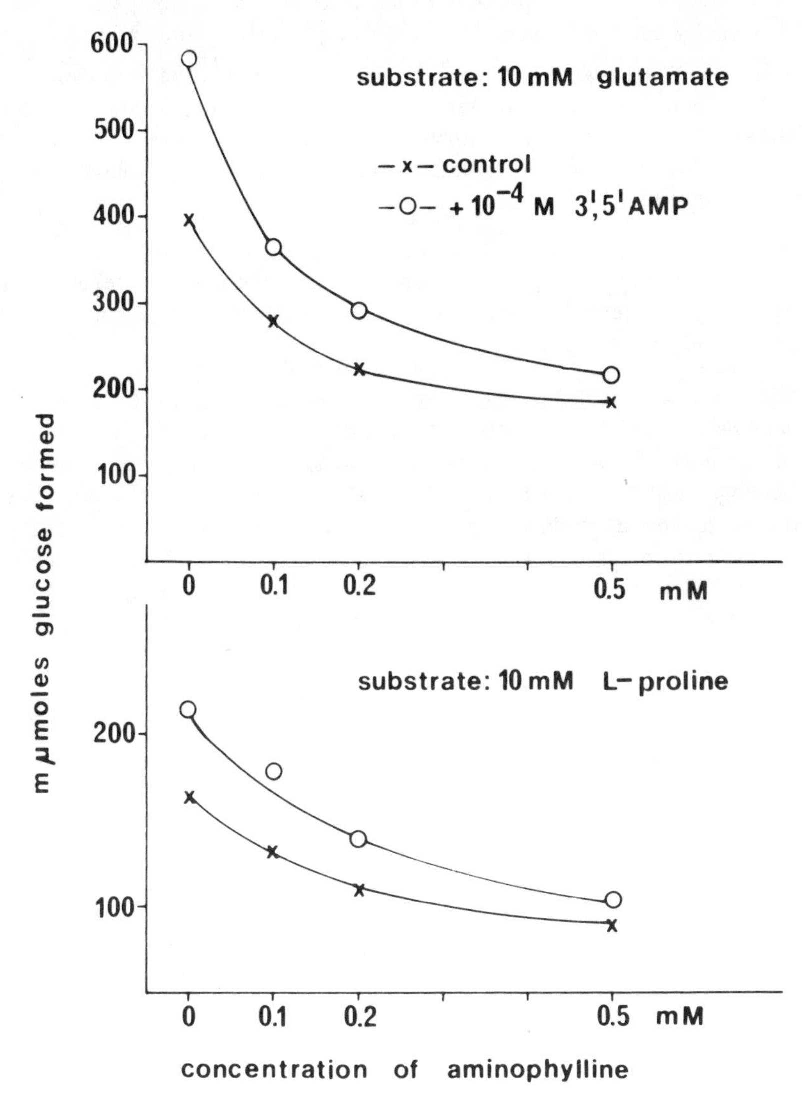

Fig. 5. Effect of aminophylline on glucose formation in rat kidney tubules.

Glucose formation within 60 minutes of incubation was studied at increasing concentrations of aminophylline in the presence and absence of 10^{-4}M cyclic AMP.

Whether the effect of cyclic AMP on glucose formation is specifically directed on some reaction(s) of the gluconeogenic pathway or is part of a more general action of this compound on kidney metabolism awaits further experimentation. With respect to the specificity of the effect of 3′, 5′-cyclic AMP studies with other cyclic nucleotides were started with the following preliminary results. The dibutyryl derivative of 3′, 5′-AMP which is effective in much lower doses as compared to the parent compound in adipose tissue and in liver showed no stimulatory effect on renal glucose production when added in doses below 10^{-4}M. Stimulation of glucose formation from three representative substrates by 10^{-4}M dibutyryl-cyclic AMP was about one half that produced by the nonsubstituted 3′, 5′-AMP. This would indicate that the reasons generally mentioned to explain the higher potency of the substituted compound do not apply to kidney tubules.

With other cyclic nucleotides no clear results could so far be obtained. Cyclic CMP and UMP (10^{-4}M) were without effect, whereas cyclic GMP and IMP yielded variable results. Further experiments are needed to prove the significance of the action of these compounds in renal gluconeogenesis.

Finally, the effect of aminophylline on renal gluconeogenesis was investigated. When this compound was added, at concentrations of 1 mM or higher, strong inhibition of glucose formation was found. As may be seen from Fig. 5, aminophylline inhibits basal gluconeogenesis at a concentration of 10^{-4}M by about 40%. However, the stimulation of gluconeogenesis produced by cyclic AMP, when expressed as a percent of the basal activity is impaired to a much smaller degree by aminophylline. This effect of aminophylline was surprising to us since in all other tissues thus far investigated methylxanthins are known to potentiate the action of cyclic AMP.

Discussion

Although there is good experimental evidence from these and other studies that cyclic AMP does stimulate gluconeogenesis in the kidney in vitro, the physiological significance of this effect is not understood at present. A most important question arising now regards the nature of the "first messenger" in the regulation of gluconeogenesis in the kidney. Unlike in liver, glucagon does not seem to be involved in the kidney (Nishiitsutsuji-Uwo et al. 1967). Chase and Aurbach (1967) and Dousa and Rychlik (1968) have demonstrated an adenyl cyclase in rat kidney cortex which was sensitive to parathyroid hormone. However, no clear information about the possible relation between parathyroid hormone and glucose formation is available. Pagliara and Goodman (1969) have reported that parathyroid hormone stimulates glucose formation in cortex slices. The effects observed by these authors were however much lower if compared with those produced by cyclic AMP itself. Vasopressin, which also has been found to activate a renal cyclase, is even less likely to be involved in gluconeogenesis, since this hormone-sensitive adenyl cyclase is located mainly in the medulla (Chase and Aurbach 1968) where no gluconeogenesis does occur.

Clearly more experiments are needed in order to gain insight into the hormonal control of carbohydrate metabolism in the kidney cortex.

Acknowledgment

The excellent technical assistance of B. Stukowski and J. Mayr is gratefully acknowledged. This work was supported by the Deutsche Forschungsgemeinschaft, Bad Godesberg, (Sonderforschungsbereich 51), FRG.

References

Alleyne, G. A. O., G. H. Scullard: J. clin. Invest. 48 (1969) 364

Bowman, R. H.: Fed. Proc. 28 (1969) 411

Burg, M. G., J. Orloff: Amer. J. Physiol. 203 (1962) 327

Chase, L. R., G. D. Aurbach: Proc. nat. Acad. Sci. (Wash.) 58 (1967) 518

Chase, L. R., G. D. Aurbach: Science 159 (1968) 545

Dousa, T., I. Rychlik: Biochem. biophys. Acta 158 (1968) 484

Guder, W., E. Siess, O. Wieland: FEBS Letters 3, (1969) 31

Krebs, H. A., D. A. H. Bennett, P. de Gasquet, T. Gascoyne, T. Yoshida: Biochem. J. 86 (1963) 22

Krebs, H. A. in: Renal Transport and Diuretics, Ed. by K. Thurau, H. Jahrmärker, Springer Berlin 1969, p. 1

Nishiitsutsuji-Uwo, J. M., B. D. Ross, H. A. Krebs: Biochem. J. 103 (1967) 852

Pagliara, A. S., A. D. Goodman: J. clin. Invest. 48 (1969) 1408

Senft, G., K. Munske, G. Schultz, M. Hoffmann: Arch. Pharmak. exp. Path. 259 (1968) 344

Slein, M. W., in: Methoden der enzymatischen Analyse, Ed. by H. U. Bergmeyer, Verlag Chemie, Weinheim 1962, p. 117

Weidemann, M. J., D. A. Hems, H. A. Krebs: Biochem. J. 115 (1969a) 1

Weidemann, M. J., D. A. Hems, H. A. Krebs: Nephron 6 (1969b) 282

Weiss, L., unpublished data

Discussion to Guder and Wieland

Exton: I think you might be interested to hear the results of some experiments we did to study the effects of caffeine and theophylline on glycogenolysis and cyclic AMP levels in the perfused liver. Caffeine and theophylline inhibited phosphodiesterase and promoted the accumulation of cyclic AMP in the presence of glucagon or epinephrine. In spite of the large increase of cyclic AMP, glucose production was frequently inhibited. The reason for this is that these agents inhibit phosphorylase by stimulating the activity of phosphorylase phosphatase thus annulling the effects of cyclic AMP on the activation of phosphorylase. It occurred to me that something analogous to this may be happening with caffeine in the kidney. Does part of the glucose production you observe come from glycogen?

Guder: No. The amount of glycogen in kidney cortex is very small and glucose formation in the absence of added substrate was not changed by methylxanthines. In the presence of substrate, however, caffeine and theophylline acted in the same manner as aminophylline.

Exton: The second point I want to talk about concerns the observation that dibutyryl cyclic AMP does not stimulate gluconeogenesis from various precursors in the kidney whereas cyclic AMP does. In some tissues it therefore seems that the conversion of the dibutyryl-derivative to cyclic AMP cannot be carried out, presumably because the esterase activity is deficient.

Krebs: You may be interested that Dr. Howard Rasmussen of the University of Pennsylvania has also used isolated kidney tubules prepared with collagenase. I recently saw the drafts of two papers submitted to the Journal of Clinical Investigations. His results are very similar to yours. He also found a stimulation of gluconeogenesis with cyclic AMP. Incidentally, such an effect has also been described for the perfused kidney by Bowman (Feder. Proc. 28 (1969) 411). Rassmussen was mainly interested in the effects of calcium and parathyroid hormone. Under certain conditions he obtained a stimulation of gluconeogenesis by parathyroid hormone, but only when sub-optimal concentrations of calcium were used.

Söling: I wonder whether the effects of aminophylline could have something to do with the intracellular pH, as gluconeogenesis in kidney tubules is certainly dependant on the intracellular pH? At lower pH values gluconeogenesis increases.

Guder: Yes, gluconeogenesis of kidney cortex increases when the kidney is taken from an acidotic rat. In vitro however, you have to maintain the physiological pH to get maximal gluconeogenesis. Therefore a fall in the intracellular pH in vitro would decrease the glucogenic rate. To answer your question: all we know is that in our experiments the methylxanthines did not change the medium pH.

Seubert: I have about the same question as Dr. Exton: Have you already done some studies on the PEP-carboxykinase.

Guder: No, we haven't.

D. H. Williamson: Coming back to Dr. Söling's question: I think that Dr. Cahill (Goodman, A. D., R. E. Fuisz, G. F. Cahill: J. clin. Invest. 45 (1966) 612) has found some effects of acidosis or gluconeogenesis using rat kidney slices. Have you looked at this at all?

Guder: No, not up to now.

Effect of Proinsulin on the Metabolism of Alanine in Isolated Perfused Rat Livers

K.-H. Rudorff, R. Windeck[+] and W. Staib
Institut für Physiologische Chemie der Universität Düsseldorf, FRG

Summary

The influence of insulin and proinsulin on the metabolism of alanine and the inhibition of the effects of proinsulin by Trypsin-Kallikrein-Inhibitor (Trasylol)®[++] was investigated in isolated perfused livers of normal rats. Proinsulin inhibited just as insulin did with gluconeogenesis, the incorporation of ^{14}C in glucose and the formation of $^{14}CO_2$ from 10 mM L-Alanine-^{14}C (U). On the other hand, incorporation of L-Alanine-^{14}C (U) in liver and plasma proteins was stimulated by proinsulin and insulin. After addition of 10 mg Trasylol® 25 minutes after the beginning of the perfusion the effects of proinsulin were inhibited. But the inhibition was not significant. The effects of proinsulin on the gluconeogenesis and on the ^{14}C-incorporation in glucose were significantly inhibited after preperfusion for 60 minutes, with 10 mg Trasylol®.In these experiments neither insulin nor proinsulin showed an effect on the $^{14}CO_2$ formation and on the ^{14}C-incorporation in liver proteins.

Our results support the thesis of a rapid conversion from proinsulin to insulin by the liver (Rudorff in press). This conversion is inhibited by Trypsin-Kallekrein-Inhibitor. A rapid conversion from proinsulin to insulin by liver proteases is probable, but we are not able to analyse proinsulin.

Introduction

Proinsulin, a one-chained peptide at an early stage insulin, was found after having incubated insuloma-tissue (Steiner and Oyer 1967) as well as isolated ß-cells from the pancreas of a rat (Steiner et al. 1957). Proinsulin has a connecting chain of 33 amino acids between the amino end of the a-chain and the carboxyl end of the b-chain. Proinsulin (big insulin) consisting of 84 amino acids can be transformed into active insulin by proteolytic enzymes. An insulin-like activity of proinsulin could not be found in experiments with isolated fat cells, frog sartorius or isolated rat diaphragms. The ability of proinsulin to lower the blood glucose level could

[+] Essential part of the doctor thesis. Med.Fat. University Düsseldorf

[++] Registered trade mark BAYER Leverkusen

be shown on normal and on pancreatectomised rats (Puls and Kroneberg 1969). This effect on a molecular base was only half as strong as that from insulin and could first be established after 30 minutes. When Trypsin was added to proinsulin this delayed effect could be suppressed. It seems that proinsulin is activated in vivo by ubiquitous proteinases (Puls and Kroneberg 1969). A hypoglycemic effect with proinsulin could not be produced in eviscerated rats while with normally fed rats a rapid reduction of glucose concentration occurred (Willms et al. 1969). We presume that a rapid activation of proinsulin occurs in the liver or the gastro-intestinal tract. In the following experiments we have analyzed the influence of proinsulin on the metabolism of alanine in isolated rat livers and we compared the effect of proinsulin with the effect of insulin. Further we tried to reduce the proinsulin effect with Trasylol®.

Methods

Livers of normal rats starved for 24 hours were perfused in a modified apparatus as described by Miller et al. (1965) with 50 ml of a half synthetic medium compound of 2 g of albumin, 10 g of hemoglobin and 5000 USP of heparin in 100 ml Krebs-Ringer-Bicarbonat buffer, pH 7,4. Cyclic perfusion was continously carried out for 90 and 120 minutes.+ In the first series of experiments 10 mM L-Alanin was added 25 minutes after the beginning of the perfusion. Five minutes later 1 μC L-Alanin-^{14}C (U) and 1 IU insulin "S" (Hoechst) or 250 μg proinsulin were added. Simultaneously 1 IU insulin "S" or 250 μg proinsulin per hour were infused. In the second series of experiments, an additional 10 mg Trypsin-Kallikrein-Inhibitor (5200 KIU/mg) was added 25 minutes after beginning the perfusion. In the third series of experiments the livers were perfused for 60 minutes with 10 mg Trypsin-Kallikrein-Inhibitor only, before proinsulin, insulin and 10 mM L-Alanin-^{14}C (U) were added. Trasylol® and the gelelectrophoreticly and chemically homogeneous proinsulin was kindly given by Drs. Schmidt and Arens (Farbenfabriken Bayer, Wuppertal, FRG). Calculating the dose of proinsulin, the difference in dose effects was considered as observed by Puls and Kroneberg (1969). The activity of Trasylol in the medium decreased about 10% after one hour. The activity of Trasylol in the medium was analized by Dr. A. Arens (Farbenfabriken Bayer, Wuppertal, FRG). Glucose and Urea were analyzed in the medium after 30, 90, and 120 minutes after the beginning of the perfusion. After 90 and 120 minutes we determined the incorporation of ^{14}C from L-Alanin-^{14}C (U) in glucose and plasmaproteins of the medium and in the liver the incorporation of ^{14}C in liver proteins. The oxidation of L-Alanin-^{14}C (U) to $^{14}CO_2$ was measured from 0-30 minutes and 30-60 minutes after addition of L-Alanin-^{14}C (U). The $^{14}CO_2$ was absorbed by ethanolamin and ethylenglycoll-monomethylether (1:2). Blood glucose and urea in medium were determined by Biochemica-Test-Combination (Firma Boehringer, Mannheim GmbH, FRG, TC-M II 15983 TBAD Blutzucker und TC-UR-I 15954 THAB Harnstoff). Plasma and liver proteins were measured according to McClean and Cohn (1958). The ^{14}C

+ Definition of the KIE see K. E. Frey, H. Kraut u. E. Werle "Das Kallikrein-Kinin-System und seine Inhibitoren", S. 11 Stuttgart, 1968 glucose was isolated as phenylosazon (Simon and Steffens 1962). After purification and lyophilisation, the osazons were burned in a 1 l Schoeninger Druckflasche and the $^{14}CO_2$ was absorbed (Rudorff et al. in press).

Results

Proinsulin inhibits, like insulin, the gluconeogenesis ($p < 0.025$), the ^{14}C-incorporation in glucose ($p < 0.025$) and the formation of $^{14}CO_2$ ($p < 0.025$) from 10 mM L-Alanin-^{14}C (U) (Tables 1, 2). On the other hand, incorporation of L-Alanin-^{14}C (U) in liver ($p < 0.025$, $p < 0.005$) and plasma proteins (n.s., $P < 0.01$) is stimulated by proinsulin and insulin (Table 2). After addition of 10 mg Trasylol 25 minutes after the beginning of the perfusion of the effects of proinsulin are inhibited. But the inhibition is not significant (Tables 3, 4). The effects of proinsulin on the gluconeogenesis and on the ^{14}C-incorporation in glucose is significantly inhibited ($p < 0.05$) after preperfusion with 10 mg Trasylol from 60 minutes (Table 5). In these experiments neither proinsulin nor insulin show an effect on $^{14}CO_2$-formation or on the ^{14}C-incorporation in liver proteins.

Discussion

It has been shown by Rudorff et al. (in press) that insulin directly inhibits the gluconeogenesis from 10 mM L-Alanin-^{14}C (U) in the isolated perfused liver of normal and alloxan-diabetic rats. This choice of series of experiments seemed to be suitable for us to examine the influence of proinsulin on the gluconeogenesis and to compare the effects of proinsulin with the effects of insulin. On a molecular base proinsulin shows the same effects as insulin (Table 1-4). On isolated fat cells, on frog sartorius, on isolated rat diaphragms and on eviscerated rats, an insulin like activity could not be shown (Puls and Kroneberg 1969, Willms et al. 1969). On the other hand, proinsulin rapidly lowers the blood glucose level in normal (Puls and Kroneberg 1969, Willms et al. 1969) and pancreatectomised rats (Puls and Kroneberg 1969). For this reason we suppose a rapid conversion from proinsulin to insulin by liver cell proteases (Rudorff et al. in press). This thesis is supported by experiments from Rees and Madison (Rees and Madison 1969) which show an insulin like activity of proinsulin on the hepatic release of glucose in dogs. Our results in the isolated perfused rat liver do not prove a conversion from proinsulin to insulin by liver cell proteases because we have no suitable method to distinguish between insulin and proinsulin. To elucidate the question of a proteolytic activation of proinsulin, we tried to suppress the proinsulin effect with Trypsin-Kallikrein-Inhibitor (Trasylol®). Our experiments show that Trasylol® inhibits the activation of proinsulin in the liver (Table 3-5). With these results we believe we have proved that proinsulin is activated in the liver. A rapid conversion from proinsulin to insulin by liver proteases is probable, but we cannot prove it because we are not able to analyze proinsulin. It is possible that high-molecular fragments of proinsulin are left by proteolytic splitting, which have an insulin-like activity. Trasylol® has no immediate effect, but a delayed one. Further our experiments point out an influence of Trasylol on the protein metabolism.

	Glucose Production μ Mole/g Liver/h	P	^{14}C-Glucose cpm/ g liver/h	P	Urea Production μ Mol/g Liver/h	P
Control (C) n = 13	35.80 ± 3.71	C/PI $p < 0.025$	667 ± 63	C/PI $p < 0.0025$	11.55 ± 1.31	C/PI n.s.
Proinsulin n = 6 (PI)	25.08 ± 3.47	PI/I n.s.	350 ± 28	PI/I $p < 0.01$	10.41 ± 1.67	PI/I n.s.
Insulin (I) n = 13	23.88 ± 2.58	C/I $p < 0.005$	494 ± 38	C/I $p < 0.025$	10.19 ± 0.97	C/I n.s.

Table 1. Effects of Proinsulin and Insulin on Glucose and Urea Production and on ^{14}C-incorporation into Glucose from 10 mM L-Alanin ^{14}C (U). $X \pm$ s.e.m. p = T-Test

	$^{14}CO_2$ cplo'/g Liver/h 0-30'	P	$^{14}CO_2$ cplo'/g Liver/h 30-60'	P	$^{14}CO_2$ cplo'/g Liver/h 0-30' 30-60'	P	Plasma protein %D/h	P	Liver protein %D/h	P
Control (C) n = 13	115,271 ± 7,435	C/PI $p < 0.025$	213,367 ± 10,069	C/PI $p < 0.005$	328,638 ± 11,165	C/PI $p < 0.025$	5.00 ± 0.17	C/PI $p < 0.005$	0.83 ± 0.17	-
Proinsulin n = 6 (PI)	80,484 ± 11,906	PI/I n.s.	160,853 ± 10,307	PI/I n.s.	241,337 ± 20644	P/I n.s.	7.41 ± 0.83	PI/I n.s.	0.92 ± 0.17	-
Insulin n = 13 (I)	68,182 ± 4,966	C/I $p < 0.005$	162,833 ± 19,205	C/I $p < 0.005$	230,625 ± 9,649	C/I $p < 0.005$	6.33 ± 0.42	C/I $p < 0.0025$	1.00 ± 0.17	-

Table 2.
Effects of proinsulin and insulin on the conversion from L-Alanin-^{14}C (U) to $^{14}CO_2$ and on incorporation into liver and plasmaproteins.
X ± s.e.m.   p = T-Test

	Glucose production μMol/g Liver/h	P	^{14}C-Glucose cpm/g Liver/h	P
Control (C) n = 7	47.34 $\pm$ 6.80	C/I $p < 0.025$	1,135 $\pm$ 83	C/I $p < 0.025$
Insulin (I) n = 6	28.71 $\pm$ 4.24	I/PI n.s.	840 $\pm$ 84	I/PI n.s.
Proinsulin (PI) n = 6	36.60 $\pm$ 5.88	C/PI n.s.	1,005 $\pm$ 130	C/PI n.s.
Proinsulin without Trasylol® (PI^+) n = 6	25.08 $\pm$ 3.47	PI^+/PI $p < 0.05$	350 $\pm$ 28	PI^+/PI $p < 0.005$

Table 3. Effect of proinsulin and insulin on glucose production and on ^{14}C-incorporation into glucose from 10 mM L-Alanin-^{14}C (U). Rat liver perfusion with 1,040 KIE Trasylol®/ml. Trasylol®, insulin and proinsulin respectively were added to the medium simultaneously. X $\pm$ s.e.m. P = T-Test

	$^{14}CO_2$ cplo'/g Liver/h 0-30'	P	$^{14}CO_2$ cplo'/g Liver/h 30-60'	P	$^{14}CO_2$ cplo'/g Liver/h 0-30' 30-60'	P	Liver protein %D/h	P
Control (C) n = 7	90,410 ± 12,480	C/I $p < 0.05$	230,300 ± 24,890	C/I $p < 0.025$	320,700 ± 34,036	C/I $p < 0.025$	3.83 ± 0.75	C/I $p < 0.005$
Insulin (I) n = 6	50,169 ± 12,520	I/PI n.s.	188,132 ± 28,170	I/PI n.s.	228,184 ± 42,180	I/PI n.s.	9.33 ± 1.25	I/PI n.s.
Proinsulin (PI) n = 6	74,450 ± 18,324	C/PI n.s.	222,672 ± 50,425	C/PI n.s.	297,212 ± 64,487	C/PI n.s.	7.16 ± 1.08	C/PI $p < 0.025$
Proinsulin without Trasylol® n = 6 (PI^+)	80,484 ± 11,906	PI^+/PI n.s.	160,853 ± 10,207	PI^+/PI n.s.	241,337 ± 20,644	PI^+/PI n.s.	7.41 ± 0.83	PI^+/PI n.s.

Table 4.

Effect of proinsulin and insulin on the conversion from 10 mM L-Alanin-^{14}C (U) to $^{14}CO_2$ and on incorporation into liver protein. Rat liver perfusion with 1,040 KIE Trasylol®/ml. Trasylol®, proinsulin and insulin respectively were added simultaneously.
X ± s.e.m. P = T-Test

	Glucose production µ Mol/g Liver/h	P	^{14}C-Glucose cpm/g Liver/h	P
Control (C) n = 6	39.62 ± 3.90	C/I $p < 0.025$	559 ± 56	C/I $p < 0.005$
Insulin (I) n = 6	29.52 ± 2.74	I/PI $p < 0.05$	378 ± 56	I/PI $p < 0.05$
Proinsulin (PI) n = 6	38.56 ± 3.0	C/PI n.s.	554 ± 70	C/PI n.s.
Proinsulin without Trasylol® (PI^+) n = 6	25.08 ± 3.47	PI^+/PI $p < 0.025$	350 ± 28	PI^+/PI $p < 0.005$

Table 5. Effect of proinsulin and insulin on glucose production and on ^{14}C-incorporation into glucose from 10 mM L-Alanin-^{14}C (U). The livers were perfused for 60 minutes with Trasylol® only before alanin, proinsulin or insulin were added to the medium.

References

McLean, R., G. L. Cohn: J. biol. Chem. 233 (1958) 657

Miller, L. L., W. T. Burke, D. E. Haft: Fed. Proc. 14 (1955) 107

Puls, W., W. Kroneberg: Naunyn-Schmiedeberg's Arch. exp. Path. 264 (1969) 295

Rees, K. O., L. L. Madison: Abstr. 29th Annual Meeting of Amer. Diab. Ass. (1969) New York. Diabetes 18, Suppl. 1 (1969) 341

Rudorff, K.-H., G. Albrecht, W. Staib: Hormone and Metabolic Research, in press

Rudorff, K. -H., G. Albrecht, W. Staib, in press

Schmidt, D. D., A. Arens: Hoppe Seylers Z. physiol. Chem. 349 (1968) 1157

Simon, H., J. Steffens: Chem. Ber. 95 (1962) 358

Steiner, D. F., D. Lunningham, L. Spiegelman, B. Aten: Science 157 (1957) 697

Steiner, D. F., P. E. Oyer: Proc. nat. Acad. Sci. (Wash.) 57 (1967) 473

Willms, B., A. Appels, H. D. Söling, W. Creutzfeld: Hormone and Metabolic Research 1 (1969) 199

Discussion to Rudorff, Windeck and Staib

Wieland: In your experiments you get an inhibition of the $C^{14}O_2$- production from C^{14}-alanine by insulin. What is your explanation for this?

Rudorff: We think that this is due to an increased incorporation of the alanine into protein. The only thing we can say is: When gluconeogenesis is increased, we get an increase of $C^{14}O_2$-formation: when gluconeogenesis is decreased, we get a decrease also in $C^{14}O_2$-formation.

Wieland: This is not clear to me. If insulin inhibits glucose formation from C^{14}-alanine, how should it at the same time inhibit the oxidation of alanine?

Exton: It is puzzling, Dr. Wieland, but glucagon does a large number of things to amino acid metabolism in the liver. We know from the studies of Dr. Mallette in our laboratory that glucagon stimulates gluconeogenesis from alanine, but it also increases the transport of alanine into the cell. The action of glucagon can be mimicked by cyclic AMP. If insulin acts by lowering cyclic AMP this may slow the entry of alanine into the cell. Of course, this is purely a hypothesis. Insulin seems to have an additional effect on protein synthesis.

Krebs: Is it possible to identify the proteins which are synthesized at an increased rate on addition of insulin? These proteins are likely to be enzymes. Has any increase in enzyme activities been observed in these short-term insuline experiments?

Staib: We tried to do this, but not up to now.

Ruderman: Did you use 1-C^{14}-alanine or uniformly labeled alanine?

Rudorff: Uniformly labeled alanine.

Ruderman: Then part of your decrease in $C^{14}O_2$ production could be accounted for by the decrease in gluconeogenesis. Using uniformly labeled alanine, you should theoretically produce one molecule of $C^{14}O_2$ for every two molecules of alanine converted to glucose. Therefore any reduction in gluconeogenesis would also reduce $C^{14}O_2$ production.

I also have a question. Did you measure immunoreactive insulin in your perfusate? Specifically I would like to know whether the increase in insulin-like activity might have been due to the conversion of proinsulin to insulin by the liver.

Söling: I think we have at least some secondary evidence that the liver itself is involved in this conversion. We have published (Willms, B. et al.: Horm. Metab. Res. 1 (1969) 199) that in eviscerated rats there is nearly no effect of proinsulin on the uptake of glucose. Under these experimental conditions a hepatic conversion of proinsulin to insulin is excluded. There are naturally still other proteases present in other tissues, that means that proinsulin can be converted to insulin in extrahepatic tissues, but at a much lower rate. When the liver function was left intact, then the effect of proinsulin on glucose uptake was similar to that of insulin. So it seems that a rather high activities of proinsulin-converting enzymes are present in the liver.

General Discussion

Exton: I was very happy to hear Dr. Fröhlich's paper and I wish to point out that the is an additional dissociation between the gluconeogenic and ketogenic actions of glu cagon. This is the difference in the sensitivity of the two processes, and also ureogenesis, to stimulation by glucagon. One observes a hierachy of effects as one gradually increases the level of glucagon in the medium. The most sensitive process is phosphorylase activation leading to increased glucose production by glycogenolysis. Next is gluconeogenesis which is slightly less sensitive than glycogenolysis in the fed rat. Possibly, gluconeogenesis is more sensitive to glucagon in the starved rat. Then a ten-fold increase in the concentration of glucagon is required before the ketogenic and ureogenic actions of the hormone are observed. These observations are made in livers from fed animals perfused without lactate. When lactate is added, the dose of glucagon has to be increased to 100 times the gluconeogenic dose before one observe a stimulation of ketogenesis. So one observes a very large dissociation between these two processes which means the elevation of cyclic AMP in the cell has to be very hig to produce a ketogenic response.

Fröhlich: We did not test lower doses, because we wanted to investigate maximum effects, and so we used high doses of glucagon.

Exton: Perhaps I might point out that the observations I described might explain a lot of discrepancies in the literature concerning the effects of glucagon on ketogenesis. We observed in our early experiments very little effect of glucagon on ketogene sis, but we were using lactate and relatively low doses of glucagon.

Wieland: I just wanted to mention, we have not looked very carefully at different dose relationships, but five years ago Dr. Struck (unpublished) made some comparisons between ureogenesis and gluconeogenesis, and at that time we had the impression that ureogenesis is more sensitive against glucagon stimulation than gluconeogene sis. This would disagree with your experiments, but I am sure you have done this mu more carefully.

Exton: One has to be careful when one makes statements about sensitivity because th presence of a single substrate, lactate, makes a lot of difference. Perhaps in the pre sence of appropriate concentrations of alanine or other amino acids, or maybe in the presence of fatty acids, the ureogenic and ketogenic responses may occur with lower glucagon levels.

J. R. Williamson: Could I ask whether glycodiazine is metabolized in any way?

Wieland: I do not think it is metabolized, but I'm sure Dr. Hasselblatt can give us more information on this point. To Dr. Menahan's studies: There is a difference between his experiments and Dr. Fröhlich s. Dr. Menahan added glycodiazine at the same time together with oleic acid or glucagon whereas Dr. Fröhlich added it

18 minutes before, thus allowing for equilibration of the drug. It appears that it takes a certain time until the drug has reached an effective concentration within the liver cell, and this may be a main difference in these experiments.

Weiss: Dr. Hasselblatt is not present now. He first used that compound. He started with studying ketogenesis in liver slices under the influence of glycodiazine. The first metabolite of this drug which is still capable of inducing hypoglycemia in vivo has no effect on ketogenesis in the in vitro system (A. Hasselblatt, Naunyn-Schmiedebergs Arch. Pharmak. exp. Path. 262 (1969) 152-164).

J. R. Williamson: What is the metabolite?

Weiss: I do not know exactly, I think it is hydroxylated.

Krebs: The metabolism of glycodiazine has been extensively studied by Gerhards, published mainly in Arzneimittelforschung (e.g. vol. 14, (1964) 394), but this work may not be relevant in the present context because the concentrations of glycodiazine in your experiments were rather high.

Fröhlich: Ten mM.

Krebs: At high concentrations of glycodiazine the changes in the concentrations of the drug in relatively short-term experiments are likely to be slight. In this respect the metabolism of the drug may therefore be unimportant. Glycodiazine is demethylated and oxidized. Another question: Was the albumin used in these experiments defatted?

Fröhlich: The albumin used was not defatted. From the amount of ketogenesis and the time of preperfusion which did last about 40 minutes, one can calculate that all fatty acids initially present in the medium should have been oxidized. Thus they probably do not play a role during the test period.

J. R. Williamson: I think that it is highly important to know the metabolism of drugs used in studies with perfused liver because, as we will hear from Dr. Thurman, any reactions going through cytochrome P450 can have a very profound effect on gluconeogenesis.

Krebs: My comments referred to short-term experiments.

J. R. Williamson: The problem is, that if it is metabolized, metabolites can have possible effect on enzymes or on pathways of metabolism, even though they do not accumulate in large amounts.

Söling: I would like to hear a comment from Dr. Exton or from Dr. Fröhlich on the following problem: Would you agree that it is tempting to assume that glucagon and fatty acids act at the same side? There is a stimulation of gluconeogenesis by cyclic AMP, by fatty acids and by glucagon in rat liver, but there is no effect by any of these agents on gluconeogenesis in guinea pig liver. So to me it seems improbable that when glucagon should act on a completely different site than does increased fatty acid oxidation, that the two species should behave differently with respect to both sites, especially in view of the fact that glycogenolysis and ureogenesis are affected by glucagon and cyclic AMP in the same way in both species.

Exton: Well, my comment to that would be that you would not see any effect of glucagon on phosphoenolpyruvate formation if pyruvate carboxylation was limiting. A block at the first step would make it impossible to see any effect of glucagon at the site of PEP-carboxykinase.

Fröhlich: One argument against any direct action of glycodiazine or its metabolites on gluconeogenesis is the fact that in the presence of this compound oleate stimulates gluconeogenesis to the same extent as in the absence of glycodiazine. This would not be expected if there would be an interference of glycodiazine with gluconeogenesis besides its effect on lipolysis.

J. R. Williamson: But you did not get the same rate with oleate in presence of glycodiazine?

Wieland: Oh yes, we did. We got exactly the same rate.

J. R. Williamson: We found that after administration of 2 or 4 mM 5-methylpyrazol 3-carboxylic acid (MPCA) to livers in an attempt to inhibit hepatic lipases, the tissue level of free CoA fell drastically, while the amount of acid soluble acyl-CoA derivatives other than acetyl-CoA showed an increase. Gluconeogenesis from lactate was inhibited, and we interpreted this effect to the observed marked fall of acetyl-CoA levels. The rise of the acid soluble acyl-CoA fraction in liver may indicate activation of the pyrazole carboxylic acid to the CoA ester. Thus, although this compound is a lipase inhibitor in adipose tissue, it has quite different metabolic effects in liver.

Ruderman: Glycodiazine, I think, acts in the same manner as tolbutamide. It is of interest that Boshell and his colleagues (Metabolism 9 (1960) 21) reported as far back as 1960 that tolbutamide can inhibit ketogenesis in liver slices. It has been pointed out that the beneficial effects of chronic sulfonylurea treatment on carbohydrate metabolism cannot be readily explained by its insulinogenic effect (J. M. Feldman and H. E. Lebovitz: Arch. Intern. Med. 123 (1969) 317). Do you have any data which suggest that the antilipolytic actions of glycodiazine might contribute to its effect on carbohydrate metabolism in vivo? For instance could you demonstrate an anti-lipolytic effect at physiological concentrations and, if so, was this effect longer in duration than the reported stimulation of insulin secretion?

Fröhlich: No, glycodiazine was not added at lower concentrations than 10 mM, as it was only used as a tool in order to suppress lipolysis and to look for simultaneous changes in gluconeogenesis.

Ballard: In reference to Dr. Söling s question to Dr. Exton on whether glucagon acts at the same site in guinea pig liver. This might not be the case if gluconeogenesis occurs mainly via the mitochondrial PEP-carboxykinase. Dr. Garber in our laboratory has shown that ß-hydroxybutyrate can be used to alter the mitochondrial redox in isolated mitochondria from guinea pig liver. He finds that an increase in the ß-hydroxybutyrate to acetoacetate ratio from 0.5 to 1 results in a sharp depression of PEP synthesis from pyruvate with diversion of oxaloacetate carbon to malate. This is analogous to Dr. Willms' effects of fat-feeding on redox changes in vivo as well as your results with hexanoate, and is perhaps an indication that mitochondrial PEP synthesis is particularly important in guinea pig liver.

Söling: I still would have this question: Why is there no action of glucagon in absence of exogenous fatty acids in guinea pig liver? This could not be explained, at least not on this basis.

Schäfer: I want to make a comment to the very important point mentioned by Dr. J. R. Williamson. I think it is really a very stimulating concept to connect between the hydroxylation of the drug and the inhibition of gluconeogenesis. Up to my knowledge it is the case, indeed, that glycodiazine is hydroxylated. The product also forms a carboxylic acid in a second step. But there is nothing known whether this carboxylic acid can be activated. It can penetrate the mitochondrial membrane, but I don't think that there has been identified any CoA-derivative. If this would occur one would expect an accumulation of such a derivative which is not further metabolized.

Guder: I would like to make a comment on the differences described by Dr. Söling between the guinea pig and the rat. Some years ago the Nashville group has described that in livers from adrenalectomized rats glucagon no further stimulates gluconeogenesis (N. Friedmann, J. M. Exton and C. R. Park: Biochem. biophys. Res. Comm. 29 (1967) 113). Additionally, J. R. Williamson did not find any effect of fatty acids, when the liver donors were adrenalectomized. If so we here have the same situation in the rat as you described for the guinea pig. My question is: Is there anything known about the role of glucocorticoids in the regulation of gluconeogenesis in the guinea pig liver? How are the cortical steroid levels in the guinea pig compared with the rat?

Söling: There are data available concerning the action of glucocorticoids on glycogen synthesis and gluconeogenesis from alanine, and as far as I recall there was only a very slight effect of corticoids in guinea pig liver as compared with rat liver.

J. R. Williamson: Have you measured the oxygen consumption in guinea pig liver? It would be interesting to calculate the energy cost of gluconeogenesis, and determine whether there is any recycling of carbon between pyruvate and PEP.

Söling: We have not done this up to now, but we are just starting to do this.

Walter: I have one question to Dr. Söling's species differences. Dr. Söling brought up the point that the pyruvate carboxylase may be fully activated in guinea pig liver in order to count for the rate of gluconeogenesis. I wonder whether there is anything known about the rate of gluconeogenesis in man?

Hanson: Did not Dr. G. Cahill do work of this type?

Ruderman: Numerous investigators, including Dr. Cahill, have used tracer techniques to measure glucose turnover in fasting man (G. P. Cahill et al.: J. clin. Invest. 45 (1969) 1751). Reported values have generally ranged between 120-200 g/day. The majority of this glucose appears to be utilized by the brain. It is of interest that in a group of obese subjects fasted for six weeks, glucose turnover fell to about half the basal value. In this situation, ketone bodies were the main fuel for the brain and hence blood glucose levels could be maintained (O. Owen et al.: J. clin. Invest. 46 (1967) 1589).

Walter: I am familiar with this work, but I wonder whether somebody has measured maximum rates, the maximum capacity of human liver?

Schäfer: I wonder whether Dr. Guder has made some measurements of oxygen consumption, because he has a very simple model for looking for the cost of energy for gluconeogenesis.

Guder: We are just doing this.

Krebs: Dr. Söling mentioned in passing that he had carried out experiments on pigeon liver. We tried to perfuse pigeon liver, but we ran into difficulties and gave it up. The difficulties arose from anatomical circumstances connected with the fact that birds have no diaphragm. As soon as one opens the abdomen the respiration stops and this made it impossible to insert the cannula quickly enough to avoid temporary anoxia. How did you overcome this difficulty?

Söling: We naturally had the same difficulties, but Dr. Willms managed this problem in that he took the liver out of the body within a few seconds. Then he inserted the catheter into the portal vein outside into the removed organ. This gave an anoxia period of about two minutes, and then we started perfusing the liver immediately before transferring the liver into the perfusion apparatus. This was the only way to handle this problem.

D. H. Williamson: Is there an intramitochondrial malic enzyme in guinea pig liver, and if so what are its properties?

Söling: We haven't yet measured it.

D. H. Williamson: I am just wondering whether there is an alternative mechanism to get from pyruvate to phosphoenolpyruvate in guinea pig liver.

Söling: We measured the total malic enzyme activity in guinea pig liver, this was very, very low, lower than in rat liver, but we did not measure the distribution.

Hanson: We have measured NADP-malic dehydrogenase in guinea pig liver and find values of 0.64 units/g (37^0C) for fed animals. This was in the cytosol with no detectable activity in the mitochondria.

Weiss: With respect to the problem of the anoxia period with pigeon livers: We in our laboratory and also Dr. Scholz in his laboratory use with much success a preperfusion system. We insert a cannula connected to a perfusion flask into the portal vein. Dr. D. H. Williamson told us this morning that he is able to remove the rat liver within six seconds. If one needs 10 to 20 seconds to bring in the cannula, then

I think this anoxia period should be allowed.

Söling: This is the way we have done it already for a long time in rat and guinea pig livers, but you cannot do it this way in pigeon liver for anatomical reasons.

Gabrielli: I have a question to Dr. Söling concerning the differences between rat and guinea pig livers: You showed the percents of gluconeogenesis inhibition by quinolinic acid. Is the rate of gluconeogenesis from alanine in guinea pig and rat livers about the same?

Söling: With 10 mM alanine in the medium the rate of gluconeogenesis in absence of quinolinic acid is a little bit higher in guinea pig liver compared with rat liver. It is in the order of 0.5 µmoles/g/min. in the rat and about 0.6 to 0.7 µmoles/g/min. in the guinea pig liver.

Gabrielli: We found that in perfused rat liver the quinolinate block on glucose production from alanine was depending from the concentration added to the perfusate. With 20 mM alanine the inhibition was 100% and with 2 mM alanine it was about 60%.

Söling: We used exactly the same concentrations as Veneziale (Veneziale, M., et al.: Biochemistry 6 (1967) 2129).

Weiss: The problem is that we do not know enough about gluconeogenesis itself. The pigeon, for example, produces glucagon and has a very special kind of gluconeogenic system. To my knowledge the only homogenate which shows a substrate dependent gluconeogenesis can be made from pigeon liver. Perfused rat and guinea pig livers besides the different location and activities of some enzymes have about the same capacity for gluconeogenesis. The former seems to be glucagon dependent, the latter not. In contrast to the pigeon both tissues loose their ability for glucose synthesis after homogenizing the cells. If guinea pig experiments had been done in the beginning of the research of hormonal control of gluconeogenesis we probably would assume now glucagon does not act in that field. There may be distinct differences in the hormonal regulation of gluconeogenesis in different species, which till now cannot be explained.

Krebs: You are right in saying that there is a lack of information. We succeeded in obtaining gluconeogenesis in pigeon liver homogenates after having tried numerous other species, but we still do not know why the pigeon liver experiments were successful. It may be connected with the compartmentation of pyruvate carboxylase and PEP-carboxykinase.

Seubert: I want to draw your attention to the PEP-carboxykinase. This enzyme is usually assayed in the presence of magnesium, but in the presence of manganese the activity is about doubled as compared with magnesium alone. The relationship between the enzyme amount and the enzyme activity shows in the presence of manganese no straight line. According to unpublished experiments of Dr. L'Age, the enzymatic activity can be increased by substrates, for instance malate or by epinephrine (3×10^{-4}M). It may be that there is a primary effect at the level of PEP-carboxykinase by a change of the enzymatic activity. I really think one should do enzymological studies again in order to clarify a possible control of PEP-carboxykinase. I still would like to ask: The activity you showed in your slides are too low, and you presumably have assayed the PEP-carboxykinase in the presence of only magnesium?

Kleineke: No, with manganese, too. We performed the assay in exactly the way you are doing it.

Guder: I have a question concerning the localization of the rate-limiting enzymes in kidney: Is anything known about species differences with respect to the localization of PEP-carboxykinase in kidney? I think, Dr. Seubert has done some studies.

Seubert: With respect to PEP-carboxykinase and kidney we have not done such studie

Utter: Dr. Ballard, did you measure the activity of pyruvate carboxylase in human liver?

Ballard: We find a pyruvate carboxylase activity of 3 to 4 μmoles/g/min. at 37^0C in livers from normal humans, but as high as 20 μmoles/g/min. in liver from a chil with high blood lactate.

J. R. Williamson: I have a general question whether there is any control by the guan nucleotides. In rat liver they are about 10% or less of the adenine nucleotides. I do have any knowledge about GTP/GDP ratios in guinea pig liver. Does anybody have d

Ballard: We have not measured GDP but Dr. Garber in our laboratory finds that the r of PEP synthesis in guinea pig liver mitochondria is very critically related to the leve of GTP.

J. R. Williamson: You said you have not measured the GDP?

Ballard: No, not GDP. GTP was measured by substracting the nucleotide triphosphat measured by the hexokinase assay from that determined with phosphoglycerate kinase

Hanson: Perhaps we should raise the question of the source of the GTP; whether it co from substrate level phosphorylation or from transphosphorylation. I think it might be key difference in trying to understand this process and its regulation.

J. R. Williamson: My understanding is that they are equally effective.

Krebs: The activity of nucleoside diphosphokinase is rather high in relation to the ge eral turnover of the nucleotides. This means that the various nucleoside di- and tri-phosphates are at near-equilibrium, and further that GTP can be formed in two ways, either when α-oxoglutarate is oxidized, or by transphosphorylations.

Hanson: It may be appropriate to amplify a comment made at this mornings session and to ask the participants if they can suggest new approaches to this problem of the regulation of gluconeogenesis. By that I mean alternatives to perfusion studies, liver slices and the other, by now familiar, approaches.

J. R. Williamson: I think Dr. Wieland's and Dr. Guder's approach with the isolated tubule cell is a most important one. This preparation seems to offer some advantage over isolated liver cells, which tend to have leaky cell membranes.

Krebs: I think only the use of a variety of different approaches can take us further. We have to continue using the perfused organ because it is the only way in which we can study an intact organ performing its functions under approximately physiological conditions. But this has to be combined with every other possible biochemical approach -- studies on pure enzymes, homogenates, slices and especially on freeze-clamped tissue, as Dr. D. H. Williamson has emphasized. New methodological ideas would be most welcome.

J. R. Williamson: Kinetic and direct readout approaches are very valuable.

Exton: I agree completely that every approach should be utilized. It is interesting to consider how many approaches we have: 1) D. H. Williamson and others studying gluconeogenesis in vivo. 2) Numerous groups perfusing livers. 3) Drs. Utter, Lardy and others working at the enzyme level. 4) We have the groups of Lardy, Walter, Haynes, etc. working with isolated mitochondria. Thus it seems that what we are lacking at the moment is the homogenate, i. e. the whole cell without its permeability barriers. I wonder whether anyone knows why the rat liver homogenate does not make glucose? Is it a question of the activity of phosphatases or of reversal of pyruvate kinase?

Krebs: I cannot claim that I know. I suspect one of the factors is the dilution of cofactors which occurs on preparing homogenates. Dilution is unavoidable because without it it is not possible to oxygenate the material adequately. Another factor may be the stability of adenine nucleotides. We recently found that liver slices invariably lose a great deal of their adenine nucleotides just by incubating. About 70% of the total adenine nucleotides are lost by liver slices within minutes (Krebs: Advances in Enzyme Regulation 8 (1970) 335). This is probably the major reason why rates of gluconeogenesis and of ketogenesis are much lower in liver slices than in the perfused organ .

Exton: Do you think Dr. Krebs, that if you blocked the enzyme degrading the adenine nucleotides and added cofactors in appropriate concentrations the system would go?

Krebs: A method which prevents dilution of cofactors may prove successful.

Wieland: I think, it is very difficult to define exactly what a homogenate is. If you would look at different homogenates prepared in different laboratories microscopally, I am sure you would find very different types of preparations. This may result also in differences in the ability to perform special metabolic reactions. So I think, one should better standardize what we call an homogenate. There is another point I would like to make: Talking on future approaches to study metabolic control I should like to mention quite another aspect and this is the morphological one. I believe that the development of methods which combine histochemistry and electronmicroscopy could help us at least in so far as to clarify the localization of key enzymes in the cells. For instance the problem of localization of pyruvate carboxylase, which has been discussed so extensively during this meeting should be theoretically accessible by a method allowing, with the electronmicroscope, for localizing the enzyme by a histochemical reaction. This of course would hardly provide quantitative data, but it would at least enable to exclude or establish the presence of an enzyme within a certain compartment. This kind of an approach in addition to other techniques, could perhaps help to get more insight into the details of metabolic regulations.

Krebs: The approach which you suggest is certainly desirable but it seems to me difficult. Have you any suggestions on how to locate pyruvate carboxylase or oxaloacetate histochemically?

Wieland: One could think on several ways using, chemical reactions, which probably would be very difficult. The immunological approach seems more promising. If you have a pure antibody against the enzyme, then you can label the antibody with an electrondense substance, for instance ferritin. Thus it seems possible to detect the ferritin-marked antibody specifically at the site of its precipitation with the antigen in the cell.

Krebs: Good luck!

Wieland: It looks indeed very complicated, but in a pilot study my coworker, Dr. Siess, had at least managed the first part of the story. As a model for enzyme localization, we took liver ADH. ADH is a very potent antigen and Dr. Siess got very good antibodies. He purified the antibody by precipitation with the antigen and subsequent dissociation. Dr. Siess developed a simple method for coupling the antibody with ferritin and now we can enter the second stage and try to localize ADH in the liver cell. As I pointed out this is just a model study but if it works with ADH it may be working with other enzymes too.

Söling: This type of approach has been used successfully for the localization of intracellular gastrin, using rabbit antigastrin antiserum and fluorescein-conjugated hog anti-rabbit globulin (Lomsky et al.: Nature 223 (1969) 618).

J. R. Williamson: Another approach requiring good courage is microinjection of single cells. If you are interested in a particular effector of an enzyme directly one actually can inject small amounts of the effector into the cell. The problem is to have a direc readout, such as the pyridine nucleotide fluorescence.

Ballard: I have a comment on this question of why gluconeogenesis is measurable only in pigeon liver homogenates. This tissue is characterized not only by having PEP-carboxykinase localized only in the mitochondria, but by having the highest activity in all livers tested of both PEP-carboxykinase and pyruvate carboxylase. These enzymes would be present in mitochondria at concentrations of about 10^{-4}M and 10^{-5}M respectively and as such there are many more molecules of these enzymes than of oxal acetate.

The Role of Pyruvate Carboxylase and P-enolpyruvate Carboxykinase in Rat Adipose Tissue +

R. W. Hanson, M. S. Patel, Lea Reshef and F. J. Ballard
Fels Research Institute and Department of Biochemistry, Temple University Medical School, Philadelphia, Pennsylvania, U.S.A., the Department of Biochemistry, Hadassah Medical School, Jerusalem, Israel and the Division of Nutritional Biochemistry, Commonwealth Scientific and Industrial Research Organization, Adelaide, South Australia

Summary

Recent studies have shown that two enzymes, pyruvate carboxylase and P-enolpyruvate carboxykinase, normally considered to play a key role in the regulation of hepatic gluconeogenesis, are involved in the metabolism of adipose tissue, "a non-gluconeogenic" tissue. Pyruvate carboxylase is present in isolated adipose tissue mitochondria at a specific activity three times that found in liver mitochondria. This high activity is consistent with an important role of pyruvate carboxylase in providing oxaloacetate to support the rapid formation of citrate for lipogenesis. The overall rate of pyruvate carboxylation by mitochondria isolated from rat adipose tissue as measured by the incorporation of ^{14}C-bicarbonate into acid stable intermediates is of the same general magnitude as that found in mitochondria from rat liver and is somewhat higher than in rat kidney and guinea pig liver mitochondria. Citrate and malate were the main radioactive products, with citrate containing about two-thirds of the radioactivity fixed.

Furthermore, pyruvate carboxylase and P-enolpyruvate carboxykinase of adipose tissue are part of a metabolic sequence for the synthesis of glyceride-glycerol (glyceroneogenesis). P-enolpyruvate carboxykinase is present in adipose tissue cytosol and has similar immunochemical properties to the enzyme in rat liver cytosol. Adipose tissue P-enolpyruvate carboxykinase is increased in activity by fasting, during diabetes, and unlike the liver enzyme, by adrenalectomy. Evidence is presented indicating that hormonal and dietary regulation of this enzyme in adipose tissue is closely correlated with the rate of glyceride-glycerol synthesis from pyruvate and that this pathway may play a role in the regulation of fat mobilization.

Gluconeogenesis has been intensively studied over the past ten years and has been found to be confined in mammals, to liver and kidney cortex (Scrutton and Utter

---

\+ Unusual Abbreviations: P-enolpyruvate = phosphoenolpyruvate, FFA = free fatty acids

Enzyme Code Numbers: P-enolpyruvate carboxykinase (EC 4.1.1.32), pyruvate carboxylase (EC 6.4.1.1), NADP-malate dehydrogenase (EC 1.1.1.40), NAD-malate dehydrogenase (EC 1.1.1.37), ATP-citrate lyase (EC 4.1.3.8), glucose-6-phosphatase (EC 3.1.3.9), fructose-1, 6-diphosphatase (EC 3.1.3.11) and glycerol kinase (EC 2.7.1.30), citrate synthase (EC 4.1.3.7).

1968, Krebs 1963). The integrative nature of this process as well as the specific controls regulating the interrelationship of gluconeogenesis with fatty acid oxidation in the mitochondria and with lipogenesis in the cytosol are, at the present time, only partly understood. One of the most fundamental aspects of the regulation of gluconeogenesis concerns those factors which influence what have been called the key gluconeogenic enzymes, glucose-6-phosphatase, fructose-1, 6-diphosphatase, pyruvate carboxylase and P-enolpyruvate carboxykinase. As was pointed out by Scrutton and Utter (1968) the maximal capacities of both pyruvate carboxylase and P-enolpyruvate carboxykinase in rat liver are only two to three-fold greater than the observed maximal rates of gluconeogenesis from various 3 and 4-carbon precursors. If these two enzymes are of prime importance in regulating hepatic glucose synthesis, we would expect that their intracellular location, kinetic and physical properties and adaptive behavior should provide information regarding their role in this process. But what is the function of these two enzymes in tissues which do not synthesize glucose? Our studies on adipose tissue have indicated that both pyruvate carboxylase and P-enolpyruvate carboxykinase are not only present in that tissue, but are also key enzymes in processes unrelated to the synthesis of glucose.

Pyruvate Carboxylase

Tissue Distribution and Properties of the Adipose Tissue Enzyme

The activity of pyruvate carboxylase in a number of rat tissues is shown in Table 1. When expressed on the basis of wet weight of tissue, the activity of pyruvate carboxylase is greatest in liver and kidney with appreciable activity in brown adipose tissue, adrenal glands and lactating mammary gland. The activity of this enzyme is highest in tissues with a high capacity for lipogenesis or gluconeogenesis. White adipose tissue has about one-seventh the pyruvate carboxylase activity of liver, but when expressed as milliunits per mg protein the activity in this tissue is as great as found in liver. In our earlier studies on the sub-cellular distribution of adipose tissue pyruvate carboxylase (Reshef et al. 1969), we reported that this enzyme was primarily mitochondrial with 20-30% of the total tissue activity in the cytosol. However, more recent studies with antibodies to rat liver pyruvate carboxylase have suggested that the cytosol activity may be due to contamination with mitochondrial enzyme (Ballard et al. in press).

Adipose tissue pyruvate carboxylase is a biotin containing enzyme, since it can be blocked by the addition of avidin. It is activated by acetyl-CoA or propionyl-CoA at concentrations of 0.75 and 1.5 mM, respectively. The assay for this enzyme generally used in our laboratory is that described by Utter and Keech for liver pyruvate carboxylase (1963), in which the fixation of ^{14}C-labeled $NaHCO_3$ is measured in the presence of ATP, pyruvate, $MgCl_2$ and acetyl-CoA. Pyruvate carboxylase in adipose tissue is dependent upon pyruvate, ATP and acetyl-CoA for activity. The radioactive product of this reaction in adipose tissue was isolated and found to be citrate when acetyl-CoA was present as the activator and oxaloacetate when propionyl-CoA was added. The formation of citrate by adipose tissue extracts under the

Tissue	Units/g	milliunits/mg protein
Liver	15.5 ± 2.1 (5)	103 ± 11 (5)
Kidney	14.1 ± 1.0 (7)	90.0 ± 8.0 (4)
Brain	0.88 ± 0.18 (4)	9.7 ± 2.0 (4)
White adipose	1.92 ± 0.29 (5)	103 ± 15 (5)
Brown adipose, (newborn animal)	4.06 ± 0.59 (5)	38.0 ± 5.5 (5)
Adrenal gland	5.30 ± 1.30 (4)	41.0 ± 7.0 (4)
Lactating mammary gland	5.00 ± 0.51 (8)	50 ± 4.0 (5)
Intestinal mucosa	0.14 ± 0.05 (4)	1.7 ± 0.4 (4)
Testis	0.61 ± 0.04 (4)	9.0 ± 0.04 (4)
Heart	0.59 ± 0.09 (4)	4.1 ± 0.1 (4)
Skeletal muscle	0.12 ± 0.02 (4)	1.0 ± 0.2 (4)

Table 1. Tissue Distribution of Pyruvate Carboxylase. Tissues were homogenized in 0.25 M sucrose, freeze dried and extracted with a solution containing 50 mM Tris, 5 mM ATP, 5 mM $MgSO_4$ and 0.5 mM EDTA at pH 7.0 to stabilize the enzyme (Ballard et al. in press). Blank values in the absence of pyruvate or acetyl-CoA have been substracted from the total rate of incorporation. All values are the mean ± s.e.m. for the number of animals shown in parantheses.

assay conditions used reflects the high level of citrate synthase (0.83 units/mg mitochondrial protein (Patel and Hanson 1970) present in this tissue.

We have isolated pyruvate carboxylase from rat liver mitochondria and prepared antibodies against it (Ballard et al. in press). These antibodies have been used for titration studies with pyruvate carboxylase from various rat tissues. As shown in Table 2, antibodies to pyruvate carboxylase from rat liver give an identical inactivation pattern with pyruvate carboxylase of rat kidney, mammary gland, brown adipose tissue and white adipose tissue. The findings demonstrate that pyruvate carboxylase is not only present at high activity in adipose tissue but also has similar immunochemical properties to the enzyme in rat liver mitochondria.

The Metabolic Role of Pyruvate Carobxylase in Adipose Tissue

The citrate cleavage pathway. Fatty acid synthesis from acetyl-CoA in adipose tissue and liver occurs in the cytosol of the cell but uses acetyl-CoA formed in the mitochondria (Wakil et al. 1958, Srere 1965). Since the diffusion of acetyl-CoA out of the mitochondria is too slow to meet the demands of rapid lipogenesis, it is widely held that citrate efflux and cleavage in the cytosol supplies the necessary acetyl-CoA for lipogenesis (Spencer and Lowenstein 1962). Young et al. (1964) and Wise and Ball (1964) proposed that oxaloacetate, formed by citrate cleavage, is reduced to malate by NAD-malate dehydrogenase with subsequent decarboxylation to pyruvate via the NADP-linked malate dehydrogenase (Fig. 1). An important feature of this proposal is the extra-mitochondrial transhydrogenation of NADH to NADPH which would augment the available NADPH to support lipogenesis.

We pointed out, however, that unless there is a reaction sequence to regenerate intra-mitochondrial oxaloacetate there will be a net loss of this intermediate and a subsequent shut-down of both citric acid cycle activity and of citrate formation for lipogenesis (Ballard and Hanson 1967). Of the various reactions capable of supplying intra-mitochondrial oxaloacetate, only pyruvate carboxylase is active enough to fit such a role. A calculation of the carbon flow in adipose tissue during lipogenesis supports this view. Flatt and Ball (1964) have reported that 0.125 μ moles of glucose per g per minute are converted to fatty acids by adipose tissue in the presence of insulin. Since this glucose would be converted to pyruvate before incorporation into lipid there would be 0.125 x 2 or 0.25 μ moles of pyruvate formed per minute. To support this rate of pyruvate conversion to fatty acid, pyruvate must be converted at a similar rate to acetyl-CoA. This in turn will form citrate which passes out of the mitochondria. Thus oxaloacetate is removed from the mitochondria at the rate of 0.25 μ moles/minute/g tissue. As shown in Table 1, there is in rat adipose tissue mitchondria the capacity to synthesize 1.92 μmoles of oxaloacetate per minute per g via pyruvate carboxylase, so that this enzyme is sufficiently active to fit the key metabolic role of generating oxaloacetate in support of the citrate cleavage path way.

The carboxylation of pyruvate by isolated rat adipose tissue mitochondria. The important position of oxaloacetate at a metabolic crossroads in adipose tissue is underlined by the results of our studies of the carboxylation of pyruvate by isolated rat adipose tissue mitochondria. The experimental evidence which supports the various functions of pyruvate carboxylase in adipose tissue has been largely based on studies of the intracellular compartmentation of various enzymes and incorporation of certain specifically labeled intermediates into end-products such as fatty acids, CO_2 and glycerol. While recognizing the importance of studies of this type, it seemed to us necessary to isolate functional adipose tissue mitochondria and to test directly the various parameters of mitochondrial metabolism that are assumed to occur.

Since adipose tissue mitochondria have not been previously studied in vitro, it is important to prove the functional integrity of mitochondria isolated by our procedures.

The detailed isolation procedure for the preparation of functionally intact adipose tissue mitochondria has been reported in detail by Patel and Hanson (1970) but there were several minor, but important modifications of the standard mitochondrial isolation procedure that should be emphasized. 1) Young animals were used as a source of adipose tissue since the protein to fat ratio is greater than in older rats. 2) Adipose tissue cells were prepared by the collagenase method of Rodbell (1964) and then ruptured by shaking in a glass centrifuge tube as originally suggested by Angel and Sheldon (1965). All attempts to homogenize the tissue directly resulted in severely disrupted mitochondria. 3) Care must be taken to exhaustively remove all lipid from the mitochondria during the subcellular fractionation procedure. Mitochondria from epididymal adipose tissue from a typical isolation, using the precautions outlined above, are shown in Fig. 2. These mitochondria appear to be largely intact and some degree of swelling is apparent. Two types of mitochondria were evident (Fig. 2D); one has a more strongly condensed matrix with swollen cristae, whereas the other type appear similar to mitochondria in fixed, unfractionated adipose tissue, having cristae which extend all the way across the organelle and a less dense matrix. The respiratory control ratio for succinate plus pyruvate for these mitochondria was 3.5, with a P:O ratio of 2.4. Since mitochondria can synthesize ATP from ADP by either oxidative phosphorylation of ADP when coupled, or by adenylate kinase when uncoupled, we have also measured the change in adenine nucleotide levels in isolated adipose tissue mitochondria (Fig. 3). No significant change in AMP concentration was observed over the 20 minute incubation period, whereas ATP was synthesized from ADP thus indicating that the ATP was generated by oxidative phosphorylation rather than by adenylate kinase.

Tissue	milliunits inactivated/20 μl antibody
liver	64.9 ± 5.1 (8)
kidney	62.1 ± 4.1 (8)
mammary gland	69.7 ± 8.4 (6)
brown adipose tissue	60.8 ± 2.7 (4)
white adipose tissue	55.3 ± 5.1 (7)

Table 2. Inactivation of Pyruvate Carboxylase from Different Tissues by a Constant Amount of Antibody. Enzyme from the particulate fractions of various tissues was extracted as outlined in Table 1. Each tissue extract was diluted with the Tris-ATP-$MgSO_4$-EDTA solution to give an approximate pyruvate carboxylase concentration of 100 milliunits per 0.5 ml. Twenty μl of antibody or control γ-globulin were added and after two hours the solutions were centrifuged and the supernatant fraction assayed for pyruvate carboxylase. Values are the mean ± s.e.m. for the number of animals shown in parentheses. Statistical analysis indicated no significant differences the the 5% probability level between the various tissues. Complete details of the experimental procedures are given in Ballard et al. in press.

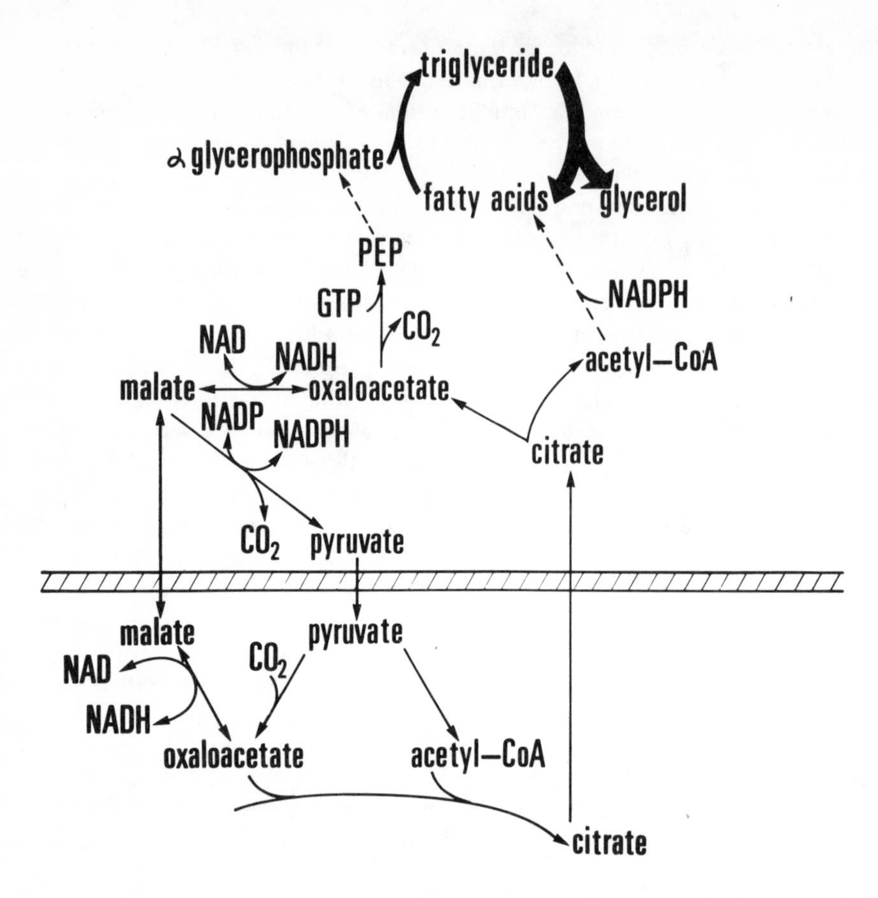

Fig. 1. Metabolic sequence showing the interaction of enzymes in the cytosol and mitochondria (below the hatched line) of rat adipose tissue. The citric acid cycle is abbreviated to show only the synthesis of citrate and wherever reactions in a sequence are omitted they are indicated with a dotted line.

The optimal concentrations of substrates and cofactors for pyruvate carboxylation by rat adipose tiddue mitochondria are 8 mM ATP, 20 mM $MgCl_2$, 40 mM $KHCO_3$ and 8 mM pyruvate. The carboxylation of pyruvate at these concentrations at 37^0C was linear with time to 30 minutes and with mitochondrial protein to 0.5 mg. The effec of varying pyruvate concentration over a concentration range of from 0 to 20 mM is shown in Fig. 4. With pyruvate present in the incubation medium the mitochondrial concentration of acetyl-CoA is probably adequate to fully activate pyruvate carboxylase (Walter et al. 1966) and the increased rate of pyruvate carboxylation with increasing pyruvate concentration seen in Fig. 4 is probably due, at least in part, to an increase in mitochondrial acetyl-CoA formation. It is interesting to note that the apparent "K_m" for the carboxylation of pyruvate by adipose tissue mitochondria calculated from Fig. 4 is about 2 mM. The actual concentration of pyruvate in the intracellular water of adipose tissue is about 0.5 mM (Ballard and Hanson 1969). The pyruvate levels are thus within the range where a small change in the concentration of pyruvate can significantly alter its rate of carboxylation by the mitochondria. It is significant to point out that lipogenesis from pyruvate by adipose tissue incubated in vitro is directly related to the concentration of pyruvate in the incubation medium (Reshef et al. 1969). In fact the normally observed block in lipogenesis in adipose tissue from starved animals can be largely overcome by increasing the

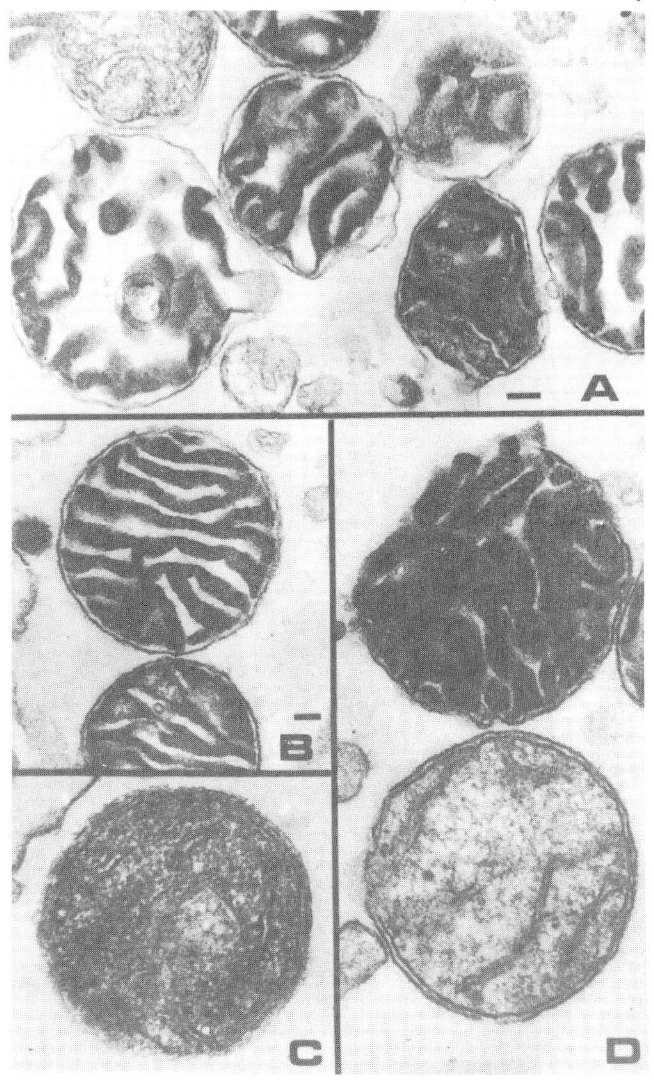

Fig. 2. Electron micrographs of mitochondria isolated from rat epididymal adipose tissue. The mitchondria were isolated and prepared for electron microscopy as decribed (Patel and Hanson 1970). The bar shown in the figure represents 0.1 μ .

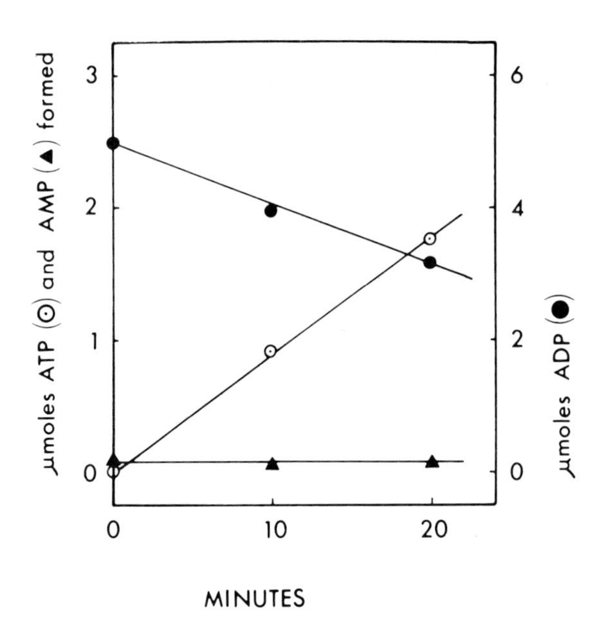

Fig. 3. ATP synthesis from ADP and phosphate by rat epididymal adipose tissue mitochondria. The reaction mixture contained 0.25 M sucrose, 10 mM potassium phosphate buffer pH 7.4, 10 mM potassium pyruvate, 10 mM malate, 5 mM ADP, 20 mM $MgCl_2$ and 15 mg defatted bovine serum albumin. After preincubation, the reaction was initiated by adding 0.5 mg of mitochondrial protein followed by incubation for varying times at 37^0C. Further details are given by Patel and Hanson (1970).

concentration of pyruvate. It is most likely that these changes in lipogenesis with increasing pyruvate concentration are related to the increased carboxylation of pyruvate by adipose tissue mitochondria.

Rognstad and Katz (1969) reported that the incubation of adipose tissue with dinitrophenol stimulated pyruvate oxidation to CO_2 but inhibited its conversion to fatty acids. These results were interpreted by the authors as a dinitrophenol-induced decrease in ATP concentration within the fat pad which in turn reduced the level of pyruvate carboxylation by the mitochondria. As seen in Fig. 3, increasing concentrations of ATP up to 8 mM increases the fixation of bicarbonate into organic acids. In previous studies (Patel and Hanson 1970), we have also shown that pyruvate carboxylation with ADP, even in the presence of ATP, was inhibited. This indicates that the ATP to ADP ratio directly influences the rate of pyruvate carboxylation to oxaloacetate in adipose tissue mitochondria.

Although isotope studies have provided indirect evidence that citrate formed within the mitochondria in adipose tissue is transferred into the cytosol, there has previously been no direct evidence to support this concept. In our experiments with isolated

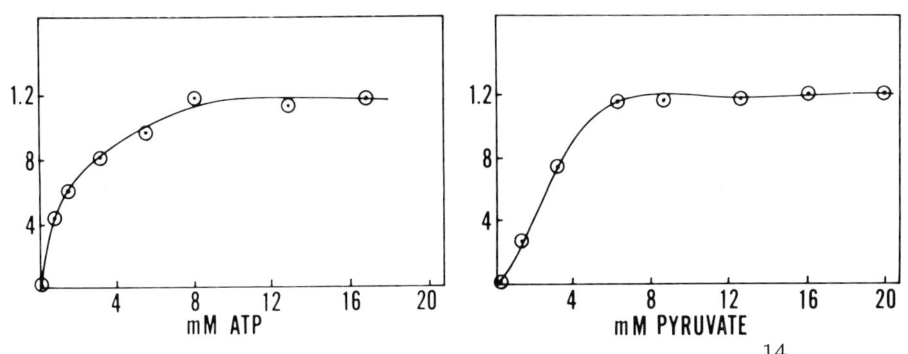

Fig. 4. Effect of ATP and pyruvate concentration on the fixation of $H^{14}CO_3^-$ by rat adipose tissue mitochondria. The reaction mixture contained, in a final volume of 1 ml; 0.25 M sucrose; 6.6 mM potassium phosphate, pH 7.4, 6.6 mM triethanolamine, pH 7.4, 20 mM $MgCl_2$ and 40 mM $NaH^{14}CO_3$. Unless otherwise indicated, the concentrations of pyruvate and ATP were 10 mM and 8 mM, respectively. Incubations were carried out for 20 minutes at 37^0C with about 0.5 mg of mitochondrial protein in each tube. Further details are given by Patel and Hanson (1970).

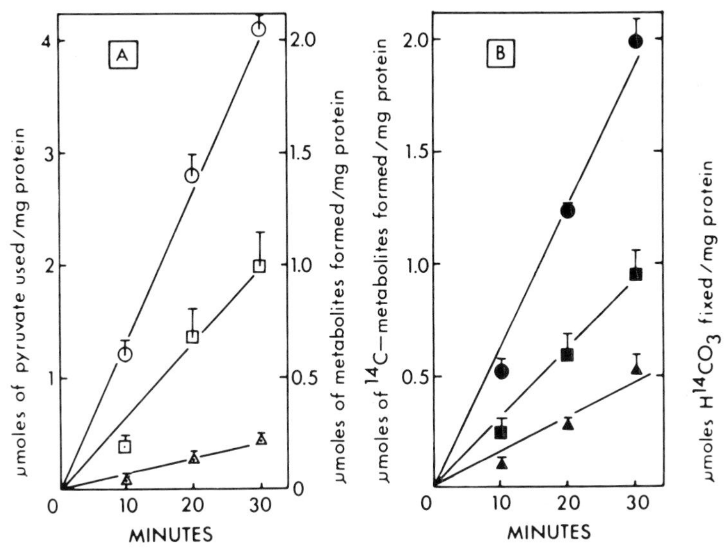

Fig. 5. Formation of organic acids from pyruvate and labeled bicarbonate by rat adipose tissue mitochondria. The reaction was described in Fig. 4 and the incubation time, at 37^0C, varied as indicated. Labeled organic acids were separated on high voltage electrophoresis as described in detail by Patel and Hanson (1970). Values are the means ± s.e.m. of four experiments. A: ○—○—○ , pyruvate utilized; □——□——□ , citrate formed; △——△——△ , malate formed. B: ●——●——● , $H^{14}CO_3$ fixed; ■——■——■ , ^{14}C-citrate formed and ▲——▲——▲ , ^{14}C-malate formed.

adipose tissue mitochondria we have measured the products formed by the carboxylation of pyruvate in vitro. Mitochondria incubated with 10 mM pyruvate and $KH^{14}CO_3$ were rapidly filtered using the Millipore filtration device described in detail by Garber and Ballard (1966). This procedure involves the filtration of an aliquot of incubation medium through a syringe-mounted Millipore filter (0.65 μ) followed by rapid washing of the mitochondria with one ml of 0.25 mM sucrose and extraction with one ml of 6% perchloric acid introduced from syringes mounted on the assembly. Using citrate synthase as a marker enzyme for mitochondrial breakage we calculated that about 10% of the mitochondria were ruptured during filtration and since this method is rapid it seemed ideally suited for our studies of the efflux of products of pyruvate carboxylation. As shown in Fig. 5, the main radioactive products formed by pyruvate carboxylation were citrate and malate and these were released into the incubation medium. The incorporation of $H^{14}CO_3$ into acid-soluble intermediates by adipose tissue mitochondria was linear over 30 minutes of incubation and about 2 μmoles of $H^{14}CO_3^-$ per mg of protein was fixed under the conditions used. Of this, 0.95 μmoles/mg protein was incorporated into citrate and 0.53 μmoles into malate Measurements of pyruvate utilization indicated that 4.1 μmoles of pyruvate were use and 1 μmole of citrate and 0.21 μmoles of malate were found in the medium after 30 minutes of incubation.

A comparison of the rate of pyruvate carboxylation by adipose tissue mitochondria with that reported by other investigators for gluconeogenic tissues is revealing. Under optimal conditions, rat liver mitochondria carboxylate 56 mumoles of pyruvate per mg mitochondrial protein per minute (Walter, Paetkau and Lardy, 1966), guinea pig liver mitochondria carboxylate 32 mμmole/min. (Somberg and Mehlman 1969) and rat kidney, 28 mμmoles per minute (Mehlman 1968), while 66 mμmoles of pyruvate was carboxylated per minute per mg protein by adipose tissue mitochondria. The ma nitude of pyruvate carboxylation emphasizes the important role of pyruvate carboxylase in adipose tissue.

<u>The role of pyruvate carboxylation in "glyceroneogenesis in adipose tissue</u>. An alternative function of pyruvate carboxylase in adipose tissue is its part in a sequence term ed "glyceroneogenesis." This pathway, shown in Fig. 1, is essentially an abbreviate form of gluconeogenesis from pyruvate with the main end product being α-glycerophosphate. The experimental evidence underlying the existance of this pathway as well as its role in adipose tissue triglyceride metabolism will be considered in detail in the following section. It should be emphasized at this point however, that any three carbon intermediate that is metabolized through pyruvate must be carboxylated via pyruvate carboxylase prior to conversion to α-glycerophosphate. In this regard, the information presented in Fig. 5 indicates that adipose tissue mitochondria can synthesize and release malate when pyruvate, ATP, $MgCl_2$ and HCO_3^- are provided. These mitochondria can form three or four times more citrate than malate; a finding which contrasts with gluconeogenic tissues such as rat liver (Walter, Paetkau and Lardy, 1966) and kidney (Mehlman 1968) where pyruvate conversion to malate is equal to, or greater than, its conversion to citrate. This difference reflects the importance of citrate synthesis to the overall process of lipogenesis in a tissue which has a predom

inant lipogenic capacity, but also implies that malate can be formed from pyruvate in adipose tissue mitochondria and released into the cytosol at rates that are sufficient to support glyceroneogenesis.

P-enolpyruvate Carboxykinase

Verification of Its Presence in Adipose Tissue

The presence of P-enolpyruvate carboxykinase in adipose tissue was predicted by the results of in vitro incubation experiments employing specifically labeled pyruvate (Ballard et al. 1967). Our own studies indicated that twice as much carbon-2 of pyruvate-^{14}C was incorporated into glyceride-glycerol by adipose tissue than was the carbon-1 of pyruvate. When the glycerol was chemically degraded, over 90% of the pyruvate carbon-1 and 43-49% of the pyruvate carbon-2 incorporated were in the α-carbons (Ballard et al. 1967). These results, which have recently been refined and extended by White et al. (1968) support the existence of the dicarboxylic acid shuttle in adipose tissue. For this sequence to function, two key enzymes, pyruvate carboxylase and P-enolpyruvate carboxykinase, must be present. Pyruvate carboxylase has been found in adipose tissue mitochondria while the application of a sensitive radio-chemical assay for P-enolpyruvate carboxykinase, developed by Chang and Lane (1968), has allowed us to also demonstrate the presence of this second enzyme in the cytosol of rat epididymal adipose tissue. P-enolpyruvate carboxykinase has an activity of about 30-40 milliunits per g tissue (37^0C) in rat adipose tissue and this activity can be increased five fold by starving the animals for 24 hours (Fig. 6). Refeeding the animals laboratory chow after various periods of fasting caused a marked reduction in P-enolpyruvate carboxykinase activity. One surprising finding concerning the adaptive nature of this enzyme in adipose tissue is its marked age dependency. In four-to-five week-old rats, starvation for 24 hours caused a five-fold increase in adipose tissue P-enolpyruvate carboxykinase activity whereas fasting caused only a 50% increase in tissue from 20-week-old animals. No such age effect was noted in the adaptive response of the liver enzyme (Reshef et al. 1969).

One of the most direct and convincing methods of demonstrating the existence of P-enolpyruvate carboxykinase in the cytosol of rat adipose tissue is to test its antigenicity against antibody prepared against the liver enzyme. P-enolpyruvate carboxykinase was purified from rat liver cytosol to a final specific activity of 16 units/mg protein by the procedure outlined in Table 3 (Ballard and Hanson 1968). The purified rat liver P-enolpyruvate carboxykinase was found to have a constant specific activity in all protein-containing fractions upon chromatography on a Sephadex G-100 column while polyacrylamide disc gel electrophoresis indicated a single protein band. Antibody against rat liver cytosol P-enolpyruvate carboxykinase was prepared by injection of the purified enzyme into a rabbit followed by the isolation of the serum γ-globulin. When this antibody was titrated against P-enolpyruvate carboxykinase from rat liver cytosol and adipose tissue cytosol, identical immunological titration curves of enzyme inactivation were observed (Fig. 7). Ouchterlony double diffusion analysis of the enzymes from adipose tissue and liver shows a single continuous precipitation line

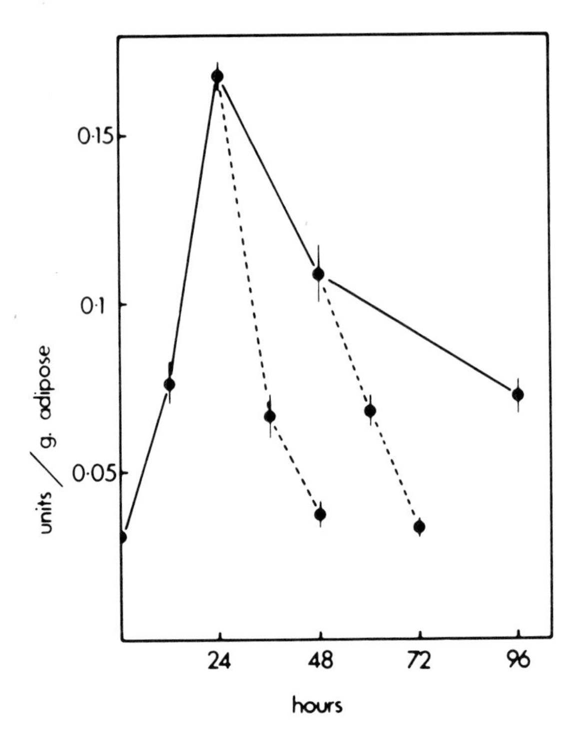

Fig. 6. Activity of P-enolpyruvate carboxykinase in adipose tissue during starvation (————) and when animals are refed laboratory chow (--------). Activity is expressed as units per g tissue at 37^{0}C where a unit is 1 μmole of $H^{14}CO_3^-$ fixed per minute. The bars represent the mean ± s.e.m. (Reshef et al. 1969).

Stage	Protein	Total activity	Specific activity
	mg	units	units/mg
1. 100,000 xg supernatant	22,500	950	0.042
2. $(NH_4)_2SO_4$ precipitate (45-65%)	5,150	708	0.138
3. Sephadex G-100	1,720	492	0.286
4. DEAE-cellulose	159	336	2.11
5. $Ca_3(PO_4)_2$ gel adsorption	38	201	5.3
6. Electrofocusing, pH 5-8	6.4	102	16

Table 3. Purification of P-enolpyruvate Carboxykinase From Rat Liver Cytosol. Livers from 20 rats, starved for 24 hours to induce P-enolpyruvate carboxykinase, were used in this typical purification. Detailed methods are given by Ballard and Hanson (1968).

against P-enolpyruvate carboxykinase antibody (Fig. 8). This evidence provides a firm experimental basis for the existence of a distinct P-enolpyruvate carboxykinase in the cytosol of epididymal adipose tissue which is immunologically identical to the enzyme from liver cytosol.

Factors Regulating P-enolpyruvate Carboxykinase Activity in Rat Adipose Tissue

Young et al. (1964), Foster et al. (1964) and Walter et al. (1966) have thoroughly documented the effects of dietary and hormonal manipulations on the activity of rat liver P-enolpyruvate carboxykinase. The adipose tissue enzyme shares many of the same controls but it differs in certain important ways. Figure 9 shows the response of both adipose tissue and liver P-enolpyruvate carboxykinase to adrenalectomy and to diabetes. When a rat is adrenalectomized, P-enolpyruvate carboxykinase activity increases three to four-fold, but the same enzyme in liver is unaffected (Reshef et al. 1969). If a normal animal is made diabetic by the injection of alloxan, both hepatic and adipose tissue P-enolpyruvate carboxykinase activities are elevated (Reshef et al. in press). But if an alloxan diabetic rat is adrenalectomized the increase in adipose tissue P-enolpyruvate carboxykinase is additive, being greater than the increase due to either adrenalectomy or diabetes alone. In liver the combination of adrenalectomy and diabetes reduces the activity of the enzyme to a level similar to the normal rat.

A direct reversal of the response of P-enolpyruvate carboxykinase to adrenalectomy and diabetes is evident in adipose tissue when the animals are given triamcinolone, a synthetic adrenal cortical steriod hormone, or insulin (Fig. 9). Triamcinolone injected into adrenalectomized-diabetic rats causes a drop in the activity of adipose tissue P-enolpyruvate carboxykinase but doubles the activity of the hepatic enzyme. Insulin given together with triamcinolone does not change the activity of adipose tissue P-enolpyruvate carboxykinase but reduces the hepatic enzyme to below the normal level of activity. The additive effect of triamcinolone and insulin on adipose tissue P-enoloyruvate carboxykinase is most clearly shown when these hormones are administered to diabetic adrenalectomized animals. In this case the drop in enzyme activity is greater than seen with either hormone alone. The results of these experiments indicate that adipose tissue and liver P-enolpyruvate carboxykinase are subject to a different pattern of hormonal regulation. They also show distinctly separate effects of triamcinolone and insulin on the control of adipose tissue P-enolpyruvate carboxykinase.

It is probable that the adrenal hormones and insulin exert their effect on adipose tissue P-enolpyruvate carboxykinase by regulating the relative rates of enzyme synthesis and degradation. We have shown in a previous study (Reshef et al. 1969) that both actinomycin D and cycloheximide will markedly reduce the activity of this enzyme in adipose tissue from both fasted adrenalectomized and diabetic-adrenalectomited rats (Reshef et al. in press). A time curve of decay of enzyme activity caused by actinomycin D and cycloheximide administration indicated a more rapid turnover induced by cycloheximide (Ballard and Hanson 1968). It is of interest to note that the P-enolpyruvate carboxykinase decay rate caused by cycloheximide is similar to that with insulin and is more rapid than with either triamcinolone or actinomycin D.

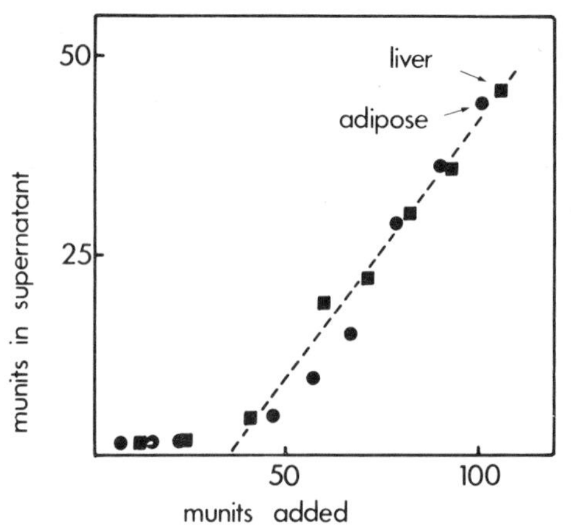

Fig. 7. Titration of an antibody preparation of rat liver cytosol P-enolpyruvate carboxykinase against the same enzyme from rat liver and adipose tissue cytosols. Constant amounts of antibody (30 μl) were incubated with cytosol fractions for one hour and the antibody-antigen complex separated by centrifugation. The activity of P-enolpyruvate carboxykinase remaining in the supernatant was measured and expressed as munits of bicarbonate fixed per minute at 37^{0}C. A constant activity of P-enolpyruvate carboxykinase was precipitated by the antibody, and beyond this amount of enzyme the P-enolpyruvate carboxykinase appeared quantitatively in the supernatant. Further details are given by Ballard and Hanson (1968).

One might speculate that the depression in the activity of rat adipose tissue P-enolpyruvate carboxykinase occurs at either the level of protein synthesis as is exemplified by the insulin-cycloheximide system, or at the level of RNA synthesis as in the case of the triamcinolone-actinomycin D system. The independence of these two systems is firmly supported by the fact that they result in an additive induction in enzyme activity (Fig. 9). However, a more definitive statement on the effect of various hormones or dietary alterations on the regulation of P-enolpyruvate carboxykinase of rat adipose tissue must await immunological analysis of enzyme turnover. What these hormonal studies do indicate, however, is the presence of a finely tuned regulatory balance of the steady state levels of this enzyme in adipose tissue. This is in contrast to the effect of the same hormonal treatment on adipose tissue pyruvate carboxylase, which retains its high level of activity and is unaltered by various treatments tested (Reshef et al. in press). P-enolpyruvate carboxykinase would therefore seem the key enzyme in the dicarboxylic acid shuttle in adipose tissue.

Physiological Role of P-enolpyruvate Carboxykinase in Adipose Tissue

If, as the available evidence indicates, P-enolpyruvate carboxykinase is active in adipose tissue, can we assign it a role in the overall metabolism of the fat cell? Our original assumption, based on the labeling studies described above, was that this enzyme functions together with pyruvate carboxylase as part of a dicarboxylic acid shuttle to provide α-glycerophosphate, which in turn is used for the re-esterification of fatty acids

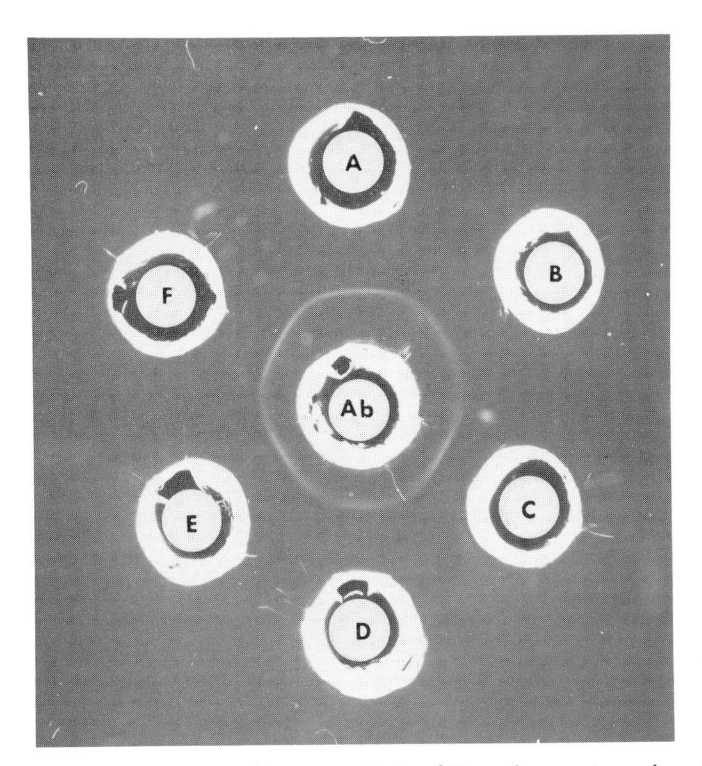

Fig. 8. Ouchterlony double diffusion analysis of P-enolpyruvate carboxykinase antibody of rat liver cytosol. The center well contained 0.02 ml of P-enolpyruvate carboxykinase antibody (Ab) and the outer wells contained 0.02 ml of the following: A) liver cytosol; B) adipose tissue cytosol; C) liver cytosol; D) adipose tissue cytosol; E) pure liver cytosol P-enolpyruvate carboxykinase; F) liver cytosol. The preparation of the tissue extracts and the treatment of the antibody are fully described by Ballard and Hanson (1968).

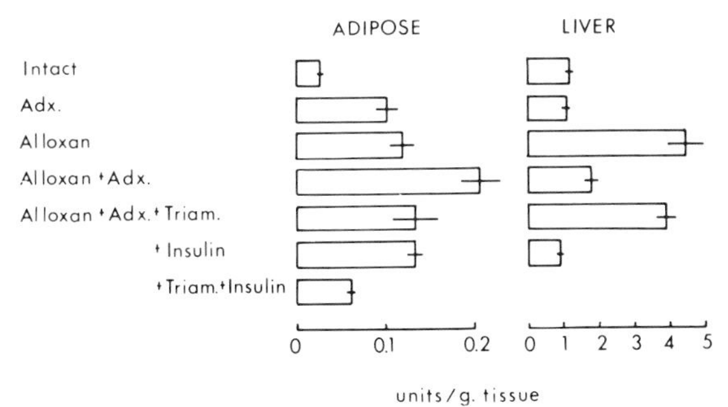

Fig. 9. The effect of diabetes and adrenalectomy on the activity of rat adipose tissue and liver P-enolpyruvate carboxykinase. Animals were made diabetic by the injection of alloxan and the activity of the enzyme was measured after five days. When both adrenalectomized diabetic animals were used the animals were adrenalectomized three days after alloxan treatment. Triamcinolone (5 mg/100 g body weight) was injected subcutaneously and insulin (6 units/100 g body weight) injected intramuscularly.

That glucose itself is not a product is well established due to the absence in adipose tissue of two of the gluconeogenic enzymes, fructose-1, 6-diphosphatase and glucose 6-phosphatase (Weber et al. 1965), as well as the negligible incorporation of pyruvate carbon into adipose tissue glycogen. In support of the glyceroneogenic pathway we found that pyruvate at high concentration (25 mM) was converted to glyceride-glycerol in adipose tissue from starved animals at three times the rate found in tissue from fed rats (Reshef et al. 1969, Jomain and Hanson 1969). However, at lower and presumably more physiological concentrations of pyruvate (0.25 - 1 mM), there was no change in glyceroneogenesis from pyruvate with diet. It is possible that factors other than pyruvate concentration influence the conversion of pyruvate to glycerol. Reshef, et al. (1967) in an earlier study, reported that butyrate and propionate stimulated the incorporation of pyruvate-^{14}C into glyceride-glycerol at a pyruvate concentration of 0.25 mM. As shown in Table 4, the addition of 0.25 mM butyrate at low pyruvate concentrations caused an increase in the labeling of glyceride-glycerol in tissue from both intact and adrenalectomized rats. This agreed closely with the increase in the activity of adipose tissue P-enolpyruvate carboxykinase. Since both butyrate and propionate are metabolized within the mitochondria, the results of this type of experiment suggest that some factor such as the activation of adipose tissue pyruvate carboxylase by butyryl-CoA or propionyl-CoA or a shift in the intramitochondrial redox state favoring malate formation from oxaloacetate is important in the regulation of glyceroneogenesis.

Treatment in vivo	P-enolpyruvate carboxykinase	Pyruvate-2-^{14}C converted to glyceride-glycerol	
	units/g	mμmoles/g/hr	
			butyrate
Intact			
Fed (7)	0.027 ± 0.002	60 ± 7	85 ± 2.5
Starved 24 hours (6)	0.160 ± 0.02	104 ± 28	340 ± 36
Adrenalectomized			
Fed (6)	0.101 ± 0.008	280 ± 49	304 ± 34
Starved 24 hours (5)	0.220 ± 0.10	330 ± 17	930 ± 38

Table 4. Glyceroneogenesis Compared With the Level of P-enolpyruvate Carboxykinase in Adipose Tissue: Effect of Adrenalectomy and Starvation. Values are the mean ± s.e.m. for the number of animals shown in parentheses. Epididymal adipose tissue was incubated in Krebs-Ringer bicarbonate buffer, pH 7.4, containing 0.25 mM pyruvate-2-^{14}C (0.5 μCi) and 0.25 mM butyrate where indicated. The glyceride-glycerol fraction was isolated and the radioactivity determined as described by Reshef et al. 1969.

Dietary status	Addition to the medium	FFA esterified μmoles/g/3 hrs
Fed	-	1.45 ± 0.52
	25 mM pyruvate	3.35 ± 0.90
Starved 24 hours	-	8.86 ± 0.57
	0.25 mM pyruvate	7.50 ± 0.41
	25 mM pyruvate	15.36 ± 0.54
	0.25 mM pyruvate + 0.25 mM butyrate	10.25 ± 0.89

Table 5. Effect of Starving and of Pyruvate and Butyrate on the Esterification of FFA in Adipose Tissue in Vitro. Each value is for the mean ± s.e.m. for five animals. FFA esterification was calculated by the following method: (net glycerol x 3) — net FFA. Both glycerol and FFA were measured in the epididymal fat pad before and after incubation in 5 ml of Krebs-Ringer bicarbonate buffer, pH 7.4, containing 3% defatted bovine serum albumin and appropriate substrates. FFA and glycerol release into the medium was also measured after the three hour incubation period. The procedures for measuring both glycerol and FFA as well as the assumptions upon which the calculations of FFA esterification are based are given in detail by Reshef et al. in press.

It seems likely that the synthesis of α-glycerophosphate during periods of starvation when net lipolysis is occurring, may provide a physiological check on fatty acid release. During diabetes, the efflux of fatty acids from adipose tissue is widely held as one of the causative factors in ketosis (Landgon 1960). Consequently, a restraint in the output of fatty acids from adipose tissue during diabetes would cause a diminution in fatty acid flow to the liver which would reduce the synthesis of ketone bodies. This is consistent with our finding of an elevated P-enolpyruvate carboxykinase activity in adipose tissue from diabetic rats together with an increase in the conversion of pyruvate to glyceride-glycerol. A direct evaluation of the effect of pyruvate on net re-esterification of fatty acids in adipose tissue is presented in Table 5. This study was carried out essentially as described by Vaughn (1962). For the purpose of the calculations in Table 5 we have assumed the following: a) glycerol is not re-esterified to triglyceride by adipose tissue due to the absence of glycerol kinase (Morgolis and Vaughn (1962); b) that glycerol is derived mainly from triglycerides and not from di- or monglycerides (Vaughn 1962) and; c) that fatty acids are not oxidized to any great extent by adipose tissue (Flatt and Ball 1964). On this basis three times the glycerol release is taken to represent the amount of fatty acid formed via lipolysis. If the measured FFA release is subtracted from this figure the rate of fatty acid re-esterification can be calculated. The results indicate that 25 mM pyruvate stimulates the net re-esterification of fatty acids in adipose tissue from fed animals with the effect being markedly greater in tissue from fasted rats. This finding agrees well with our earlier report (Reshef et al. 1969) that pyruvate-2-^{14}C incorporation into glyceride-glycerol is greater in adipose tissue of starved animals than in tissue from fed rats. It is also consistent with an increased activity of P-enolpyruvate carboxykinase in adipose tissue from starved rats. At a lower pyruvate concentration (0.25 mM) we found a re-esterification of fatty acids by adipose tissue from starved rats, which was increased if 0.25 mM butyrate was added to the incubation medium.

Conclusions

Both pyruvate carboxylase and P-enolpyruvate carboxykinase play an important role in adipose tissue metabolism. Pyruvate carboxylase is emerging as one of the key regulatory enzymes in lipogenesis because of its anaplerotic function in the replenishment of four-carbon intermediates used for biosynthesis. In this regard, it is of interest that pyruvate carboxylase is absent in ruminant adipose tissue. This tissue does not synthesize fatty acids from glucose and hence does not deplete intra-mitochondrial oxaloacetate levels for citrate formation (Hanson and Ballard 1967). It is also becoming apparent that in adipose tissue the mitochondrial carboxylation of pyruvate is subject to many of the same controls characteristic of gluconeogenesis in mammalian liver. The dependence of pyruvate carboxylase on the availability of ATP and of pyruvate within the adipose tissue cell is consistant with several studies (Reshef et al. 1969, Patel and Hanson 1970, Rognstad and Katz 1969) which indicate that variations in the levels of these two compounds strongly affect lipogenesis. We are only beginning to understand the potential importance of this uniquely situated enzyme in the regulation of a number of vital biosynthetic processes.

During periods of starvation, pyruvate carboxylase and P-enolpyruvate carboxykinase participate in a metabolic sequence termed "glyceroneogenesis." P-enolpyruvate carboxykinase is a unique adipose tissue enzyme in that it is increased by starvation, which usually diminishes the metabolic functioning of this tissue. Since both adrenalectomy and diabetes also induce adipose tissue P-enolpyruvate carboxykinase this enzyme probably is involved in opposing the adverse effects created by the deletion of adrenal hormones and of insulin. By acting to increase the intracellular synthesis of α-glycerophosphate via glyceroneogenesis, an increase in P-enolpyruvate carboxykinase could effectively decrease FFA release. Our studies on the re-esterification of fatty acids by adipose tissue from starved rats in the presence of pyruvate are consistent with such a concept. The fine points of the control of the glyceroneogenic pathway in adipose tissue remain uncertain since one of the key enzymes in this process, P-enolpyruvate carboxykinase, is subject to both hormonal and dietary regulation at the level of protein synthesis and degradation. The regulatory controls are further complicated because in rat adipose tissue the synthesis of glyceride-glycerol from compounds such as pyruvate, lactate, alanine or serine involves pyruvate carboxylation which is an intra-mitochondrial process. Factors such as a shift in the mitochondrial redox state, the availability of acetyl-CoA and the rate of ATP production also undoubtedly regulate the flow of carbon over the glyceroneogenic sequence. With the increasing emphasis on the importance of fatty acids in influencing hepatic gluconeogenesis it would seem reasonable to state that a process such as glyceroneogenesis in adipose tissue, which may act to control the rate of FFA release from the depots to the liver, could be a major factor in the metabolic interaction between liver and adipose tissue.

Acknowledgment

This work was supported in part by U.S. Public Health Service grants AM-11279, HD-02758, CA-10916, CA-10439, CA-05097; an institutional grant IN-88 from the American Cancer Society; and by grants from the Research Fund of the Hebrew University-Hadassah Medical School and the Queen Elizabeth II Fellowships Committee. The electron microscopy of adipose tissue mitochondria in Fig. 2 was carried out by Dr. Michael Higgins, Temple University Medical School.

References

Angel, A., H. Sheldon: Ann. N. Y. Acad. Sci. 131 (1965) 157
Ballard, F. J., R. W. Hanson: J. Lipid Res. 8 (1967) 73
Ballard, F. J., R. W. Hanson, G. A. Leveille: J. biol. Chem. 242 (1967) 2746
Ballard, F. J., R. W. Hanson: J. biol. Chem. 244 (1968) 5625
Ballard, F. J., R. W. Hanson: Biochem. J. 112 (1969) 195
Ballard, F. J., R. W. Hanson, L. Reshef: Biochem. J., in press
Chang, H. C., M. D. Lane: J. biol. Chem. 241 (1968) 5625
Flatt, J. P., E. G. Ball: J. biol. Chem. 239 (1964) 675
Foster, D. O., P. D. Ray, H. A. Lardy: Biochemistry 2 (1964) 555
Garber, A. J., F. J. Ballard: J. biol. Chem. 244 (1966) 369
Hanson, R. W., F. J. Ballard: Biochem. J. 105 (1967) 529
Jomain, M., R. W. Hanson: J. Lipid Res. 10 (1969) 674
Krebs, H. A., In: Advances in Enzyme Regulation, Vol. I. Pergamon, New York 1963, p. 385
Landgon, R. G., In: Lipid Metabolism, Ed. by K. Bloch, John Wiley, New York 1960, p. 247
Mehlman, M. A.: J. biol. Chem. 243 (1968) 3289
Morgolis, S., M. Vaughn: J. biol. Chem 237 (1962) 44
Patel, M. S., R. W. Hanson: J. biol. Chem. 245 (1970) 1302
Reshef, L., J. Niv, B. Shapiro: J. Lipid Res. 8 (1967) 688
Reshef, L., F. J. Ballard, R. W. Hanson: J. biol. Chem. 244 (1969) 5577
Reshef, L., R. W. Hanson, F. J. Ballard: J. biol. Chem. 244 (1969) 1994
Reshef, L., R. W. Hanson, F. J. Ballard: J. biol. Chem., in press
Rodbell, M.: J. biol. Chem. 239 (1964) 375
Rognstad, R., J. Katz: Biochem. J. 111 (1969) 431
Scrutton, M. S., M. F. Utter: Ann. Rev. Biochem. 37 (1968) 249
Somberg, E. W., M. A. Mehlman: Biochem. J. 112 (1969) 435
Spencer, A. F., J. M. Lowenstein: J. biol. Chem. 237 (1962) 3640

Srere, P. A.: Nature 205 (1965) 766
Utter, M. F., D. B. Keech: J. biol. Chem. 238 (1963) 2603
Vaughn, M.: J. biol. Chem. 237 (1962) 3354
Wakil, S. J., E. B. Tilchener, D. M. Gibson: Biochem. biophys. Acta. 29 (1958) 225
Walter, P., V. Paetkau, H. A. Lardy: J. biol. Chem. 241 (1966) 2523,
Weber, G., H. J. Hird, N. B. Stamm, D. S. Wagle, in: Handbook of Physiology, Ed. by A. E. Renold and G. F. Cahill, Jr. Section 5 American Physiological Society, Washington, D.C. 1965, p. 255
White, L. W., H. R. Williams, B. R. Landau: Arch. biochem. biophys. 126 (1968) 552
Wise, E. M., Jr., E. G. Ball: Proc. nat. Acad. Sci. (Wash.) 52 (1964) 1255
Young, J. W., E. Shrago, H. A. Lardy: Biochemistry 3 (1964) 1687

Discussion to Hanson, Patel, Reshef and Ballard

Krebs: As for the role of pyruvate carboxylase, is there not another point to be considered, i.e. its possible role in the supply of reducing equivalents in the cytoplasm? The citrate cleavage enzyme pathway regenerates the oxaloacetate needed for the synthesis of citrate. But unless the oxaloacetate is lost by side reactions there is no need for its resynthesis. What may be relevant in this context is the concept developed by Ball (E. M. Wise and E. G. Ball: Proc. nat. Acad. Sci. 52 (1964) 1255; M. S. Kornacker and E. G. Ball: Proc. nat. Acad. Sci. 54 (1965) 899) that a special mechanism is needed for the generation of NADPH in the cytoplasm. The suggestion is that this is formed by the malic enzyme converting malate and $NADP^+$ to pyruvate + CO_2 + NADPH. The pyruvate can be regenerated to oxaloacetate within the mitochondria under the influence of pyruvate carboxylase to subsequently undergo reduction to malate. Would this not give a satisfactory explanation for the relatively high activity of pyruvate carboxylase?

Hanson: The data we have presented here are in essential agreement with the findings of Ball and his associates. What we are suggesting is that since citrate efflux from the mitochondria is obligatory its cleavage in the cytosol would supply both acetyl-CoA (for lipogenesis) and oxaloacetate. With glucose as substrate, NADH formed by triosedehydrogenation reactions would be available to convert oxaloacetate to malate. The enzyme NADP-malate dehydrogenase could then convert the malate to pyruvate thereby effectively transhydrogenating NADH to NADPH. This means that there is a cycle, we call it "citrate clevage pathway" others term it "Malate pathway" or "pyruvate cycle." The net result of such a pathway would be the depletion of mitochondrial oxaloacetate. Therefore pyruvate carboxylase would be required to replenish this oxaloacetate.

Krebs: I agree; on balance it comes to the same.

Hanson: When pyruvate is substrate the situation is very different since there is a shortage of reducing equivalents. Pyruvate by-passes the pentose and glycolytic pathways and all of the necessary NADH must come from mitochondrial reactions. Dr. Ball has published on this subject but all of the details of lipogenesis from pyruvate in adipose tissue are still not clear.

Krebs: To sum up, there is a very satisfactory explanation for the high activity of pyruvate carboxylase and for the relatively low activity of PEP carboxykinase.

Hanson: Our values for P-enolpyruvate carboxykinase must be multiplied by seven since we are measuring the enzyme in the direction of carboxylation using the $^{14}CO_2$ fixation assay of Chang and Lane (J. biol. Chem. 241 (1966) 2413). If one corrects for the protein content of adipose tissue the activity of P-enolpyruvate carboxykinase in the two tissues is not too far different. What we do not know, and I would invite your comments on, is whether our proposed role for P-enolpyruvate carboxykinase, as part of "glyceroneogenesis" is a reasonable one from a physiological point of view.

Krebs: Is FDPase definitely absent?

Söling: There is a small but consistent activity of FDPase, measurable in adipose tissue, and even in isolated fat cells.

Hanson: Who had done that?

Söling: Me.

Hanson: We have attempted to measure this enzyme in adipose tissue and could not find it.

Krebs: But is glucose actually formed?

Hanson: If you incubate adipose tissue with labeled pyruvate you do not get label in glycogen. The enzymes (FDPase and G-6-Pase) are missing, or are so low in activity to be almost negligible.

Wieland: You had an increased rate of re-esterification in the starved animals. Does this mean that there is an increase in lipogenesis?

Hanson: No, in the fed animals there is a large extracellular water space relative to metabolizable space in the tissue. According to some calculations we did, there is something like 1 mM glucose in the extracellular water in fat tissue. So there is a tremendous store of available precursors, which complicate any metabolite studies with whole fat pads.

Exton: I was interested in your interpretation of the pyruvate effect shown at this symposium. Did you mention the possibility that this might be a sparing effect on pyruvate dehydrogenase?

Hanson: Yes, that is a possibility. Although we initially tried to explain it on the basis of redox changes.

Ruderman: In recent years Bjorntrop and his colleagues as well as others have demonstrated that ketone bodies can inhibit adipose tissue lipolysis (reviewed by N. B. Ruderman et al.: Arch Intern. Med. 123 (1969) 299). Acetate and lactate have been shown to have a similar effect. Possibly the effects of butyrate which you reported might be on this basis.

I also have a question. Did you determine whether adipose tissue pyruvate carboxylase activity is influenced by the animal's nutritional state?

Hanson: This goes into the controversy we have discussed at this meeting namely, where pyruvate carboxylase is in the cell. We have looked at the distribution of adipose tissue enzymes, using as gentle methods as we can: by treating the cells with collagenase and running them down along the side of a glass vellel to open them up without homogenizing at all. After isolating mitochondria we found roughly about 30% of the enzyme to be soluble. We assume that this was real, but J. Ballard has recently done antibody titration studies and he found the mitochondrial and cytosolic enzyme to be immuno-chemically identical. So now I am wondering whether we are really seeing a difference, whether there is really a by-model distribution. I am not sure about this.

Ruderman: Did the total activity vary according to the nutritional state?

Hanson: No, it is not adaptable but is very cold labile, so that you must be very careful how you handle it when assaying its activity.

Stimulation of Amino Acid Metabolism and Gluconeogenesis from Amino Acids in the Liver of Fasting Rats Treated with Antilipolytic Agents

A. Hasselblatt, U. Panten and W. Poser
Pharmakologisches Institut der Universität Göttingen, FRG

Summary

When starving rats were prevented from mobilizing their fat stores by the antilipolytic agent 3,5-dimethylisoxazole the following adaptive changes in hepatic metabolism occurred:

1) In fasting rats more urea was produced following dimethylisoxazole as indicated by a two-fold rise in plasma urea levels. Livers from such animals produced more urea when isolated and perfused without added substrates. Plasma urea levels were similarily elevated in fasting adrenalectomized rats. The increase in amino acid metabolism is not mediated, therefore, by the action of adrenal cortical hormones. An increase in urea formation following dimethylisoxazole was absent in fed animals which do not rely upon lipolysis to secure a sufficient supply in energy yielding substrates. It is considered, therefore, to represent an adaptive response to the shortage in substrates for energy metabolism arising when lipolysis is inhibited in starving rats.

2) Livers from starving rats which had received dimethylisoxazole produced more glucose than those of corresponding controls when perfused without added substrate. Similarily liver slices taken from these rats did incorporate more bicarbonate carbon into glucose when incubated in vitro. It is concluded from these results that gluconeogenesis from endogenous substrates has been stimulated. Adrenalectomized rats were unable to form additional glucose from endogenous amino acids and fatal hypoglycemia resulted when fasting adrenalectomized rats were injected with dimethylisoxazole.

3) Although more glucose is formed in livers of fasting rats treated with dimethylisoxazole, the capacity of such livers to utilize alanine for gluconeogenesis did not increase to a major degree. It is concluded that glucose formed in addition in livers of treated animals is derived from endogenous substrates. These are made available to gluconeogenesis by an increased proteolysis resulting in an elevated deamination of amino acids, and an increase in urea formation.

Introduction

The primary effect common to various compounds, as methylated isoxatole- or pyrazole derivatives and of nicotinic acid is to inhibit lipolysis. As they suppress the cleavage of triglycerides in the adipose tissue the amount of free, unesterified fatty acids (FFA) released into the blood is reduced. Following the injection of antilipolytic agents, FFA-levels in the plasma decline. Several metabolic effects are secondary to the diminished supply of FFA: hepatic ketogenesis is reduced and ketonemia in fasting or diabetic rats is reversed (Bubenheimer et al. 1966). Hepatic levels of acetyl-coenzyme A and long chain acyl esters of coenzyme A in the liver decline and a drop in plasma triglycerides indicates that less esterified fatty acids are released from the liver (Schwabe and Hasselblatt 1967). Especially the high efficiency of antilipolytic agents to suppress ketonemia in the rat has been considered to indicate that these drugs might be useful in the treatment of diabetic patients (Beringer et al. 1969). It should be borne in mind, however, that, in the absence of insulin, antilipolytic agents fail to increase glucose utilization to a major degree. On the other hand, they reduce the supply in FFA and ketone bodies, both of which may be of vital importance as energy yielding substrates in the diabetic patient whose glucose utilization is impaired. A similar shortage in energy yielding substrates does arise when fasting rats are prevented from mobilizing their fat stores by antilipolytic agents. Thus in starving rats injected with 3,5-dimethylisoxazole a rise in plasma corticosterone indicated that the shortage in energy yielding substrates has given rise to a state of sufficient emergency to evoke a stress response from the adrenal glands (Hasselblatt 1969). Due to the shortage in substrates derived from fatty acids more glucose is metabolized and blood glucose levels and liver glycogen stores are reduced. To prevent fatal hypoglycemia additional glucose has to be formed from body proteins (Hasselblatt 1969, Hasselblatt et al. 1970). Effects of antilipolytic agents are thus not limited to the adipose tissue. They are useful experimental tools to study the relationships existing between metabolism of fatty acids, glucose and amino acids.

Methods

A description of procedures employed in measuring blood glucose, plasma FFA and the incorporation of $1\text{-}^{14}C$ glycine into a glucose and glycogen in the rat and the details of the incubation technique for experiments on liver slices have been given by Hasselblatt (1969). The stimulatory effect of the antilipolytic agent 3,5-dimethylisoxazole on urea synthesis in the liver has been reported by Hasselblatt et al. (1970). For perfusion experiments livers of male rats (Wistar, W 64) of 180-240 gms body weight starved for 36 hours were used. A dosage of 3,5-dimethylisoxazole was injected subcutaneously at a dose of 1 mg/kg three hours prior to removal of the live

Livers were perfused in a system in which a constant temperature of 35^0C was maintained by a temperature controlled water mantling. A thin layer dialysator supplied by Eschweiler, Kiel, FRG, served as an oxygenator of high efficiency. It was capable to raise oxygen pressure in the medium employed from 150 to 700 mm Hg in a single passage at a flow rate of 35 ml/min. The perfusion medium was made up according to Hems et al. (1966). It contained 2.5% bovine albumin, heparin (2 mgs 100 ml) and bovine erythrocytes to give a final concentration of 2.5% hemoglobin i Krebs-Henseleit bicarbonate buffer. To prevent bacterial growth a mixture of carbenicilline, ampicilline and oxacilline was added.

Oxygen consumption was determined by polarographic measurement using membrane stabilized platin wire electrodes. The flow rate was adjusted to 32 ml/min for the initial ten minutes of perfusion and to 20-24 ml/min. for the rest of the perfusion time. Venous oxygen pressure was approximately 200 mm Hg under these conditions when livers of control animals were perfused. The total initial volume of the perfusion fluid was 60 ml. For analysis samples of 1 ml were withdrawn and deproteinised by addition of ice cold perchloric acid. In the supernatant glucose was determined by the hexokinase test (Boehringer test), urea according to Archibald (1945) and α-amino nitrogen by the method of Stegemann (1960).

Results

Following a one-day starvation period, lipolysis is activated in adipose tissue of rats. This is shown by the elevated levels of plasma FFA in fasting rats as compared to those of fed animals (Fig. 1). A dosage of 3,5-dimethylisoxazole lowered plasma FFA levels in fasting rats when injected at a dose of 1 mg/kg three hours prior to dea In fed rats lipolysis was low and scarcely inhibited by dimethylisoxazole. It might b expected, therefore, that any response which is secondary to the antilipolytic action will be most prominent in starving rats, whereas it is unlikely to be of significance i fed animals. As seen here plasma urea levels increase in starving rats while there is no such response in fed animals. This effect is therefore not likely to arise from a direct stimulatory action of dimethylisoxazole on urea synthesis in the liver. It is a secondary response to antilipolysis that hepatic amino acid metabolism is stimulated resulting in a two-fold rise in plasma urea levels and a four-fold increase in urea ex cretion (Hasselblatt 1970). The effect on plasma urea levels apparently does not depend on functioning adrenal glands as quite a similar response was present in rats whi had been adrenalectomized a week prior to the experiment. Blood glucose levels we not significantly altered in fed rats, and slightly reduced in starving animals. In adr alectomized rats a severe hypoglycemia with low mean blood glucose levels of 17 mgs/100 ml resulted from the treatment. Part of these animals succumbed to fatal hypoglycemic convulsion. Severe hypoglycemia does not occur in intact starving rats as they are able to maintain a sufficient level of blood glucose by an increased formation of glucose from body proteins. Adrenalectomized rats are known to be unable to increase gluconeogenesis to meet an elevated demand for glucose as present in diabetes (Wagle and Ashmore 1961). Similarily in our experiments adrenalectomized rats did not form sufficient glucose from precursors gained from amino acid metabolism to prevent hypoglycemia.

When a tracer dose of 1-^{14}C-glycine was injected intraperitioneally into starving rats (5 μC/100 g body weight) the specific activity of blood glucose and liver glycogen was higher in animals having received dimethylisoxazole (Fig. 2). Thus a larger part of the glucose dissolved in the blood and stored in the liver is derived from the ^{14}C-amino acid injected. As the rates of glucose formation are difficult to assess from experiments on whole animals gluconeogenesis has been measured employing in vitro systems as liver slices and the isolated perfused rat liver.

Liver slices taken from fasting rats form glucose when incubated in vitro in a potassium enriched buffer in the presence of L-alanine, L-lactate or pyruvate (Fig. 3). The amount of newly formed glucose has been calculated in these experiments from the difference in glucose liberated by acid hydrolysis before and after the incubation. Liver slices taken from fasting rats which had received dimethylisoxazole formed more glucose when incubated in the presence of L-alanine while no significant difference was found when lactate or pyruvate served as precursors. Similar results were obtained when the incorporation of bicarbonate-^{14}C into glucose was measured (Fig. 4). More labeled glucose was formed when liver slices of treated rats were incubated in the presence of L-alanine. A similar difference was found when no substrate had been added and glucose formation proceeds from endogenous sources. As was expected no such difference was found when liver tissue from fed rats was incubated. Adrenalectomized rats proved unable to increase hepatic gluconeogenesis when injected with antilipolytic agents. This does explain that in these animals dimethylisoxazole caused severe hypoglycemia. In these as in many additional experiments we were unable to demonstrate that glucose formation from pyruvate or lactate was stimulated in fasting rats injected with the antilipolytic agent dimethylisoxazole. It is not likely, therefore, that any of the metabolic steps involved in glucose formation from pyruvate has been activated. Also gluconeogenesis from L-fructose was not altered by the treatment. Apparently the shortage in energy yielding substrates induced in fasting rats by antilipolytic agents does not change the capacity of the gluconeogenetic pathway in the liver. More glucose is formed in the livers of treated rats as additional substrates are made available for gluconeogenesis. These substrates are derived from amino acids which are liberated by an increased breakdown of endogenous proteins and which are deaminated at an accelerated rate. Thus more glucose and more urea is formed in livers of fasting rats injected with dimethylisoxazole.

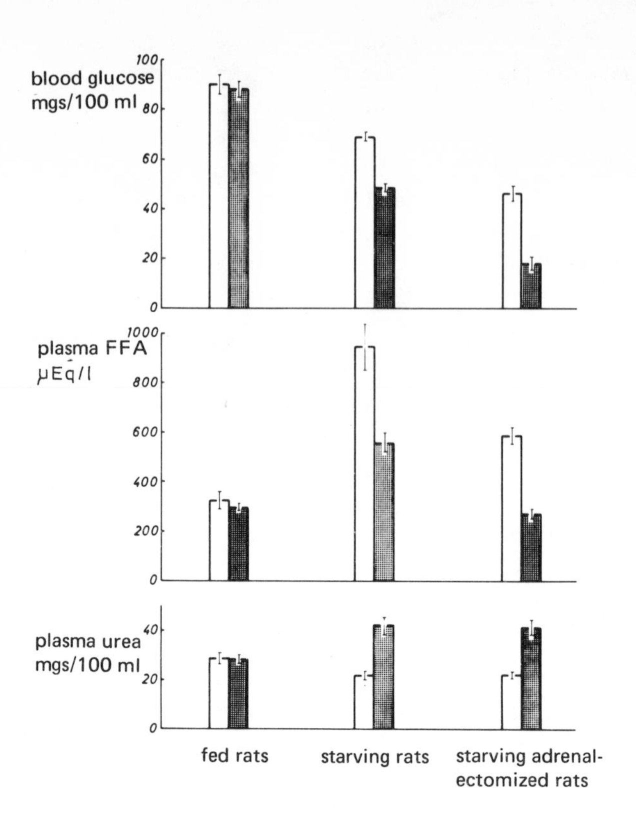
blood glucose
mgs/100 ml
100
80
60
40
20
0
plasma FFA
μEq/l
1000
800
600
400
200
0
plasma urea
mgs/100 ml
40
20
0
fed rats
starving rats
starving adrenal-
ectomized rats

Fig. 1

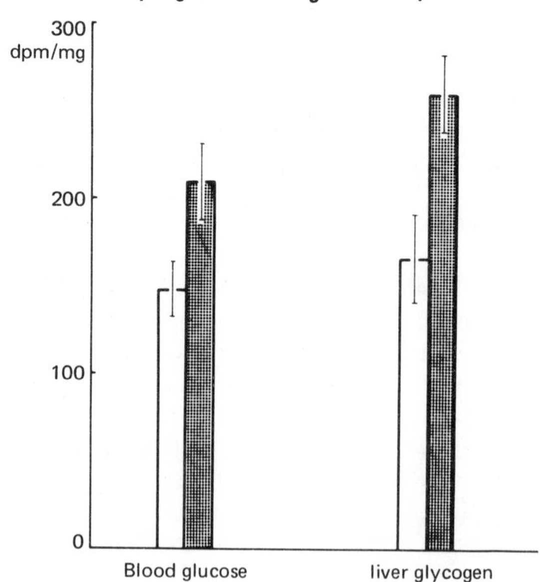
Specific Activity of Blood Glucose and
Liver Glycogen Following 1-14C-Glycine in vivo
300
dpm/mg
200
100
0
Blood glucose
liver glycogen

Fig. 2

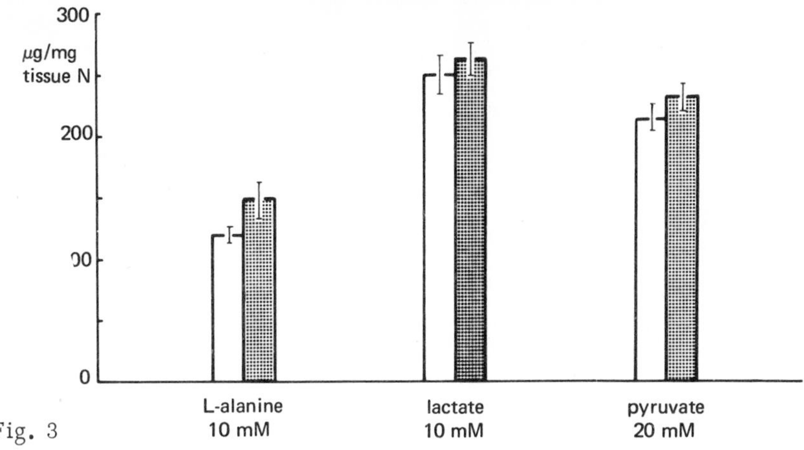
Glucose Formed by Liver Slices when Incubated in vitro
(Incubation for 90 minutes)
300
µg/mg
tissue N
200
00
0

L-alanine
10 mM
lactate
10 mM
pyruvate
20 mM

Fig. 3

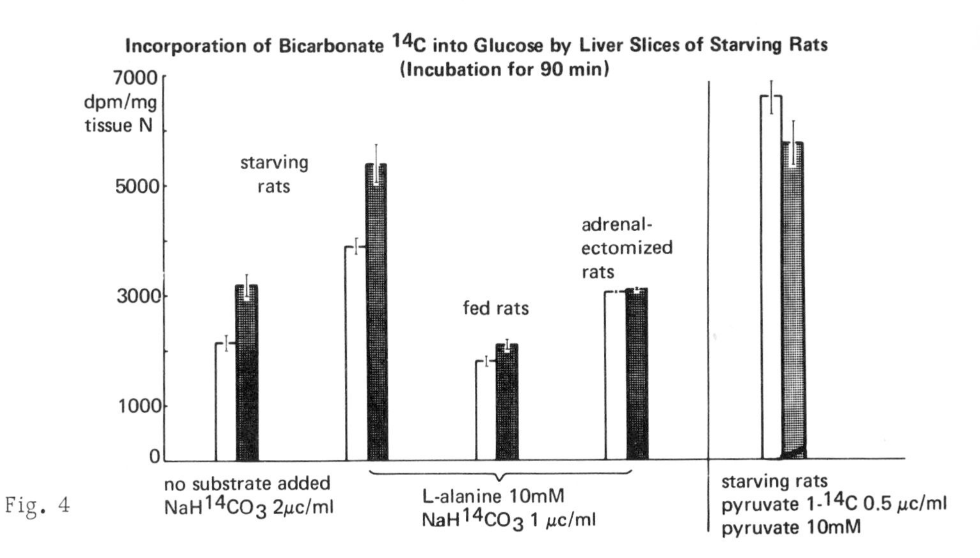
Incorporation of Bicarbonate 14C into Glucose by Liver Slices of Starving Rats
(Incubation for 90 min)
7000
dpm/mg
tissue N
5000
3000
1000
0
starving
rats
fed rats
adrenal-
ectomized
rats
no substrate added
NaH14CO3 2µc/ml
L-alanine 10mM
NaH14CO3 1 µc/ml
starving rats
pyruvate 1-14C 0.5 µc/ml
pyruvate 10mM

Fig. 4

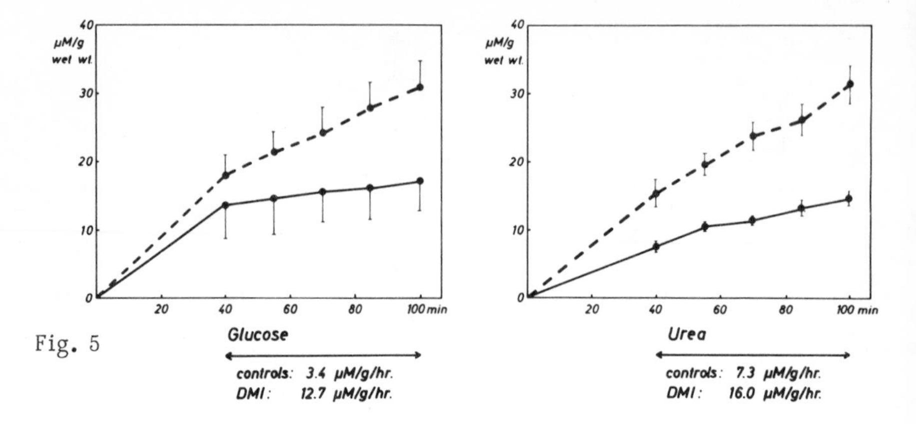
Glucose and Urea Formed by the Perfused Rat Liver without Added Substrate
(— controls - - donor rats treated with 1 mg/kg dimethylisoxazole 3 hrs. before sacrifice)
μM/g
wet wt.
40
30
20
10
0
20
40
60
80
100 min
Glucose
controls: 3.4 μM/g/hr.
DMI: 12.7 μM/g/hr.
Urea
controls: 7.3 μM/g/hr.
DMI: 16.0 μM/g/hr.

Fig. 5

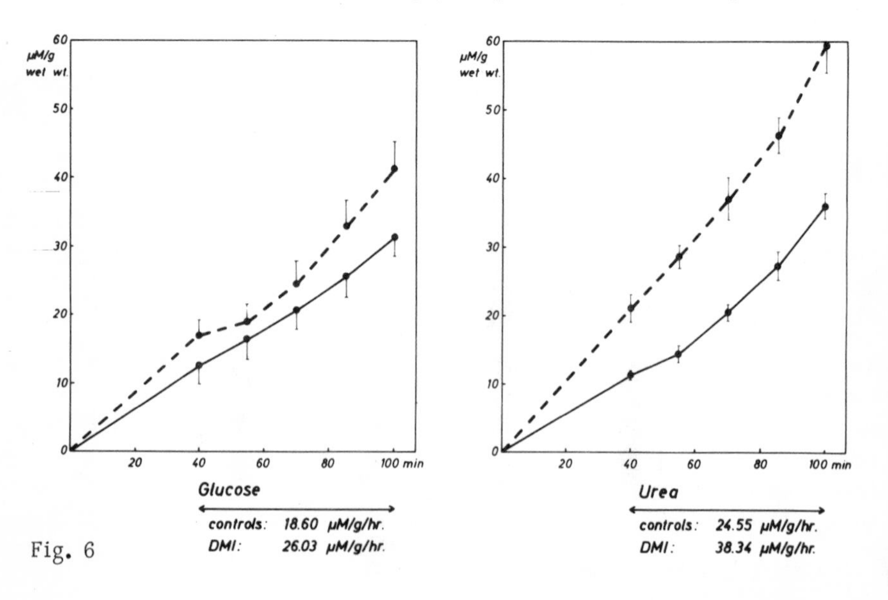
Glucose and Urea Formed by the Perfused Rat Liver with L-Alanine (20 mM)
(— controls - - donor rats treated with 1 mg/kg dimethylisoxazole 3 hrs. before sacrifice)
μM/g
wet wt.
60
50
40
30
20
10
0
20
40
60
80
100 min
Glucose
controls: 18.60 μM/g/hr.
DMI: 26.03 μM/g/hr.
Urea
controls: 24.55 μM/g/hr.
DMI: 38.34 μM/g/hr.

Fig. 6

Results of experiments on isolated perfused rat livers were in agreement with these conclusions. When livers of starving rats which had received dimethylisoxazole three hours prior to death were perfused without added substrate they formed more glucose and more urea than those of corresponding controls (Fig. 5). As substrates were lacking relatively small amounts of glucose (3.4 μM/g/hr) and of urea (7.3 μM/g/hr) were produced by livers of untreated controls. Respiration was low, the mean oxygen consumption being 2.35 μM/g/min within the last hour of the perfusion. Considerably more glucose (12.7 μM/g/hr) and urea (16.0 μM/g/hr) were produced by livers of treated rats. Also oxygen consumption was higher (2.80 μM/g/min) than in control experiments.

It can be concluded from these experiments that endogenous proteolysis in the liver has been activated. Thus more amino acids are made available and the gluconeogenetic pathway is supplied with additional substrates. This does explain that more glucose is formed in the absence of added substrates. In further experiments it has been checked whether the capacity of the liver to also utilize amino acids for glucose formation has increased. Following a preperfusion without substrate at 37 minutes enough L-alanine was added to obtain a final concentration of 20 mM in order to obtain maximal gluconeogenesis from alanine (Fig. 6). In the presence of alanine 5.5 times more glucose was produced by livers of starving rats (18.6 μM/g/hr) than in experiments were no substrate had been added. Urea production rose 3.4 fold to 24.55 μM0g/hr. Moreover, a decline in the α-amino nitrogen levels in the perfusion which was 16.84 mM when calculated as alanine at 40 minutes and 9.00 mM at the end of the perfusion, indicates that alanine is taken up by the liver. Oxygen uptake rose following the addition of alanine from 2.28 to 3.13 μM/g/min. Livers of treated rats produced more glucose (26.03 μM/g/hr) and more urea (38.34 μM/g/hr) than those of control animals. Oxygen consumption rose from 3.13 to 4.17 μM/g/min in these experiments. The net increase in glucose production observed in the presence of alanine however, did not surpass the amount of additional glucose formed in the absence of added substrate (Fig. 7). Livers of treated rats perfused without substrate produced 9.3 μM of glucose/g and hour in addition to untreated controls while the difference was 7.4 μM/g/hr in the presence of alanine. These data do not support the assumption that in starving rats subjected to antilipolysis the capacity to utilize alanine for glucose formation has increased to a major degree. If this were the case a more pronounced effect on gluconeogenesis would be expected to result from the exposure to saturating concentrations of alanine.

The amount of urea produced in addition by livers of treated rats in the presence of alanine was higher than in perfusion experiments without added substrates (13.8 as compared to 8.7 μM/g/hr). Moreover more lactate plus pyruvate was found to be released from livers of treated rats when perfused with alanine (19.8 μM/g/hr) than from those of control animals (13.6 μM/g/hr). Thus additional glucose may be derived in these experiments from alanine which is deaminated at an accelerated rate.

Glucose Produced in the Last 60 Minutes of Perfusion

	no substrate added	*L-alanine (20 mM) added*
DMI:	*12.7 μM/g*	*26.0 μM/g*
controls:	*3.4 μM/g*	*18.6 μM/g*
effect of DMI:	*+ 9.3 μM/g*	*+ 7.4 μM/g*

Fig. 7

Discussion

The increase in hepatic glucose formation and urea synthesis induced by antilipolytic agents does represent a secondary and adaptive response to the shortage in energy yielding substrates arising in starving rats which are prevented from mobilizing their fat stores. This has been shown in previous experiments where isoxazole derivatives failed to stimulate gluconeogenesis when added to liver tissue in vitro (Hasselblatt 1969). It is also evident from the present data obtained from fed rats. As animals having free access to food do not actively mobilize their fat stores lipolysis in adipose tissue and plasma levels of FFA are low. Dimethylisoxazole is scarcely effective in lowering plasma FFA levels in fed rats. Changes in hepatic metabolism induced by antilipolysis are not likely to occur, therefore, in fed animals. Thus in fed rats plasma urea levels did not increase and hepatic gluconeogenesis was not stimulated following an injection of dimethylisoxazole.

Experiments on liver slices and on perfused livers of fasting rats have shown that additional glucose formed by livers of starving rats subjected to antilipolysis is derived from substrates made available to the gluconeogenetic pathway by an increased breakdown of body proteins. As this adaptive increase in proteolysis and the corresponding rise in plasma urea levels was also present in adrenalectomized rats it is not mediated by endogenous adrenal cortical hormones. Unlike intact rats, adrenalectomized rats were unable to utilize the substrate gained from deamination of amino acids to form additional glucose. Thus starving adrenalectomized rats did succumb to fatal hypoglycemia when injected with dimethylisoxazole.

Although glucagon may have been released in starving rats treated with the antilipolytic agent and its effects may persist in the perfused liver a different response would be expected to this hormone. Glucagon is known to stimulate glucose formation from pyruvate by activating a metabolic step between pyruvate and phosphoenolpyruvate in the gluconeogenetic pathway (Exton and Park 1968). An increase in permeability of the mitochondrial membrane to pyruvate has been discussed as the possible site of action common to glucagon, adrenaline and cortisol (Adam and

Haynes 1969). There was no evidence in our experiments that glucose formation from pyruvate is increased in starving rats treated with dimethylisoxazole. In livers perfused with high concentrations of alanine a two to three-fold increase in glucose formation was observed in the presence of glucagon (Malette et al. 1969). In our experiments the capacity of perfused livers of starving rats injected with dimethylisoxazole to form glucose from saturating amounts of alanine was not increased to a major degree.

The adaptive proteolysis in livers of starving rats treated with antilipolytic agents and resulting in an increase in glucose and urea formation from endogenous amino acids may not necessarily involve hormonal regulation but could well represent a more direct response to changing metabolite patterns in the liver cell. These results demonstrate that glucose formation can be increased sufficiently to prevent severe hypoglycemia in the rat without elevating the capacity of the gluconeogenetic pathway.

References

Adams, P. A. J., R. C. Haynes: J. biol. Chem. 244 (1969) 6444

Archibald, R. M.: J. biol. Chem. 157 (1945) 507

Beringer, A., K. H. Tragl, H. Mösslbacher, B. Sokopp, G. Geyer, W. Waldhäusl: Diabetologia 5 (1969) 125

Bubenheimer, P., A. Hasselblatt, U. Schwabe: Klin. Wschr. 44 (1966) 713

Exton, J. H., C. R. Park: J. biol. Chem. 243 (1968) 4189

Hasselblatt, A.: Naunyn-Schmiedebergs Arch. exp. Path. Pharmak. 262 (1969) 441

Hasselblatt, A., U. Panten, W. Poser: Life Sci. 9 (1970) 21

Hems, R., B. D. Ross, M. N. Berry, H. A. Krebs: Biochem. J. 101 (1966) 284

Malette, L. E., J. H. Exton, C. R. Park: J. biol. Chem. 244 (1969) 5713

Schwabe, U., A. Hasselblatt: 12. Symposion der Deutschen Ges. für Endocrinologie, Springer, Berlin (1967), p. 226

Stegemann, H.: Z. physiol. Chem. 219 (1960) 102

Wagle, S. R., J. Ashmore: J. biol. Chem. 236 (1961) 2868

Discussion to Hasselblatt, Panten and Poser

Seubert: How do you explain the different effects of pretreatment with antilipolytic agents on gluconeogenesis from lactate and alanine? As far as I understood, you postulate a mobilization of amino acids by proteolysis as the cause of the increased rat of glucose synthesis. But how do you explain that in the isolated liver gluconeogenesis from alanine is increased but not in lactate? At least you must assume an additional control point between alanine and pyruvate.

Hasselblatt: Well, when you add alanine to the perfused liver, more glucose is formed by livers from treated rats than by those of control animals. This difference is less pronounced, however, than in experiments where the livers were perfused without added substrate. Thus some -- if not all -- of the additional glucose formed by livers of treated rats in the presence of alanine is derived from substrates liberated by endogenous proteolysis and not from alanine. I think that the permeation of alanine into the liver cell might be a limiting factor in gluconeogenesis from alanine, and if this were so the gluconeogenetic pathway would not be saturated and thus endogenously liberated amino acids would increase glucose formation even in the presence of alanine.

Guder: I have two questions to Dr. Hasselblatt: 1) What happens to glucose formation if you add your substance to the perfusion system? 2) Is anything known about the effect of dimethylisoxazole on insulin secretion? This would be an important factor within the whole animal.

Hasselblatt: To answer your first question: There is no direct effect of isoxazole derivatives on hepatic gluconeogenesis when these agents are added to the liver in vitro. The same is true for nicotinic acid. As for insulin levels, it is well known that insulin secretion is stimulated by fatty acids. Lowering of plasma FFA might therefore lower insulin levels. This has not been measured, however, in these experiments and thus remains a theoretical possibility.

Schoner: In connection with the presentation of Dr. Hasselblatt I would like to show you that alanine in addition to its role as a substrate for gluconeogenesis may also act as an inhibitor of the glycolytic enzyme pyruvate kinase. As you have already been told by Dr. Söling, the activities of pyruvate kinase in rat liver exceed by about 4-5 times those of the PEP-synthesizing enzymes pyruvate carboxylase and PEP-carboxykinase. Since PEP-carboxykinase and pyruvate kinase are both located in the cytoplasmic compartment, the activities of pyruvate kinase would be sufficient to degradate all phosphoenolpyruvate formed via the biosynthetic pathway. This reflux of metabolites may be prevented by metabolite control of hepatic pyruvate kinase by ATP. However, studies on the metabolite control of gluconeogenesis during the last years revealed that hepatic ATP contents do not alter during gluconeogenesis (Tarnowski and Seeman 1967). Since in contrast to diabetes and starvation hepatic pyruvate kinase is not suppressed during cortisol dependent gluconeogenesis (Weber et al. 1966) additional increase of gluconeogenesis under this hormonal state by metabolite control of liver pyruvate kinase seemed to be possible. In a study on the effects of metabolites which are or could be under the control of cortisol on the activity of hepatic pyruvate kinase we found an inhibition of the enzyme by alanine (Steubert et al. 1968).

As you may see from Fig. 1, a comparative kinetic investigation of the inhibitory effects of alanine and ATP with partially purified pyruvate kinase from rat liver (Schoner et al. in press) showed that both metabolites increase the S-shape of the PEP saturation curve. As can be seen from the increase of the slopes of the curves in the Hill graph, this is due to a heterotropic interaction of alanine and ATP on the PEP saturation curve. The inhibitory effect of alanine is restricted to the pyruvate kinase from rat liver and kidney and is specific for L-alanine; D-alanine and ß-alanine are without any inhibitory effect. A similar kinetic behavior of both inhibitors, alanine and ATP, can also be demonstrated in a study on the enzymatic activity at various inhibitor concentrations at different levels of phosphoenolpyruvate (Fig. 2). The inhibitory action of both metabolites is reduced with increasing concentrations of PEP. Hepatic contents of phosphoenolpyruvate in cortisol treated rats are 70 nmoles/g wet weight (Hornbook et al. 1966) and those of alanine 794 nmoles/g wet weight (Schoner et al. in press). A comparison of these tissue contents with the lower curve of the alanine dependent inhibition (Fig. 2, upper part) supports a physiological role of the metabolite control of pyruvate kinase by alanine. The Hill graphs of these data demonstrate a homotropic interaction of both inhibitors. The kinetic similarities between alanine and ATP in the type of inhibition could be further strengthened in an investigation of the action of both metabolites in the presence of fructose-1, 6-diphosphate. As is demonstrated in Fig. 3, the concentrations necessary for half maximal inhibition of liver pyruvate kinase by alanine and ATP are raised by the presence of the positive effector fructose-1, 6-diphosphate. However, from these data it is also evident that in the presence of 3.9 uM FDP alanine is by far less inhibitory than ATP.

The shape of the PEP saturation curve depends considerably on the pH of the reaction medium (Fig. 4.) The change of pH of the reaction medium from pH 7.95 to pH 6.25 transforms the PEP saturation curve from a sigmoidal shape (pH 7.95) to a hyperbola of the Michaelis-Menten type (pH 6.25). The increase of the Hill coefficients indicates a heterotropic action of H^+ on the PEP saturation curve. In a comparative investigation of the effect of hydrogen ion concentration on the metabolite control of hepatic pyruvate kinase (Fig. 5) the kinetic similarities in the action of alanine and ATP could be confirmed. Also in this system alanine proved to be less inhibitory than ATP.

On the base of these kinetic studies we assume in accordance with Rozengurt et al. (1969) that hepatic pyruvate kinase exists in two confirmational states. The influence of metabolites on the transition between the two different states may be described as:

$$\text{pyruvate kinase} \underset{\text{(R)} \qquad \text{FDP, } H^-}{\overset{\text{ATP, alanine, } OH^-}{\rightleftharpoons}} \text{-pyruvate kinase} \quad \text{(T)}$$

Because of the low effectivity of alanine against the activation of the enzyme by FDP (Fig. 3), alanine may act in the presence of ATP only under physiological conditions. The data available from cortisol treated rats support a metabolite control of hepatic pyruvate kinase by alanine under this hormonal state. However, more data seem to be necessary to confirm a physiological role of alanine in controlling gluconeogenesis by inhibition of pyruvate kinase in liver.

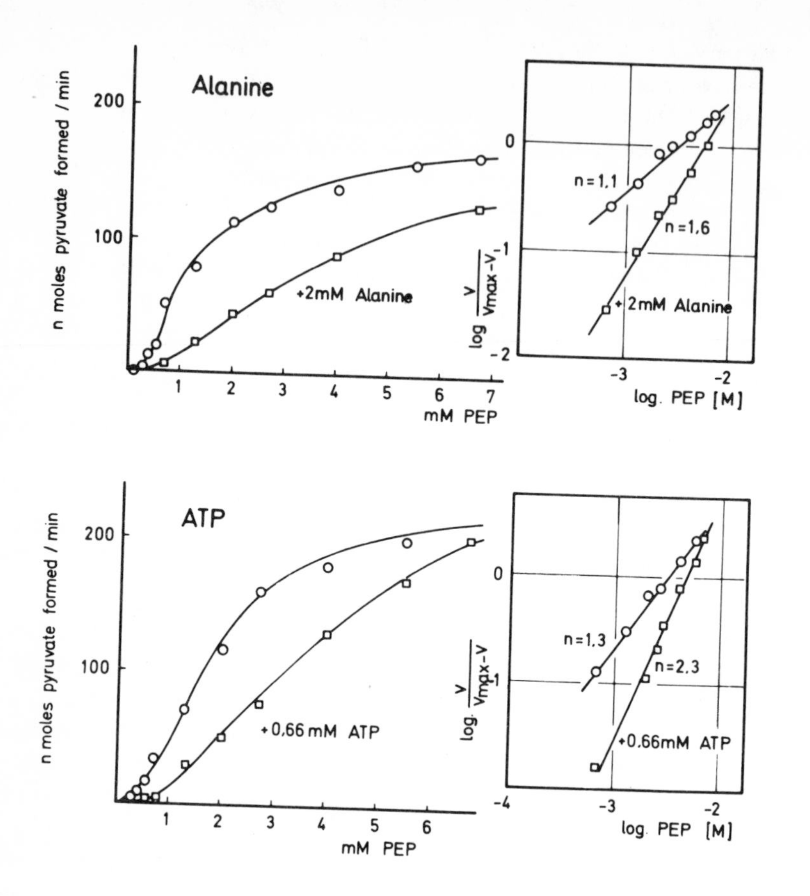

Fig. 1. Variation of the activity of liver pyruvate kinase with PEP concentration in presence and absence of alanine and ATP. The reaction mixture contained 66 mM imidazole pH 7.25; 66 mM KCl, 6.6 mM $MgSO_4$, 0.33 mM ADP, 0.33 mM NADH, concentrations of PEP, alanine and ATP as indicated. Four units lactate dehydrogenase and 30 ug pyruvate kinase (specific activity 8.0 U/mg). Temperature 30^0C.

References

Hornbrook, K. R., H. B. Burch, O. H. Lowry: Molec. Pharmacol. 2 (1966) 106

Rozengurt, E., L. J. de Asua, H. Carminatty: J. biol. Chem. 244 (1969) 3142

Schoner, W., W. Prinz, U. Haag, W. Seubert: Hoppe-Seylers Z. physiol. Chem. in press

Seubert, W., H. V. Henning, W. Schoner, M. L'Age: Advances in Enzyme Regulation, (Editor G. Weber), Vol. 6, p. 153, Pergamon Press, Oxford 1968

Tarnowski, W., M. Seemann: Hoppe-Seylers Z. physiol. Chem. 348 (1967) 829

Weber, G., M. A. Lea, G. A. Fischer, N. B. Stamm: Enzymol. biol. Chem. 7 (1966) 11

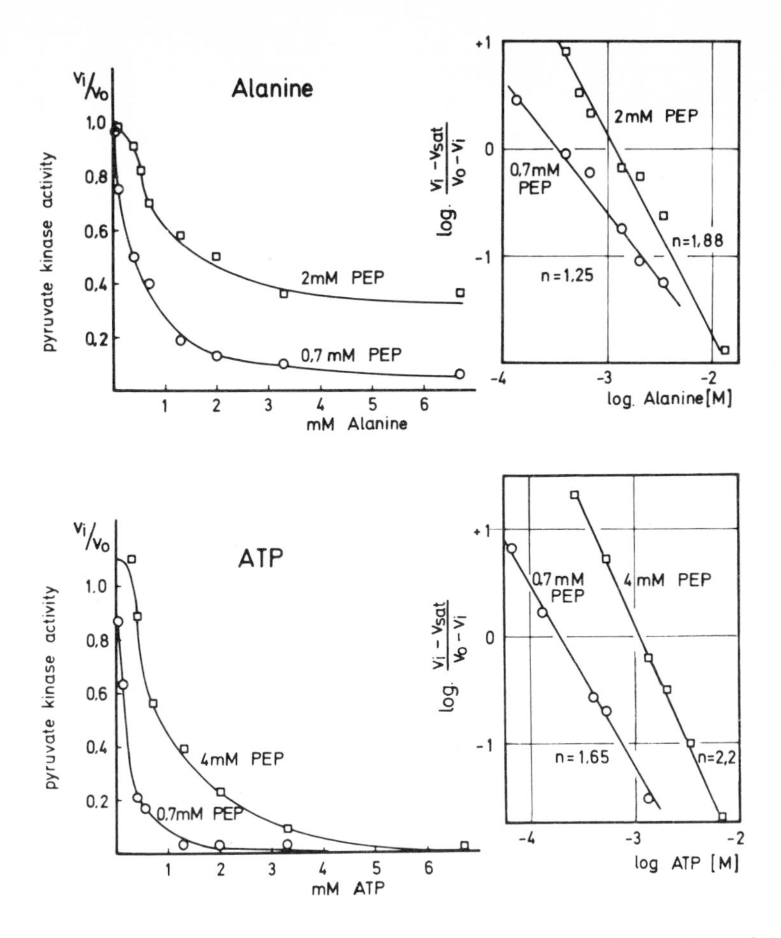

Fig. 2. Effect of the concentrations of alanine or ATP on the activity of liver pyruvate kinase at different levels of PEP. For conditions see Fig. 1.

Krebs: Dr. Hasselblatt, have you measured the concentration of free fatty acids in the liver after treatment with dimethylisoxazole? Does it fall, and if so, why are fatty acids not set free from triglyceride stores?

Hasselblatt: Liver triglycerides do decline following dimethylisoxazole. This reduction amounts to two thirds of the initial concentration and is achieved within three to six hours. The oxygen consumption of livers taken from treated rats is higher than that of control livers and exceeds the amount of oxygen necessary for the production of additional glucose and urea. We did not investigate where the substrate for the higher oxygen consumption comes from.

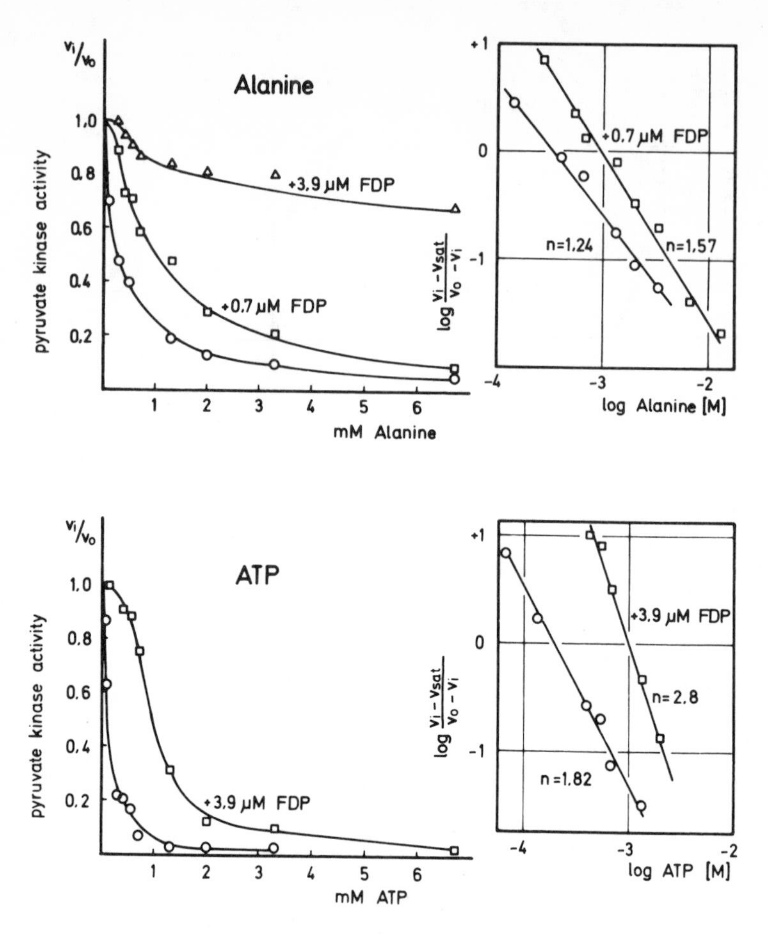

Fig. 3. Action of fructose-1, 6-diphosphate on alanine-and ATP-dependent inhibition of liver pyruvate kinase. The amount of 0.7 mM PEP, fructose-1, 6-diphosphate as indicated. For other conditions, see Fig. 1.

Krebs: You answered my question partly by saying that the liver loses triglycerides on pretreatment. What about the concentration of free fatty acids?

Hasselblatt: I did not measure free fatty acids in liver tissue. As free fatty acids have a very high turnover rate it is unlikely that free fatty acids in the liver constitute a pool of metabolic energy. Their concentration is approximately as high as in the blood.

Utter: May I return to the pyruvate and alanine effects on gluconeogenesis for a moment for another reason? As far as I recall the rats with pyruvate is higher than with alanine either with or without the analogue. So you might say that the pyruvate

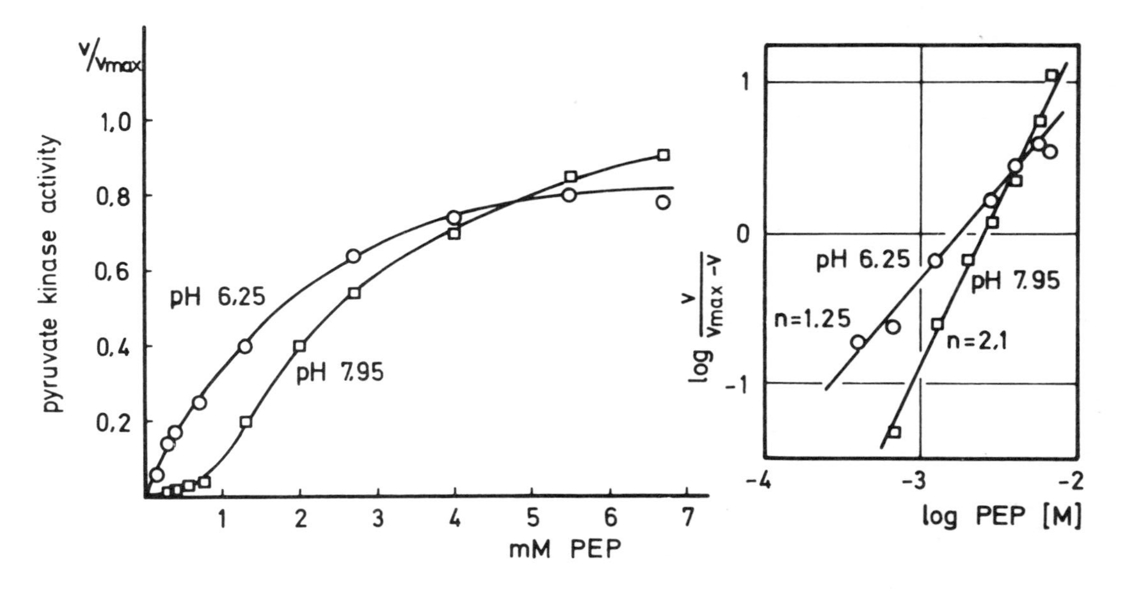

Fig. 4. Variation of the activity of liver pyruvate kinase with PEP-concentration at different pH-values. The amount of 66 mM imidazole buffer pH 6.25 or 7.95 was used. For other conditions, see Fig. 1.

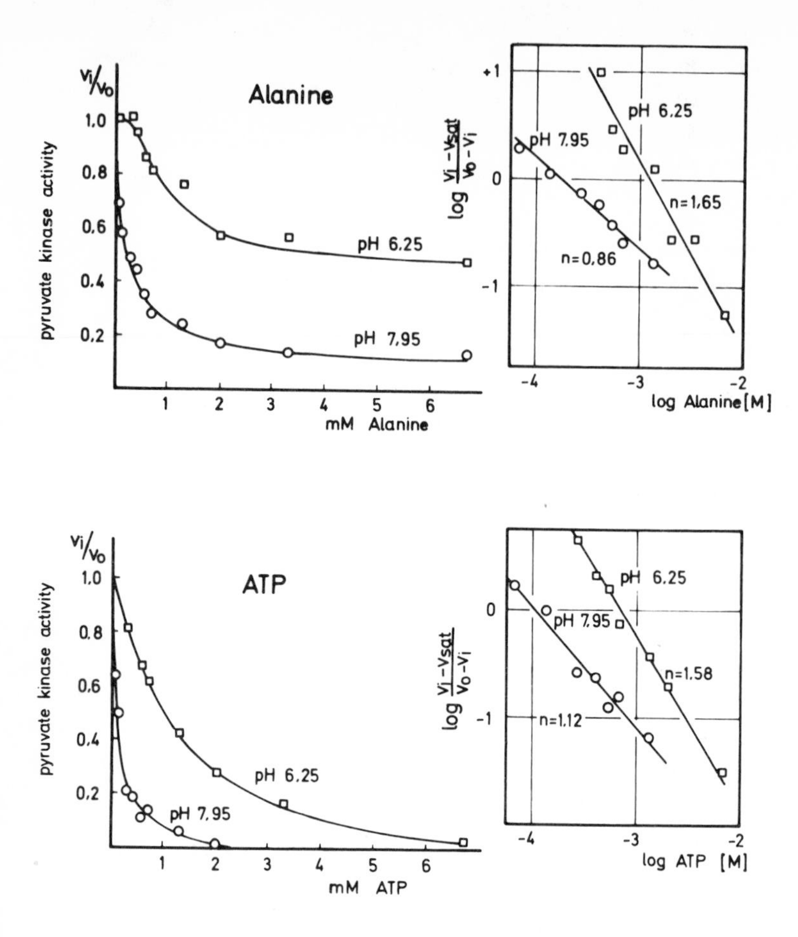

Fig. 5. Effect of the concentrations of alanine and ATP on the activity of liver pyruvate kinase at different pH-values. For other conditions, see Fig. 1 and 4.

was not saturating and this could be another possible interpretation of the differences. May I ask: What happened to the urea production in the presence of pyruvate? Did you measure it? Is there any shutting down of this particular process in the presence of the analogue?

Hasselblatt: We have not yet measured the effect of pyruvate on urea production of the perfused liver.

Utter: I am interested in what the factor is which turns on the process. Is this pyruvate or something derived from it? Alanine is not a very good choice because it is only a source for urea. I wonder whether pyruvate does something to the urea production.

Hasselblatt: We did not measure urea production in the presence of lactate and pyruvate. There is some evidence that urea formation is suppressed in the presence of pyruvate in liver slices (Metz, R., et al.: Metabolism 17 (1968) 158) and in the perfused liver, when lactate is added to the perfusion (Hems, R., et al.: Biochem. J. 101 (1966) 284.

Transfer of Carbon and Hydrogen Across the Mitochondrial Membrane in the Control of Gluconeogenesis

J. H. Anderson, W. J. Nicklas, B. Blank, C. Refino and J. R. Williamson
Johnson Research Foundation, University of Pennsylvania, Philadelphia, Pennsylvania, U.S.A.

Presented by J. R. Williamson

Summary

According to current concepts, the membrane of rat liver mitochondria is relatively impermeable to oxalacetate, and carbon or hydrogen equivalents for gluconeogenesis are thought to be transported out of the mitochondria as malate and aspartate. Present work is concerned with 1) attempts to elucidate under which metabolic conditions malate or aspartate serve as major carriers of carbon flux for gluconeogenesis in perfused rat liver; 2) the nature of the factors controlling the anion transport systems; and 3) the influence of anion concentration gradients across the mitochondrial membrane on the control of gluconeogenesis. Three different experimental approaches have been employed; 1) addition of specific inhibitors of transport or enzyme systems to the perfused liver; 2) tracer studies with (1-^{14}C) lactate, (1-^{14}C) pyruvate, and (2-^{14}C) acetate; and 3) calculations of the intracellular distribution of metabolites from assumptions of near equilibrium of glutamate-pyruvate and glutamate-oxalacetate transaminases and lactate, malate and ß-hydroxybutyrate dehydrogenases. Results of the latter calculations predict a predominantly cytosolic distribution for oxalacetate and α-ketoglutarate and a predominately mitochondrial distribution for aspartate and glutamate. Malate distributions depended on the nature of the gluconeogenic precursor. Malate concentration gradients from mitochondria to cytosol were predicted to be low with lactate but high with pyruvate as substrate. A high malate concentration gradient under the latter conditions is consistent with it being the carrier of both carbon and hydrogen atoms for gluconeogenesis. Addition of butylmalonate, which inhibits a malate:phosphate exchange in isolated mitochondria, resulted in a diminished rate of glucose production from pyruvate, an increased malate concentration gradient from mitochondria to cytosol, an oxidation of the cytosolic NAD-system and an inhibitory interaction in the gluconeogenic sequence at glyceraldehyde-P-dehydrogenase. Gluconeogenesis from lactate and lactate uptake were almost completely inhibited by 0.2 mM amino-oxyacetic acid (an inhibitor of glutamate oxaloacetate transaminase), while gluconeogenesis from pyruvate, and pyruvate uptake were not affected. Since amino-oxyacetic acid does not inhibit lactate dehydrogenase, these results establish that aspartate efflux from mitochondria is associated with gluconeogenesis from lactate while with pyruvate as substrate a transaminase step is not obligatory for gluconeogenesis.

Steady state isotope tracer studies revealed equilibration between intermediates within ten minutes. However, isotope equilibration was not achieved between the mitochondrial and cytosolic pools of malate and aspartate. Mitochondrial oxaloacetate did not equilibrate equally between mitochondrial malate and mitochondrial aspartate. The present work supports the concept that exchange of anions across the mitochondrial membrane is an integral feature of gluconeogenesis in rat liver, and that malate transport can be a rate-controlling step in the overall sequence.

Introduction

Gluconeogenesis in the rat liver requires the transport of various anions across the mitochondrial membrane. The dependency of gluconeogenesis on anion transport is determined both by the cellular location of pyruvate carboxylase and P-enolpyruvate carboxykinase and by the nature of the metabolic fuels available for gluconeogenesis. In rat liver, pyruvate carboxylase is found mainly, if not exclusively, in the mitochondria (Bottger et al. 1969) while other enzymes of the gluconeogenic sequenc are located in the soluble fraction of the cell. Since NADH and oxalacetate at physiological concentrations do not readily traverse the mitochondrial membrane (Haslam and Griffiths 1968, Chappell 1968), it has been generally accepted that the carbon and hydrogen requirements for gluconeogenesis are met by the transport of malate and aspartate out of the mitochondria (Walter et al. 1966). When the carbon source is at the same reduction state as glucose (e. g., lactate) or when reducing equivalents are generated in the cytosol by alternative dehydrogenase reactions (e. g., alcohol and xylitol dehydrogenases), the stoichiometries for carbon and hydrogen balance suggest that aspartate transport from the mitochondria should predominate over malate transport. When the gluconeogenic precursor is more oxidized than glucose (e. g., pyruvate), malate transport provides both the carbon and hydrogen requirements for glucose production. Metabolism of glucogenic amino acids, on the other hand, involves the mitochondrial export of both malate and aspartate. The central feature of these postulates is that metabolism of mitochondrial oxalacetate to malate or aspartate is dependent on the availability of reducing equivalents in the cytosol. However, the nature of the factors controlling the intramitochondrial metabolism of oxalacetate have not been clearly established. Furthermore, it is apparent that this type of control will only be operative in those organs like rat liver which have a predominantly mitochondrial distribution of pyruvate carboxylase and a predominatly cytosolic distribution of P-enolpyruvate carboxykinase. In guinea pig liver, P-enolpyruvate is produced by the mitochondria (Nordlie and Lardy 1963) so that aspartate or malate transport need not be involved in the supply of carbon for gluconeogenesis. However, transport of reducing equivalents from mitochondria will still be required when substrates more oxidized than glucose serve as precursors; transport of malate and glutamate out of and aspartate and α-ketoglutarate into the mitochondria must occur.

In this report we will explore the possibilities that compartmentation of metabolites may determine the pathway of oxalacetate metabolism in rat liver and that rates of anion transport may control the overall rate of gluconeogenesis. Three approaches to these problems have been used: 1) specific inhibitors of several of the steps in the proposed pathways have been tested as inhibitors of gluconeogenesis. A direct test of the validity of the concept that anion transport across the mitochondrial membrane is a possible control step is provided by the use of 2-n-butylmalonate, an inhibitor of the malate:phosphate exchange reaction (Robinson and Chappell 1967). The possibility of involvement of aspartate transaminases in gluconeogenesis has been examined through the use of amino-oxyacetic acid, an inhibitor of pyridoxal phosphate-linked enzymes (Roberts et al. 1964). 2) Calculations have been made on steady state concentrations of metabolites in the mitochondrial and cytosolic spaces from measurements of total tissue contents of intermediates and assumptions of near-equilibrium of dehydrogenase and transaminase enzymes. 3) Measurements have been made of the specific activities of selected intermediates of the citric acid cycle and the gluconeogenic pathway after perfusion of livers with specifically labeled ^{14}C-isotopes of lactate, pyruvate, acetate and bicarbonate. These three approaches, while not entirely agreeing quantitatively, lead to the conclusion that concentration gradients between the mitochondrial and cytosolic spaces are established for the metabolites of the malate-aspartate shuttle and other intermediates of the citric acid cycle and that changes in the intracellular distribution of these metabolites can effectively control several metabolic pathways.

Effects of butylmalonate in perfused rat liver. Gluconeogenesis from both pyruvate and lactate is inhibited by 5 mM butylmalonate (Fig. 1). Inhibition of gluconeogenesis by butylmalonate with pyruvate as the substrate is consistent with the hypothesis that malate is the carrier for both hydrogens and carbons across the mitochondrial membrane during conversion of pyruvate to glucose. With lactate as substrate, however, it at first appears paradoxical that n-butylmalonate should inhibit gluconeogenesis because with this substrate aspartate is hypothesized to be the carrier of carbon between the mitochondrial space and the cytosol. However, the cytosolic conversion of aspartate to oxalacetate via the glutamate-oxalacetate transaminase consumes α-ketoglutarate and generates glutamate while the reverse reactions occur in the mitochondrial space. To maintain metabolic balance, a movement of α-ketoglutarate from the mitochondria to the cytosol equal to the aspartate transport must occur, as well as a return of glutamate to the mitochondria. Movement of α-ketoglutarate across the mitochondrial membrane occurs, however, only in the presence of malate and studies with isolated mitochondria have indicated that this movement occurs via a malate: α-ketoglutarate antiport which functions with a 1:1 stoichiometry (Papa et al. 1969). Thus, for each µmole of aspartate utilized in the cytosol, one µmole of malate must be returned to the mitochondria. To the extent that the malate entering the mitochondria is derived from oxalacetate in the cytosol, there is no net production of 4-carbon moieties in the cytosol and gluconeogenesis cannot proceed. Gluconeogenesis can occur only if the malate exchanging with α-ketoglutarate is derived from a malate:phosphate exchange in the opposite direction. When the

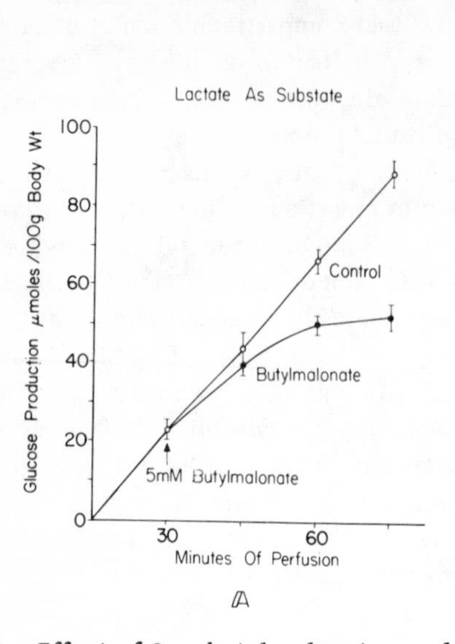

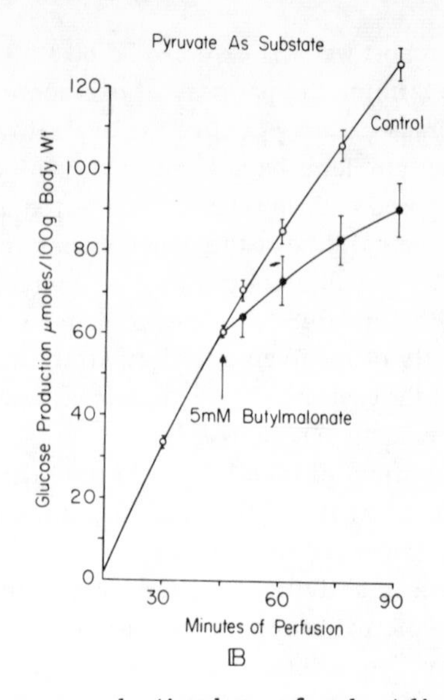

Fig. 1. Effect of 2-n-butylmalonate on glucose production by perfused rat liver. A) Livers from rats starved for 24 hours were perfused for 15 minutes with no substrate, followed by 15 minutes with 10 mM lactate, the substrate concentration being maintained by infusion. After 30 minutes of perfusion, 5 mM 2-n-butylmalonate was added to the perfusate. Small samples of perfusate were removed at 15 minute intervals and assayed as described by Williamson et al. (1970), Williamson et al. (1969b). B) Similar to A but using 2 mM pyruvate as substrate with 5 mM butylmalonate added after 45 minutes of perfusion.

malate:phosphate antiport is inhibited by butylmalonate, the malate-aspartate shuttle can perform net transport of hydrogens but not of carbons; thus it becomes a futile cycle with respect to gluconeogenesis.

The effects of butylmalonate on intracellular metabolites in livers perfused with pyruvate have been described in detail by Williamson et al. (1970) and will be summarized briefly here. An inhibition of malate transport from the mitochondria would be expected to increase the mitochondrial malate concentration relative to that in the cytosol and produce an increase of the mitochondrial pyridine nucleotide reduction state while lowering that in the cytosol. Table 1 shows that such an effect occurred. The ratio of lactate to pyruvate decreased after butylmalonate addition while the ratio of ß-hydroxybutyrate to acetoacetate increased. This differential effect of butylmalonate on the cytosolic and mitochondrial redox couples is the most direct effect of inhibition of the malate:phosphate exchange reaction. Additional effects of butylmalonate were revealed by crossover analysis.

Analysis of sites of interaction in the gluconeogenic pathway by crossover plots demonstrates three forward crossovers: 1) between pyruvate and oxalacetate, 2) between 3-P-glycerate (3PGA) and dihydroxyacetone-P (DAP), 3) between fructose-1, 6-di-P

Perfusion conditions		Ratio of lactate	Ratio of
Oleate	Butylmalonate	to pyruvate	ß-hydroxybutyrate to acetoacetate
mM			
0	0	4.9 ± 0.4	.31 ± 0.03
0	5	4.0 ± 0.1	.40 ± 0.03
difference		-0.9+	+.09+
0.52	0	7.7 ± 0.7	.73 ± 0.11
0.51	5	6.4 ± 0.4	.90 ± 0.08
difference		-1.4	+0.17
1.57	0	11.9 ± 1.1	0.72 ± 0.04
1.32	5	8.6 ± 1.1	1.07 ± 0.04
difference		-3.3+	+0.35+

Table 1. Effect of Butylmalonate on Oxidation-Reduction Couples in Perfused Rat Liver. Livers from rats starved 24-30 hours were perfused with 110 ml of Krebs-Henseleit bicarbonate medium containing 4 g per 100 ml of defatted bovine serum albumin and 2 mM pyruvate. Pyruvate was added by infusion to maintain the concentration between 1.5 and 2.0 mM. Oleate was added after 15 minutes and 2-n-butylmalonate (5mM) after 45 minutes of perfusion. The livers were frozen at 90 minutes with aluminum tongs cooled in liquid N_2. Values shown are means ± s.e.m. with six or eight animals in each group. +Determined by t test for statistical significance, $p < 0.05$.

(FDP) and fructose-6-P (F6P) (Fig. 2). Which crossover appears in a given experiment depends on the rate of fatty acid oxidation. Presumably, the same interactions occur in all experiments but their relative strengths differ. These experiments thus demonstrate that an interaction does not always lead to a crossover. The crossover appearing between 3PGA and DAP is probably attributable to a simple redox potential change subsequent to the inhibition of malate:phosphate exchange and the more oxidized state of the cytosolic pyridine nucleotides.

The crossover between fructose di- and mono-P could be caused either by a decreased activity of fructose diphosphatase or an increased activity of phosphofructokinase. Of these two possibilities we favor the latter and suggest that phosphofructokinase activity is increased as a result of a fall in the concentration of citrate (an inhibitor) in the cytosol. A redistribution of citrate would be an expected consequence of the

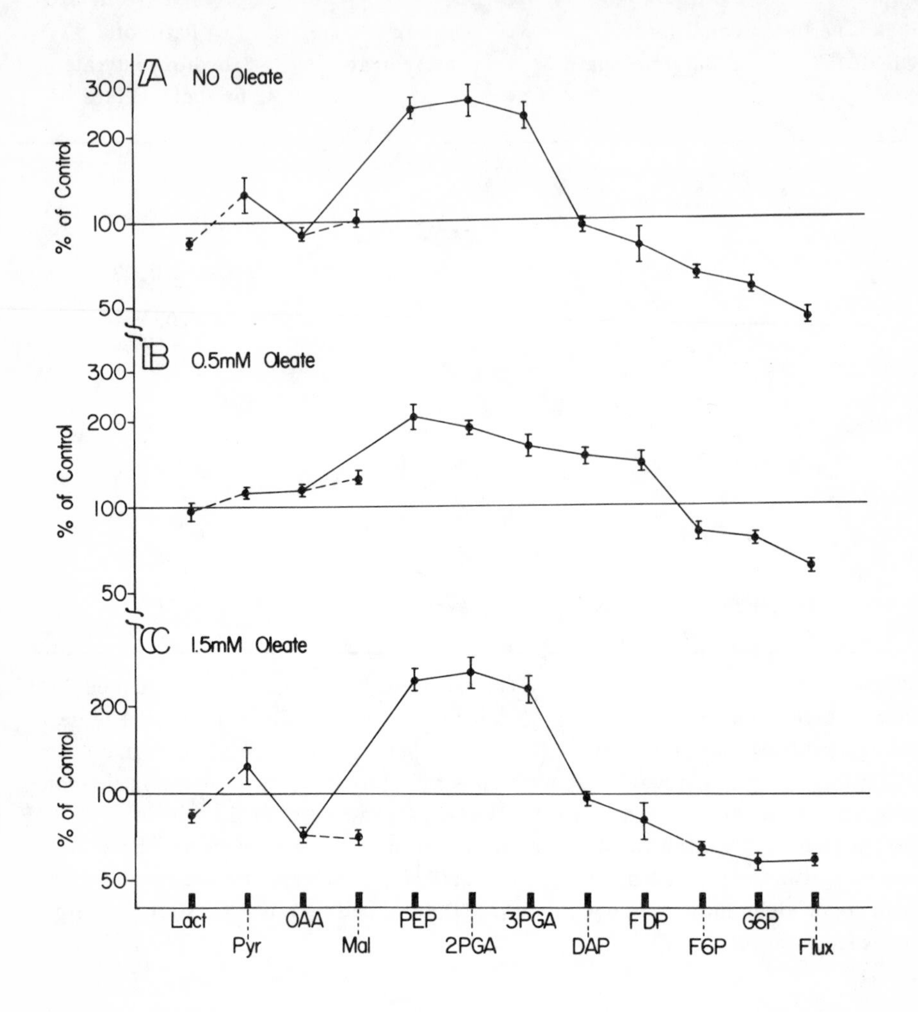

Fig. 2. Crossover plots showing the effect of butylmalonate on intermediates of the gluconeogenic sequence. Livers from rats starved for 24 hours were perfused with 2 mM pyruvate and the indicated concentration of oleate, the latter present as an albumin-oleate emulsion (Williamson et al. 1969b). After 45 minutes of perfusion, 5 mM 2-n-butylmalonate was added to the perfusate. The livers were frozen after 90 minutes total perfusion and assayed as described by Williamson et al. (1970), Williamson et al. (1969b). The broken lines (- - -) join intermediates which are partners of substrate oxidation-reduction couples. Abbreviations: Lact. (lactate), Pyr. (pyruvate), OAA (oxalacetate), Mal. (malate), PEP (P-enolpyruvate), 2PGA (2-P-glycerate), 3PGA (3-P-glycerate), DAP (dihydroxyacetone-P), FDP (fructose-1, 6-di-P), F6P (fructose-6-P), G6P (glucose-6-P).

increased malate concentration gradient since the malate:citrate exchange is not inhibited by butylmalonate (Meijer and Tager 1969). Some calculations of the magnitude of this redistribution of citrate but based on independent assumptions will be presented later in this report.

The observed oxidation of pyridine nucleotides in the cytosol cannot be responsible for a change in glucose production rate without some feedback control affecting the net rate of formation of P-enolpyruvate (PEP); otherwise, large quantities of carbon intermediates would have accumulated between PEP and 1, 3-di-P-glycerate. Likewise, the interaction between FDP and F6P is unlikely to result in a net change in steady state flux without feedback to steps influencing the net formation of PEP. Otherwise, all the intermediates prior to the hexose monophosphates would have accumulated in large quantities. Similarly, inhibition of malate transport from the mitochondria would tend to cause large increases of citric acid cycle intermediates, particularly malate, if this were the sole effect of butylmalonate. The metabolite data demonstrate an inhibitory interaction between pyruvate and PEP, which, in light of the above considerations, we interpret as reflecting two processes: 1) direct inhibition of pyruvate carboxylase, 2) an increase of pyruvate kinase activity. The direct inhibitory effect on pyruvate carboxylase might be produced by butylmalonyl-CoA, which could be formed by non-specific acyl-CoA ligases. Direct proof of such inhibition is lacking at present, but both malonyl-CoA and methylmalonyl-CoA have been shown to be effective inhibitors of pyruvate carboxylase (Utter and Scrutton 1969). However, if such a direct inhibition were the sole effect of butylmalonate, the oxygen consumption of the liver would be expected to decrease in relation to the decreased energy demands of gluconeogenesis. Such is not the case (Williamson et al. 1970); thus another explanation must be sought to account for the fact that butylmalonate caused the energy cost of gluconeogenesis to increase. Increased activity of pyruvate kinase and phosphofructokinase resulting in an increased rate of recycling of carbon between pyruvate and PEP (via pyruvate carboxylase, PEP carboxykinase and pyruvate kinase) and between FDP and F6P (via fructose-1, 6-di-phosphatase and phosphofructokinase) with concurrent ATP expenditure not linked to net glucose formation can account for this observation. As discussed above, increased activity of phosphofructokinase is thought to be mediated by the fall of citrate in the cytosol. Hepatic type L pyruvate kinase is strongly inhibited by ATP and alanine, and this inhibition can be reversed by FDP at physiological concentrations (Tanaka et al. 1967, Schoner 1970, Safer and Williamson, unpublished observations). Furthermore, the tissue PEP concentrations are within the *Km* region for the enzyme. Consequently, the necessary feedback control linking interactions at glyceraldehyde-3-P dehydrogenase or phosphofructokinase to the input flux of carbon can be mediated by elevated levels of FDP and PEP. The data thus reflect the combined effects of two sites of inhibition by butylmalonate administration, decreased malate:phosphate exchange across mitochondrial membrane and decreased net PEP formation. The differential effects of butylmalonate on the cytosolic and mitochondrial redox potential, and the intracellular distribution of malate, α-ketoglutarate and citrate (calculations to be presented below) demonstrate the inhibition of malate:phosphate exchange, while the lack of accumulation of citric acid cycle intermediates accompanied by the absence of a marked change of oxygen consumption indicates a net inhibition of PEP formation.

Calculation of metabolite distribution in rat liver. To estimate the intracellular distribution of metabolites between the cytosolic space and the mitochondrial matrix space it is assumed that if certain reversible enzymes are very active in the cell, they will probably maintain their substrate-product ratios close to the equilibrium ratio. The rat liver contains high activities of cytosolic and mitochondrial malate dehydrogenases (Berkes-Tomasevic and Holzer 1967) and glutamate-oxalacetate transaminases (Boyd 1961), and of cytosolic glutamate-pyruvate transaminase (Segal et al. 1962). None of these enzymes is known to be subject to specific metabolic control. These five enzymes form a network which should catalyze a near-equilibrium relationship between diphosphopyridine nucleotides, malate, oxalacetate, aspartate, α-ketoglutarate, glutamate, pyruvate and alanine in the cytosol and between diphosphopyridine nucleotides, malate, oxalacetate, aspartate, α-ketoglutarate and glutamate in the mitochondrial space. If the total liver content of each of the dicarboxylic acids is known, as well as the apparent cytosolic and mitochondrial pyridine nucleotide redox potentials and the cytosolic pyruvate and alanine concentrations, there remains only five unknowns; the distribution of each dicarboxylic acid between the cytosolic and mitochondrial matrix spaces. Since there are five equilibrium equations available and five unknowns, the required distributions can be solved algebraically. The algebraic equations required for this analysis are presented in Fig. 3.

The lactate-pyruvate ratio and the ß-hydroxybutyrate-acetonacetate ratios were used to estimate the apparent cytosolic and mitochondrial pyridine nucleotide redox potentials, respectively. The equilibrium constants for the several enzymic reactions involved were obtained from (Krebs and Veech 1969). The full experimental data have been published elsewhere (Williamson et al. 1970). A critical evaluation of the calculation technique will be published separately (Anderson and Garfinkel 1970).

The results of application of this analytical technique to data obtained from livers of starved rats perfused with pyruvate in the presence and absence of oleate and butylmalonate are shown in Table 2. This table lists the fraction of each dicarboxylic acid which would have to be present in the cytosol if the five cytosolic and mitochondrial enzymes were close to equilibrium. Since about 90% of the intracellular water is cytosolic (Williamson et al. 1969a), a fraction less than about 0.9 indicates that the mitochondrial concentration of that metabolite is higher than the cytosolic concentration. The analysis predicts that malate, aspartate and glutamate are at higher concentrations in the mitochondrial matrix, while oxalacetate and α-ketoglutarate are more concentrated in the cytosolic space.

The available experimental data did not include information concerning the cytosolic alanine concentration. To explore the effects of the alanine concentration on the results of these calculations the algebraic equations were coded into a simple Fortran IV digital computer program and the calculations were performed with several assumed

Perfusion Conditions		Fraction of Total Metabolites in Cytoplasmic Space				
Oleate	Butylmalonate	Malate	Oxalacetate	Aspartate	α-Ketoglutarate	Glutamate
mM	mM					
0	0	.532	.968	.124	.993	.41
0	5	.379	.964	.089	.991	.29
.52	5	.209	.917	.044	.986	.22
.51	5	.150	.920	.033	.975	.10
1.57	0	.219	.883	.070	.976	.30
1.32	5	.170	.920	.035	.982	.15

Table 2. Calculated Intracellular Distribution of Dicarboxylic Acids in Perfused Rat Liver

$$Keq_{MDH} = \frac{[OAA]_{cyto}}{[MAL]_{cyto}} \times \left(\frac{NADH}{NAD}\right)_{cyto}$$

$$Keq_{MDH} = \frac{[OAA]_{mito}}{[MAL]_{mito}} \times \left(\frac{NADH}{NAD}\right)_{mito}$$

$$Keq_{GOT} = \frac{[ASP]_{cyto} \times [\alpha KG]_{cyto}}{[OAA]_{cyto} \times [GLUT]_{cyto}}$$

$$Keq_{GOT} = \frac{[ASP]_{mito} \times [\alpha KG]_{mito}}{[OAA]_{mito} \times [GLUT]_{mito}}$$

$$Keq_{GPT} = \frac{[\alpha KG]_{cyto} \times [ALA]_{cyto}}{[GLUT]_{cyto} \times [PYR]_{cyto}}$$

Fig. 3. Equilibrium relationships used to calculate intracellular distribution of dicarboxylic acids. Abbreviations: Keq (equilibrium constant), MDH (malate dehydrogenase), GOT (glutamate-oxalacetate transaminase), GPT (glutamate-pyruvate transaminase), OAA (oxalacetate), MAL (malate), ASP (aspartate), KG (α-ketoglutarate), GLUT (glutamate), ALA (alanine), PYR (pyruvate), cyto. (cytosolic), mito. (mitochondrial).

alanine concentrations between 0.25 and 2.5 mM (the rat liver in vivo contains 0.46 mM alanine (Williamson et al. 1967), while the perfused rat liver contains about 1.7 mM alanine (Mallett et al. 1969). Although the malate and oxalacetate distributions were independent of the alanine concentration, the fraction of aspartate and glutamate in the cytosol were almost linearly proportional to the cytosolic alanine concentration. The fraction of α-ketoglutarate in the cytosol was nearly independent of the alanine concentration. Consequently to minimize the concentration gradients between the mitochondrial matrix and the cytosol for aspartate and glutamate, a value of 2.5 mM cytosolic alanine was selected. The data in Table 2 show the results of these calculations using this value (2.5 mM) for cytosolic alanine. Tables 3 and 4 show the

Perfusion Conditions		Cytoplasmic Concentration of Metabolites (mM)					
Oleate	Butylmalonate	Malate	Oxalacetate	Aspartate	α-Ketoglutarate	Glutamate	Citrate
0	0	.16	.0080	.10	1.5	2.7	.023
0	5	.12	.0074	.067	1.3	1.7	.011
.52	0	.25	.081	.067	1.0	1.3	.017
.51	5	.24	.094	.074	.64	.74	.0087
1.57	0	.61	.013	.15	1.3	2.2	.070
1.32	5	.33	.094	.087	.67	.93	.016

Table 3. Calculated Cytoplasmic Concentration of Metabolites. The data of Table 2 have been converted to concentrations by the formula: concentration in cytosol = total liver content x fraction cytoplasmic ÷ volume of cytosol. The perfused rat liver contains 2.0 ml cytosolic water per g dry weight (Williamson 1969a).

Perfusion Conditions		Mitochondrial Concentration of Metabolites (mM)				
Oleate	Butylmalonate	Malate	Oxalacetate	Aspartate	α-Ketoglutarate	Glutamate
mM	mM					
0	0	1.5	.0027	6.9	.10	39.
0	5	2.0	.0028	6.8	.11[‡]	41.
.52	0	10.	.0073	15.	.14	45.
.51	5	13.	.0080	21.	.16[‡]	65.
1.57	0	22.	.0170	19.	.31	52.9
1.32	5	16.	.0086	24.	.13	53.5[‡]

Table 4. Calculated Mitochondrial Concentrations of Metabolites. The data of Table 2 have been converted to concentrations in the mitochondria by the formula: concentration in mitochondria = total liver content x (1.-fraction cytoplasmic) ÷ volume of mitochondria. The rat liver contains 0.2 ml mitochondrial matrix water per g dry weight (Williamson 1969a). ‡ Change sensitive to alanine concentration assumption (see text).

corresponding concentrations of metabolites in the cytosol and mitochondrial matrix assuming 2.0 ml cytosolic water and 0.2 ml matrix water per gram dry weight liver (Williamson 1969a). Except where noted, the relative changes produced by butylmalonate were independent of the assumed alanine concentration.

If butylmalonate acted in vivo as it does in vitro to inhibit the malate:phosphate exchange across the mitochondrial inner membrane, it would be expected to inhibit malate egress from the mitochondria and lower the fraction of malate in the cytosol; these calculations demonstrate this effect (Table 2). With the malate:phosphate exchange inhibited, movement of malate from the mitochondria to the cytosol to supply carbons and hydrogens for gluconeogenesis would have to occur via a limited exchange of malate for cytosolic α-ketoglutarate and citrate. These effects would tend to shift the α-ketoglutarate and citrate into the mitochondria and lower their concentrations in the cytosol. This expected fall in the cytosolic concentration of α-ketoglutarate is seen in each of the three conditions examined (Table 3). The cytosolic citrate concentration which would be in equilibrium with the cytosolic α-ketoglutarate concentration can be estimated by assuming that aconitase and NADP-linked isocitrate dehydrogenase catalyze near-equilibria between citrate, isocitrate and α-ketoglutarate in the cytosol. The NADPH/NADP ratio is calculated from the pyruvate/malate ratio and the *Keq* for malic enzyme (Krebs and Veech 1969, Williamson 1969a). These very approximate estimates of the cytosolic citrate concentration are shown in Table 3. Butylmalonate in each case lowers the predicted cytosolic citrate concentration, even though a crossover between fructose-1, 6-di-P and fructose-6-P was only observed in livers perfused with 0.5 mM oleate (Fig. 2). This effect of butylmalonate in lowering the cytosolic citrate concentration was independent of the total citrate content of the liver (Williamson et al. 1970); in fact, the total liver citrate paradoxically rose on addition of butylmalonate to livers perfused with 0.5 mM oleate. The calculated changes in the cytosolic citrate concentration resolve this paradoxical rise in total citrate in the presence of an apparent deinhibition of phosphofructokinase.

These calculations demonstrate that despite the variable effects of butylmalonate on the total tissue metabolite levels, quantitative analysis of the data leads consistently to three effects of butylmalonate: decreases in the cytoplasmic concentration of malate, α-ketoglutarate and citrate. These calculated changes in intracellular distribution of anions can each be explained on the basis of the predicted effects of butylmalonate on the malate:phosphate antiport and its interaction with other dicarboxylic acid transporting systems. They thus illustrate the kinds of effects these transport mechanisms can have in control of gluconeogenesis.

Effects of amino-oxyacetic acid in perfused rat liver. To examine whether the glutamate-oxalacetate transaminases are involved in the gluconeogenic process in the rat liver, amino-oxyacetic acid was introduced into the perfused rat liver system. Amino-oxyacetic acid (AOA) binds to the pyridoxal phosphate coenzyme of transaminases (Roberts et al. 1964) and alters amino acid metabolism in the brain (Berl et al. in press) as well as inhibiting gluconeogenesis in rat kidney cortex slices in vitro (Rognstad and Katz 1970). Livers from starved rats were perfused with 8 mM

L(+) lactate or 2 mM pyruvate and the substrate utilization, oxygen consumption and glucose production were measured according to Williamson et al. 1969b. After a 15 minute perfusion with no substrate to determine the basal metabolic rates, the substrate was introduced and then maintained at a constant concentration by infusion. Fifteen minutes later, AOA was introduced to the oxygenator over a three-minute period to produce a uniform concentration of 0.2 mM AOA. AOA produced no effect on the substrate utilization, oxygen consumption and glucose production in livers perfused with 2 mM pyruvate (Fig. 4). In marked contrast, however, in livers perfused with 8 mM lactate, 0.2 mM AOA produced about 90% inhibition of lactate utilization and glucose production, while returning the oxygen consumption rate back to the basal level observed in the absence of substrates.

These data clearly demonstrate inhibition by AOA of gluconeogenesis from lactate but not from pyruvate. Since AOA does not inhibit lactate dehydrogenase (Rognstad and Katz 1970), the data strongly support the hypothesis that the conversion of lactate to glucose by the rat liver involves transaminases, presumably the glutamate-oxalacetate transaminases, and that these enzymes are not involved in any major way in pyruvate metabolism. The data thus support the concept that the carbon pathway of gluconeogenesis in liver is altered by the level of reduction of the gluconeogenic precursor; a malate shuttle carrying both carbons and hydrogens out of the mitochondria during pyruvate metabolism, and an aspartate shuttle carrying carbons out of the mitochondria during lactate metabolism. Furthermore, the marked inhibition of lactate utilization accompanied by return of the oxygen consumption rate to the basal level indicates that the aspartate shuttle is involved in lactate metabolism per se in addition to its use as a glucose precursor. These data can be taken as strong evidence in favor of the malate aspartate cycle being essentially the only effective route for hydrogen transport into the mitochondria in the rat liver.

Isotopic exchange across the mitochondrial membrane in perfused livers. To investigate the ability of the anion pools to communicate across the mitochondrial membrane, radioisotope experiments were performed using one, two and three carbon moieties as isotopic precursors. Livers from starved rats were perfused in a recirculating system with 10 mM L(+) lactate or 2 mM pyruvate for 30 minutes to achieve steady state levels of metabolites in the livers. Subsequently, 5 to 20 uC of (^{14}C) HCO_3, (2-^{14}C) acetate, (1-^{14}C) lactate or (1-^{14}C) pyruvate were introduced into the perfusate. Five to 35 minutes later, the livers were frozen, extracted with perchloric acid and the metabolites isolated by ion exchange chromatography using techniques similar to those described by LaNoue et al. (1970). A summary of these data is shown in Table 5. The relative specific activities (malate = 1) of citric acid cycle and gluconeogenic intermediates became relatively constant within five minutes after introduction of the isotope, and data obtained from longer time intervals were pooled.

The degree of labeling of the several intermediates was markedly dependent on the number of carbons in the added tracer, but remarkably independent of whether lactate or pyruvate was employed as the substrate for gluconeogenesis. Since conversion

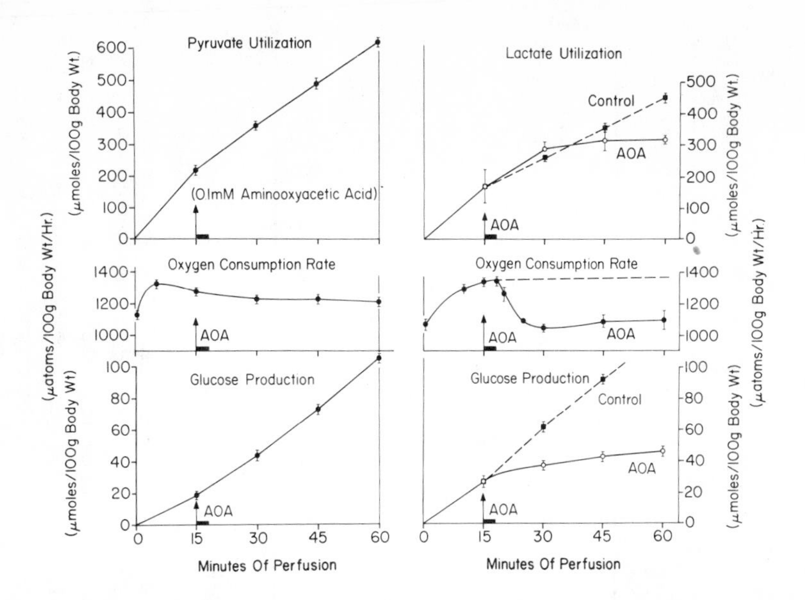

Fig. 4. Effects of amino-oxyacetic acid on perfused rat livers. Livers from rats starved 24-30 hours were perfused with 100 ml of Krebs-Henseleit bicarbonate medium containing 4 g per 100 ml of bovine serum albumin. After 15 minutes of perfusion in the absence of substrate, either 2 mM pyruvate or 8 mM lactate was added and maintained by infusion. Fifteen minutes later amino-oxyacetic acid (AOA) was added to bring the concentration of AOA to 0.2 mM. Substrate utilization and glucose production were determined enzymatically. Venous oxygen tension was measured by polarography using a modified Clark electrode (Williamson et al. 1969c). Left — pyruvate as substrate; right — lactate as substrate.

of lactate and pyruvate to glucose occurs via characteristically different routes as described above, the lack of dependence of the isotopic labeling pattern on the reduction state of the substrate suggests that the data reflect isotopic exchange more than net metabolic flux. The achievement of a relatively constant specific activity for most intermediates within five minutes supports this supposition.

The non-equilibration of malate and aspartate. Although the total liver malate and aspartate were in approximate isotopic equilibrium in livers perfused with (2-^{14}C) acetate, the total liver malate and aspartate did not achieve isotopic equilibrium with each other when one- or three-carbon moieties were employed as the isotopic precursor. Since it was found that less than 7% of the aspartate was extracellular, the low relative specific activity of aspartate cannot be attributed to an isotopically inert extracellular pool. The lack of isotopic equilibration between the total liver aspartate and malate suggests that there may be a limitation on the communication of these anions between their cytosolic and mitochondrial pools.

	Relative Specific Activity (Malate = 1.0)				
Substrate	Lactate			Pyruvate	
Isotopic precursor	$(^{14}C)HCO_3$	D,L(1-^{14}C)Lactate	(2-^{14}C)Acetate	(1-^{14}C)Pyruvate	(2-^{14}C)Acetate
Glucose	0.37	0.76	0.55	0.51	0.45
G6P	0.38	1.34	1.96	0.91	0.75
PEP	0.39	0.58	0.66	0.44	0.75
Aspartate	0.35	0.59	1.01	0.42	0.98
Citrate	1.08	0.99	4.0	1.03	3.26
Glutamate	0.36	0.34	4.09	0.36	3.7

Table 5. Isotopic Labeling of Gluconeogenic and Citric Acid Cycle Intermediates in Perfused Rat Liver. Liver perfusion and isotope administration are described in the text. The intermediates listed were isolated from acid-extracts of the frozen livers. Glucose was obtained as the neutral fraction after passage of an aliquot through Dowex 1 and AG 50 resins. The other intermediates were isolated by chromatography on Dowex 1 formate and acetate resins as by LaNoue et al. (1970). PEP, eluted from Dowex 1 formate resin, was converted enzymatically to lactate and rechromatographed before assay. The G6P data were obtained by assaying the glucose in an acid hydrolysate of the hexose phosphate peak and assuming that the hexose phosphate is 75% glucose-6-phosphate and 25% fructose-6-phosphate due to near equilibrium of phosphohexoseisomerase. The data are expressed relative to the specific activity of malate. The data for bicarbonate and acetate as precursor represent the averages of two livers per group. The data for lactate and pyruvate as precursors represent the averages of 6-8 livers per group.

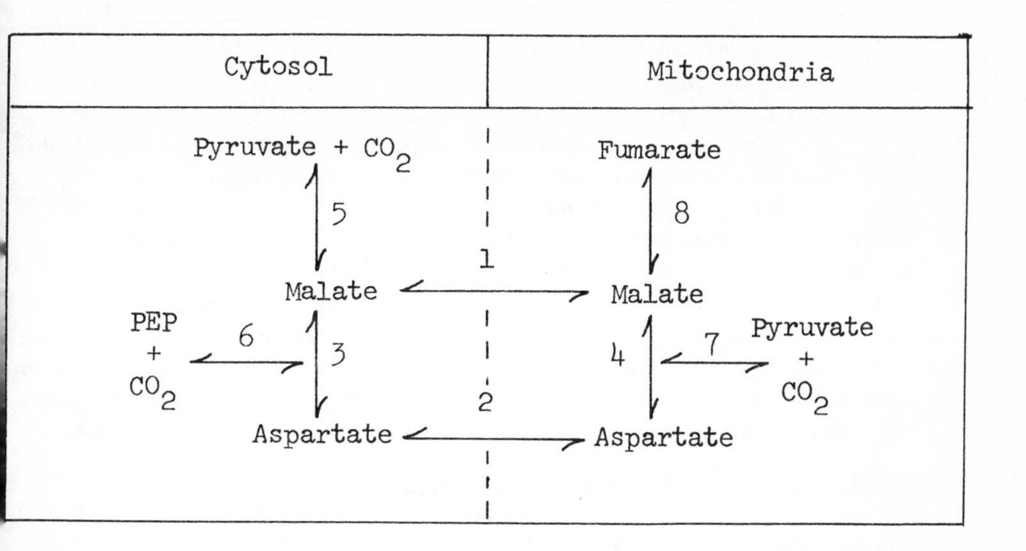

Fig. 5. Isotope exchange of malate and aspartate in rat liver. Reactions 1 and 2 represent exchanges of malate and aspartate across the mitochondrial membrane. Reactions 3 and 4 represent the cytosolic and mitochondrial malate dehydrogenases and glutamate-oxalacetate transaminases. Reaction 5 is the malic enzyme while reaction 6 is the phosphoenolpyruvate (PEP) carboxykinase. Reaction 7 represents the pyruvate carboxylase and reaction 8 the mitochondrial fumarase. The cytoplasmic fumarase and urea cycle are omitted for clarity of presentation.

The exchange reactions under consideration are shown in Fig. 5, assuming only two functional pools of malate and aspartate. Examination of this scheme reveals that the minimal necessary condition to explain the non-equilibration of total malate with total aspartate is that any two sides of the central loop of this scheme are not rapid. Even a slow isotope exchange rate for two or more of these reactions would eventually lead to isotopic equilibration unless there were also a rapid exchange of at least one member of the loop with material of a different relative specific activity, i.e., exchange or flow at pyruvate carboxylase, PEP carboxykinase, succinate dehydrogenase or malic enzyme. Thus, in addition to two slow exchange rates within the central loop there must be a fairly rapid exchange or flow into or out of the loop. Since the experiments were performed under conditions of gluconeogenesis with lactate and pyruvate as substrates, at least two of the input and output reactions were active.

A priori there is little reason to suspect any of the malate exchange rates to be slow. Malate can cross the mitochondrial inner membrane by several carriers (Chappell 1968). The malate dehydrogenases (MDH) appear to function catalytically by ordered addition of substrates and release of products with malate and oxalacetate being the central substrate pair; hence isotopic exchange between NAD and NADH might be slow in the presence of high malate concentrations but the malate-oxalacetate exchange rate should be rapid (Silverstein and Sulebele 1969).

Aspartate, however, can cross the mitochondrial inner membrane only in the presence of glutamate (Azzi et al. 1967). Although the rat liver contains relatively high quantities of glutamate, the equilibrium distributuion calculations described above predict that most of the glutamate is sequestered in the mitochondria (Table 2). If these calculations at all reflect the in vivo situation, such unequal concentrations of glutamate in the cytosolic and mitochondrial spaces could conceivably limit the ability of aspartate to cross the mitochondrial membrane and isotopically equilibrate with its other pool. The high concentrations of glutamate in the mitochondria could also saturate the mitochondrial glutamate-oxalacetate transaminase and limit the isotopic equilibration between mitochondrial aspartate and oxalacetate. Thus, significant sequestration of glutamate within the mitochondrial matrix in vivo could produce the required minimum of two slow exchanges in the central loop of Fig. 5 and account for the non-equilibration of total liver aspartate with malate.

Estimation of the relative specific activity of mitochondrial malate. With the one- and three-carbon precursors it is not possible to state where the isotope enter Fig. 5. Both pyruvate carboxylase and malic enzyme allow entry and exit of one- and three-carbon moieties to and from the central four-carbon loop. With acetate as the isotopic precursor, however, (^{14}C) effectively enters this scheme only through the citric acid cycle at mitochondrial fumarate (assuming that the $^{14}CO_2$ produced from (2-^{14}C) acetate is effectively diluted by the bicarbonate buffer employed in perfusion). Thus, in (2-^{14}C) acetate perfused livers, the most highly labeled compound in the central loop of Fig. 5 must be the mitochondrial malate.

To obtain an estimate of the relative specific activity of the mitochondrial oxalacetate, citrate isolated from livers perfused with (2-^{14}C) acetate was degraded by treatment with citrate lyase to obtain acetate and oxalacetate. The oxalacetate was converted to malate by use of malate dehydrogenase and NADH. Comparison of the citrate content of these livers with the rate of flux through the citric acid cycle (estimated from the oxygen consumption rate) indicates that the turnover time for the citrate pools would be about one minute. Thus, unless a large percentage of the citrate is metabolically inert, the citrate degradation data should reflect the recent history of the relative specific activity of the mitochondrial oxalacetate. The results indicated that in the livers perfused with lactate and (2-^{14}C) acetate the mitochondrial oxalacetate relative specific activity was 1.67, while in livers perfused with pyruvate and (2-^{14}C) acetate, the mitochondrial oxalacetate specific activity averaged 1.39 relative to total liver malate.

Since (^{14}C) from (2-^{14}C) acetate enters mitochondrial oxalacetate from mitochondrial malate and is diluted by an input of relatively low specific activity pyruvate and CO_2 via pyruvate carboxylase, the mitochondrial malate relative specific activity must be the same or higher than that of the mitochondrial oxalacetate and thus considerably higher than the relative specific activity of the total malate. It follows directly that the cytoplasmic malate must have a low relative specific activity in order that the total liver malate relative specific activity be 1.0. Thus, there must be a limitation on the isotopic equilibration between cytoplasmic malate and mitochondrial malate.

In the livers perfused with pyruvate and ($2-^{14}C$) acetate the cytoplasmic malate is derived from the mitochondrial malate. With a total malate content of about 0.6 umoles per g dry weight and a gluconeogenic flux of about 100 umoles per g dry weight per hour, the total malate should turnover completely in 0.36 minutes. Hence, it is quite unlikely that after ten minutes in the presence of isotope, the cytoplasmic malate would not have been at the same relative specific activity as the mitochondrial malate unless rapid isotopic dilution of cytoplasmic malate occurred by exchange with some pool of a relatively low specific activity.

From the scheme in Fig. 5, several possibilities for such an exchange are obvious: exchange with pyruvate and CO_2 via malic enzyme; exchange with PEP via MDH and PEP carboxykinase; and exchange with cytoplasmic aspartate via MDH and glutamate-oxalacetate transaminase. However, in livers perfused with lactate and ($2-^{14}C$) acetate the cytoplasmic aspartate is derived from the mitochondrial oxalacetate and should have a relative specific activity approaching 1.67. Hence, in these livers, exchange between cytoplasmic malate and aspartate would probably raise rather than lower the relative specific activity of cytoplasmic malate. It is most likely that in all the livers perfused with ($2-^{14}C$) acetate, the relative specific activity of cytoplasmic malate is lowered by exchange with pyruvate and CO_2 via malic enzyme.

Although we cannot firmly conclude which cytoplasmic reaction is involved in lowering the relative specific activity of the cytoplasmic malate, we are led to the conclusion that the rate of exchange between cytoplasmic malate and mitochondrial malate is not rapid compared with this other process. This conclusion is strikingly different from the suggestion by Rognstad (1970) that there is rapid isotopic communication between cytoplasmic and mitochondrial malate in kidney cortex slices.

Estimation of the relative specific activity of cytoplasmic malate. A complementary approach to the question of communication of malate across the mitochondrial membrane consists of using the PEP relative specific activity to yield a minimum estimate for the relative specific activity of its cytoplasmic precursor. With pyruvate as substrate, this precursor should be the cytoplasmic malate.

In livers perfused with pyruvate and ($2-^{14}C$) acetate the relative specific activity of PEP is approximately 0.75 (Table 5). The actual relative specific activity of cytoplasmic malate in these livers is probably higher because one carbon is lost in conversion of malate to PEP via MDH and PEP carboxykinase. The relative specific activity of the cytoplasmic malate could be more accurately estimated with knowledge of the distribution of isotope in PEP. Most of the isotope in malate derived from ($2-^{14}C$) acetate should be in the C-2 and C-3 positions, however, so the relative specific activity of cytoplasmic malate is probably not much higher than that of PEP (an upper limit would be 20% higher than the PEP). Thus, again we are led to the conclusion that there is a marked difference between the relative specific activities of the cytoplasmic and mitochondrial malate pools.

Estimation of the malate distribution in livers with pyruvate as substrate. The relative specific activity (RSA) of the total malate (1.0 by definition) reflects the weighted contributions for the cytoplasmic and mitochondrial pools:

$$1.0 = RSA_{cMAL} \times f_{cMAL} + RSA_{mMal} \times (1-f_{cMAL})$$

where RSA_{cMAL} is the relative specific activity of the cytosolic malate pool, RSA_{mMAL} is the relative specific activity of the mitochondrial malate pools, and f_{cMAL} is that fraction of the total malate which is in the cytosol. Using the RSA of PEP as a minimal estimate for the RSA of cytoplasmic malate and the RSA of mitochondrial oxalacetate as a minimal limit for the RSA of mitochondrial malate,

$$f_c = \frac{RSA_{mMAL} - 1}{RSA_{mMAL} - RSA_{cMAL}} = 0.61$$

That is, for these minimal estimates for the two relative specific activities, 61% of the malate must be cytoplasmic. This estimate is in good agreement with the value of 53% calculated by assuming near-equilibrium of the cytosolic and mitochondrial malate dehydrogenases (Table 2). Thus, the two independent methods of estimating the intracellular distribution of malate agree that the malate concentrations in the cytosol and mitochondria in pyruvate perfused livers are not equal and that a malate concentration gradient must be established from mitochondria to cytosol (Williamson et al. 1969b).

Acknowledgments

This work was supported by grants from the United States Public Health Service (GM 12202) and by a Grant-in-Aid from the American Heart Association to J. R. Williamson, who is also an Established Investigator of the American Heart Association.

References

Anderson, J. H., D. Garfinkel: Computers and Biomedical Research, in press (1970)

Azzi, A., J. B. Chappell, B. H. Robinson: Biochem. biophys. Res. Commun. 29 (1967) 148

Berkes-Tomasevic, P., H. Holzer: Europ. J. Biochem. 2 (1967) 98

Berl, S., W. J. Nicklas, D. D. Clarke: J. Neurochem., in press

Boyd, J. W.: Biochem. J. 81 (1961) 434

Bottger, I., O. Wieland, D. Brdiczka, D. Pette: Europ. J. Biochem. 8 (1969) 113

Chappell, J. B.: Brit. Med. Bull. 24 (1968) 150

Haslam, J. M., D. E. Griffiths: Biochem. J. 109 (1968) 921

Krebs, H. A., R. L. Veech, in: The Energy Level and Metabolic Control in Mitochondria, Ed. by S. Papa, J. M. Tager, E. Quagliariello and E. C. Slater. Adriatica Editrice, Bari 1969, p. 329

LaNoue, K., W. J. Nicklas, J. R. Williamson: J. biol. Chem. 245 (1970) 102
Mallett, L. E., J. H. Exton, C. R. Park: J. biol. Chem. 244 (1969) 5713
Meijer, A. J., J. M. Tager: Biochem. biophys. Acta. 189 (1969) 136
Nordlie, R. C., H. A. Lardy: J. biol. Chem. 238 (1963) 2259
Papa, S., R. D'Aloya, A. J. Meijer, J. M. Tager, E. Quagliariello, in: The Energy Level and Metabolic Control in Mitochondira, Ed. by S. Papa, J. M. Tager, E. Quagliariello and E. C. Slater. Adriatica Editrice, Bari 1969, p. 159
Roberts, E., J. Wein, D. G. Simonsen: Vitam. and Horm. 22 (1964) 503
Robinson, B. H., J. B. Chappell: Biochem. biophys. Res. Commun. 28 (1967) 249
Rognstad, R., J. Katz: Biochem. J. 116 (1970) 483,
Rognstad, R.: Biochem. J. 116 (1970) 493
Safer, B., J. R. Williamson, unpublished observations
Schoner, W.: Hoppe-Seylers Z. physiol. Chem. 351 (1970) 129
Segal, H. L., D. S. Beattie, S. Hopper: J. biol. Chem. 237 (1962) 1914
Silverstein, E., G. Sulebele: Biochem. biophys. Acta 185 (1969) 297
Tanaka, T., Y. Harano, F. Sue, H. Morimura: J. biochem. (Tokyo) 62 (1967) 71
Utter, M. F., M. C. Scrutton, in: Current Topics in Cellular Regulation, Ed. by B. L. Horecker and E. R. Stadtman. Vol. I, Academic Press, New York, 1969, p. 253
Walter, P., V. Paetkau, H. A. Lardy: J. biol. Chem. 241 (1966) 2523
Williamson, D. H., O. Lopes-Vieira, B. Walker: Biochem. J. 104 (1967) 497
Williamson, J. R., in: The Energy Level and Metabolic Control in Mitochondria, Ed. by S. Papa, J. M. Tager, E. Quagliariello and E. C. Slater. Adriatica Editrice, Bari, 1969a, p. 385
Williamson, J. R., E. T. Browning, and R. Scholz: J. biol. Chem. 244 (1969b) 4607, 4617
Williamson, J. R., J. Anderson, E. T. Browning: J. biol. Chem. 245 (1970) 1715

Discussion to Williamson

Seubert: Dr. Williamson, I saw from your slides, please correct me if I am not right, that aspartate is now going from the cytosol to the mitochondria. How does this scheme fit with the stoechiometric conversion of lactate to glucose, if it is still assumed that oxaloacetate cannot move through the mitochindrial membrane?

J. R. Williamson: The older schemes were drawn with the concept that aspartate was the main phosphoenolpyruvate mover across the mitochondrial membrane. I would not go so far to say that aspartate never moves across the mitochondrial membrane, but this is just one possibilitiy; maybe malate always moves. If we have lactate as a substrate for phosphoenolpyruvate formation, then the hydrogen is already present in the cytoplasm in qunatums sufficient for gluconeogenesis. So if malate instead of aspartate is coming across we want to get rid of this hydrogen. Then what we have to look for is some other type of shuttle. As we cannot transform this hydrogen across, we want to reduce an anion called X.

Seubert: What type of new shuttle could in your opinion, oxidize the excess of NADH in the cytoplasm?

J. R. Williamson: This is the interest of xylitol. There is a DPN enzyme in the cytosol and in the mitochondria. It is an extremely active enzyme. I think Dr. Krebs thought it might not be of significance because of the high K_M of cytoplasmic and mitochondrial sorbitol dehydrogenase.

Krebs: My main reason for not considering xylitol seriously as a hydrogen carrier is that I thought it is absent from the normal liver.

Seubert: The almost identical specific radioactivity of malate and pyruvate or lactate in the isotope experiments is, in my opinion, not at all proof of malate being a carrier of carbon across the mitochondrial membrane.

J. R. Williamson: No, it is only consistent with it.

Seubert: Identical specific radioactivities can also be obtained just by equilibration as has also been demonstrated for citrate (G. Müllhofer et al., FEBS letters 4 (1969) 33). There is no doubt that malate is involved in hydrogen transport, but I cannot see a definite proof for the transport of carbon by malate in the isotope experiments, especially in connection with lactate.

J. R. Williamson: I don´t agree with that. The specific activity of malate being higher than that of PEP as even expected from the randomization if it is almost complete is merely consistent with being a precursor. This tablet is not a proof that pyruvate is not carboxylated in the extra-mitochondrial space. I agree with you.

Hanson: I want to mention two points: 1) Crossover plots based on 5% or 30% changes in the concentration of oxaloacetate place a great stress on the absolute accuracy of your procedure for measuring oxaloacetate. My question is: Are your procedures for measuring this intermediate good enough to pick up 5% or 30% changes? 2) Sidney Weinhouse has measured, in as yet unpublished studies, the degree of recycling of pyruvate via this "futile cycle." His data suggest that although recycling does occur via pyruvate kinase in perfused livers of starved rats, it is relatively minor as compared to the overall rate of gluconeogenesis from pyruvate. Also his studies would tend to over-estimate the futile cycle since the 2-^{14}C of pyruvate can be introduced into the 3- position of lactate if any malate is recycled to pyruvate through NADP-malate dehydrogenase.

J. R. Williamson: There are numbers of publications for the rates of recycling. Sometimes it was very low at 40%. I have second-hand figures even as high as 50%. I would like to see the data and the type of label that Winehouse used. There are big problems with the interpretation of these data.

Hanson: The problem is that one cannot correct the contribution of NADP-malate dehydrogenase. Therefore, in fed animals the rate of recycling is very much higher than in starved rats due in part to the fact that two separate pathways may be stimulated.

J. R. Williamson: Surely in vivo under certain conditions it must be negligible, because teleologically it does not make sense. What interests me about it is whether it is possible for some rather abnormal states, one cannot get either in vivo or manipulate in the isolated perfused liver.

Interaction of Gluconeogenesis with Mixed Function Oxidation in Perfused Rat Liver

Ronald G. Thurman[+] and Roland Scholz
Institut für Physiologische Chemie und Physikalische Biochemie der Universität München, Munich, FRG, and Johnson Research Foundation, University of Pennsylvania, Philadelphia, U.S.A.

Summary

Mixed function oxidation in perfused rat liver depends upon the rate of generation of extra-mitochondrial NADPH (Thurman and Scholz, E. J. Biochem. 10:459, 1969). Of the intermediates necessary for this NADPH generation, malate and glucose-6-phosphate are also involved in the synthesis of glucose.

The addition of aminopyrine, a substrate for mixed function oxidation, caused slight inhibition of glucose production from lactate and dihydroxyacetone in livers from normal starved rats, but complete inhibition in livers from phenobarbital treated animals. Lactate utilization was diminished, but not to the extent to which glucose synthesis was inhibited. Moreover, uptake of added glucose was observed following aminopyrine addition to phenobarbital treated livers.

Aminopyrine oxidation is about five times faster in liver microsomes from phenobarbital treated animals due to an induction of the mixed function oxygenase system (Remmer and Merker, Ann. N. Y. Acad. Sci. 123:79, 1965). Since the inhibitory effect of aminopyrine on glucose production was also inducible, a direct action on gluconeogenic enzymes is improbable. Furthermore, an alteration of the energetic state by aminopyrine also appears to be unlikely since respiration of isolated mitochondria was not inhibited by concentrations employed in the perfusion experiments.

The surface fluorescence of reduced pyridine nucleotides decreased sharply following the addition of aminopyrine, but concomitant changes in the redox ratios of NADH linked metabolite couples were not observed. This suggests that a rapid oxidation of NADPH occurred and that the extra-mitochondrial NADH and NADPH pools are not in rapid equilibrium.

These findings lead to the conclusion that gluconeogenesis is inhibited indirectly by aminopyrine as a consequence of enhanced mixed function oxidation, i. e. drug

+ RGT is a recipient of a NATO fellowship.

metabolism. The data are consistent with increased flux through the glucose-6-phosphate dehydrogenase reaction due to increased NADPH oxidation and/or inhibition of glucose-6-phosphatase.

Introduction

The mixed function oxidation system of liver is located in the membranes of the smooth endoplasmic reticulum. It consists of a flavoprotein (NADPH-cytochrome P-450 reductase), a heat stable component, and a terminal oxidase, cytochrome P-450 (Lu and Coon 1968, Klingenberg 1958, Omura and Sato 1964). This system requires NADPH and molecular oxygen to transform a wide variety of endogenous and exogenous substances — including drugs, carcinogens, cholesterol and fatty acids — into more polar compounds (Gillette 1964). Its activity for hydroxylation and demethylation reactions is induced three-to five-fold by pretreatment of the animal for several days with phenobarbital (Remmer and Merker 1963).

In our previous work (Thurman and Scholz 1969), aminopyrine was used as a model substrate for mixed function oxidation to study the interaction of the drug metabolizing system with intermediary metabolism in perfused liver. Aminopyrine was selected because in concentrations lower than 1 mM it is not an inhibitor of mitochondrial respiration like barbiturates (Chance and Hollunger 1963, Thurman and Scholz, unpublished data, Scholz et al. 1966). It was observed that the increase in oxygen uptake of perfused liver following the addition of aminopyrine varied with the metabolic state (Thurman and Scholz 1969). For example, in livers of fed animals (high glycogen content) the addition of aminopyrine caused a larger increase in respiration than in glycogen depleted livers. In the fed state, this increase was enhanced when glycogen breakdown was stimulated following inhibition of mitochondrial respiration. Thus, it was concluded that NADPH for mixed function oxidation was provided mainly by the reactions of the pentose-phosphate shunt and that under most metabolic conditions the rate limiting step for drug metabolism in intact cells is the generation of NADPH in the extra-mitochondrial space. Furthermore, the smaller stimulation of oxygen uptake in glycogen depleted livers was completely abolished by inhibition of mitochondrial respiration. In the absence of glycogen, therefore, reducing equivalents for mixed function oxidation are derived from mitochondrial oxidations. A shuttle mechanism was postulated to provide for the egress of mitochondrial hydrogen into the extra-mitochondrial space involving pyruvate carboxylase, malate dehydrogenase, and malic enzyme (Thurman and Scholz 1969) similar to that proposed for lipogenesis (Young et al. 1964).

Of the intermediates involved in NADPH generation in the cytosol, malate and glucose-6-phosphate are also involved in gluconeogenesis. The purpose of this study was to ascertain if increased oxidation of extra-mitochondrial NADPH due to stimulated mixed function oxidation influences the ability of the liver to synthesize glucose.

Results and Discussion

Livers from starved rats were perfused with Krebs-Henseleit bicarbonate buffer (Krebs and Henseleit 1932) containing bovine serum albumin as described previously (Scholz et al. 1969). After addition of lactate as gluconeogenic precursor the rate of glucose output from the liver increased to 71 umoles/g/hr in control livers and to 40 umoles/g/hr in livers from phenobarbital pretreated rats (Fig. 1, Table 1). The addition of aminopyrine caused a rapid decrease in glucose output of about 40% in control livers. In induced livers, however, a complete inhibition and a small and transient glucose uptake was observed. Uptake of added glucose was also stimulated by aminopyrine addition.

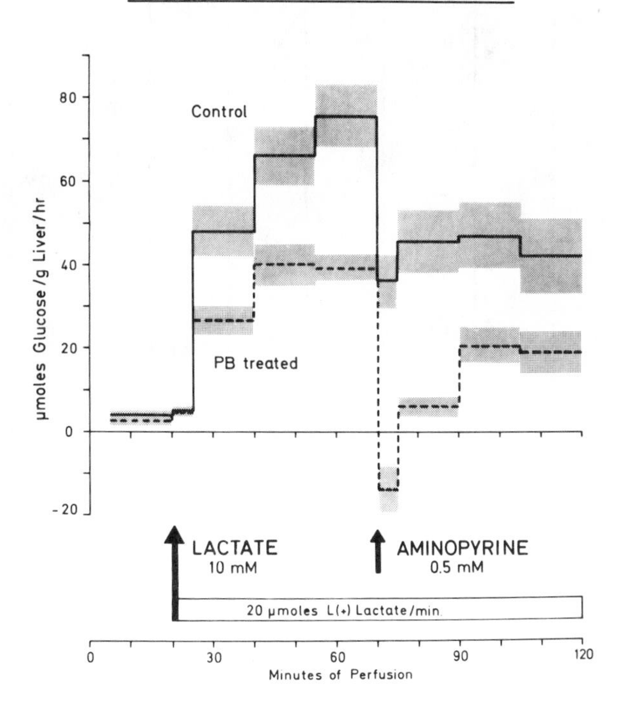

Fig. 1. Glucose production from lactate and the effect of aminopyrine in livers from control (———) and phenobarbital treated (----) rats. Livers from 24-hour-starved rats following seven days of intraperitoneal injection of phenobarbital (60 mg/kg/day) were perfused with Krebs-Henseleit bicarbonate buffer as described previously (Krebs and Henseleit 1932, Scholz et al. 1969). After a 20 minute substrate free perfusion period, lactate (10 mM) was added and maintained between 8 and 10 mM with a continous infusion. The additions of lactate and aminopyrine are indicated by arrows. The rates of gluconeogenesis are referred per gram dry weight times four, assuming 25% dry weight for liver.

µmoles/g Liver/hr or µmoles/100 g Rat/hr	Control		Phenobarbital treated	
	g Liver	100 g Rat	g Liver	100 g Rat
Glucose Production[1)] from Dihydroxyacetone	53 ± 4	201	39 ± 5	187
Glucose Production[1)] from Lactate	71 ± 7	270	40 ± 4	192
Lactate Uptake[1)] unaccounted for as Glucose or Pyruvate	151	570	106	510
Glucose-6-Phosphatase Activity[2]	2800	10700	1800	8700
g Liver equivalent to 100 g Rat	3.8 ± 0.1		4.8 ± 0.2	
Rat Weight (g)	177 ± 6		186 ± 6	

Table 1. 1) Rates calculated over the time interval 20 to 50 minutes after addition of the gluconeogenic substrate (see Fig. 1). Mean values and standard errors of the mean with seven to eight livers in each group.

2) The activity of glucose-6-phosphatase was determined by measuring the glucose production from added glucose-6-phosphate in buffered suspensions of once washed rat liver microsomes. Rates are expressed per gram liver wet weight by using 75 mg of microsomal protein per gram liver (Thurman and Scholz 1969). Mean values from three livers each.

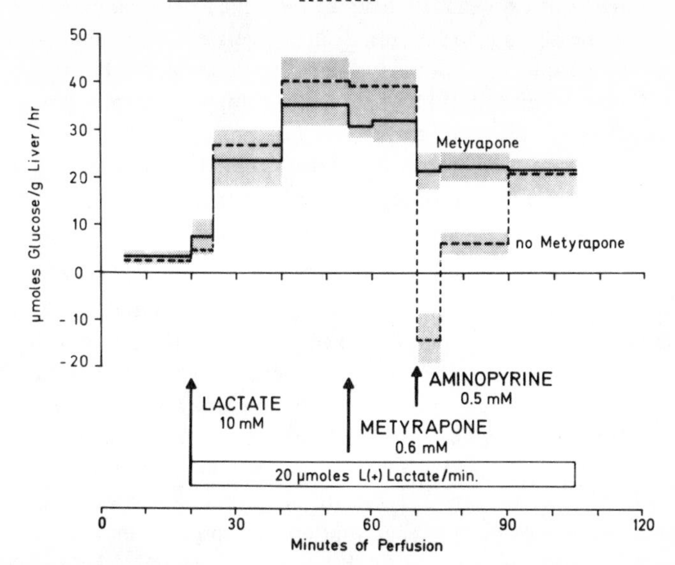

Fig. 2. Effect of aminopyrine on glucose production from lactate in the presence (———) and absence (----) of metyrapone in livers from phenobarbital treated rats. Conditions as in Fig. 1.

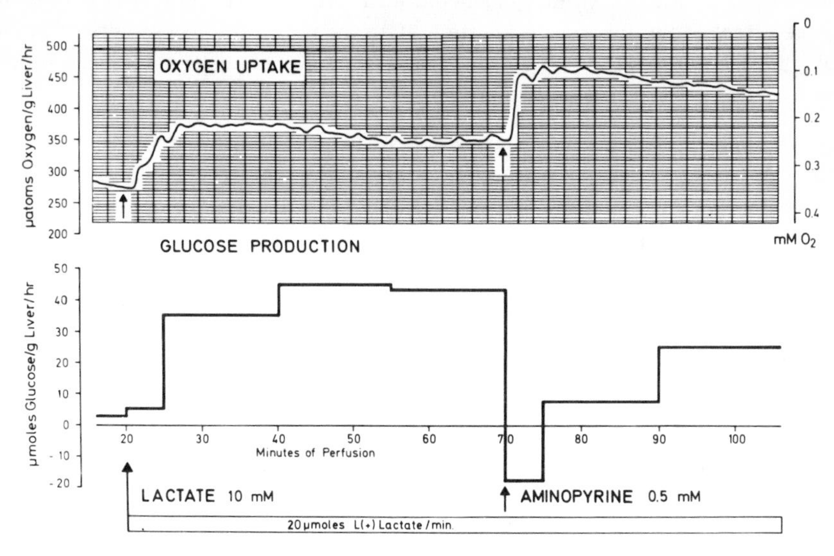

Fig. 3. Comparison of glucose production (lower panel) and respiration (upper panel) following lactate and aminopyrine additions to a phenobarbital treated rat liver. Venous oxygen concentrations were monitored with a miniature platinium electrode. Arterial oxygen concentrations were maintained constant (near 0.8 mM). Respiratory rates were calculated from the "arterio-venous" oxygen concentration differences, the flow rate, and the liver weight (gram dry x 4).

Similar observations were made when dihydroxyacetone was the gluconeogenic precursor. In the absence of exogenous substrates for mixed function oxidation, the rates of glucose output were 53 and 39 μmoles/g/hr in livers of control and pretreated animals, respectively (Table 1). Aminopyrine addition decreased the rate to 30 μmoles/g/hr in control livers, whereas in induced livers a complete inhibition was observed. With both gluconeogenic precursors employed, the aminopyrine effect on glucose output was larger in pretreated livers than in controls indicating that the effect was induced by phenobarbital. Thus, a direct effect of aminopyrine on enzymes in the sequence of glucose synthesis seems unlikely.

The induction of the aminopyrine effect on both oxygen uptake (Thurman and Scholz 1969) and on glucose output (Fig. 1) indicates that not the aminopyrine molecule itself but a metabolic consequence due to mixed function oxidation or a metabolite of aminopyrine causes the inhibition of glucose output. This conclusion was supported in experiments using an inhibitor of the mixed function oxygenase system, metyrapone. This substituted propanone, which is itself a substrate for mixed function oxidation, converts cytochrome P-450 into an inactive form (Hildebrandt et al. 1969). When aminopyrine was added to the perfused liver in the presence of metyrapone, only a slight increase in the rate of oxygen consumption was observed indicating that mixed function oxidation of aminopyrine was inhibited (Thurman and Scholz in press). This finding confirms for the whole organ experiments first performed with isolated microsomes (Metter et al. 1967). In the present experiments, metyrapone in concentrations below 1 mM did not affect gluconeogenesis from lactate (Fig. 2). On the other hand, the effect of the subsequent addition of aminopyrine on glucose output was largely abolished. Inhibition of gluconeogenesis, therefore, does not depend upon the presence of aminopyrine but on its metabolism.

The oxygen uptake of livers from starved rats was about 270 μatoms/g/hr (Fig. 3). Addition of 10 mM lactate increased this rate by 100 μatoms/g/hr due to the increased energy requirements for gluconeogenesis. The subsequent addition of 0.5 mM aminopyrine caused a further stimulation of respiration by 100 μatoms/g/hr, which is similar to the maximal stimulation observed in the absence of gluconeogenesis (Thurman and Scholz 1969). This experiment suggests, therefore, that aminopyrine metabolism did not diminish the prior rate of oxygen consumption and that the energy requirements for biosynthetic processes were unchanged during enhanced mixed function oxidation despite the fact that glucose output was abolished.

About four molecules of lactate were used per molecule of glucose formed in livers from both control and pretreated rats (Fig. 4). Thus, one-half of the lactate used was not accounted for as glucose or pyruvate. When aminopyrine was added, the rate of lactate uptake remained unchanged although glucose output was suppressed. Therefore, the rate of lactate uptake unaccounted for as glucose or pyruvate increased to 210 and 190 μmoles/g/hr in control and induced livers, respectively. The perplexing question is: Where have all the carbons gone? The following possibilities exist: First, it is attractive to postulate that glucose-6-phosphate is still synthesized from lactate. This glucose-6-phosphate could be used either for glycogen formation or enter the pentosephosphate shunt. However, glycogen was not formed in these experiments and most of the carbon entering the shunt pathway would re-enter the gluconeogenic pathway so that the problem of disposing of the lactate carbon remains. Moreover, it is unlikely that intermediates accumulate in the tissue if one considers the high rate of lactate uptake. On the other hand, lactate may be oxidized to acetyl-CoA. However, a subsequent combustion to carbon dioxide would consume three to four times more oxygen than the total respiration of the liver. Furthermore, the rate of acetoacetate and ß-hydroxybutyrate formation was unchanged following aminopyrine. Finally, a possibility consistent with the present data is a preferential oxidation of lactate to acetyl-CoA with the subsequent synthesis of fatty acids. The necessary NADPH for this biosynthesis as well as for mixed function oxidation could be provided by the primary reactions of the pentose-phosphate shunt. The possible interaction between mixed function oxidation and triglyceride synthesis is presently under investigation.

The extra-mitochondrial $NADPH-NADP^+$ system is involved in mixed function oxidation and lipogenesis. Via common intermediates it is also connected with gluconeogenesis. $NADPH-NADP^+$ redox changes could play an important role in the interaction between these metabolic systems. Insight into intracellular redox changes can

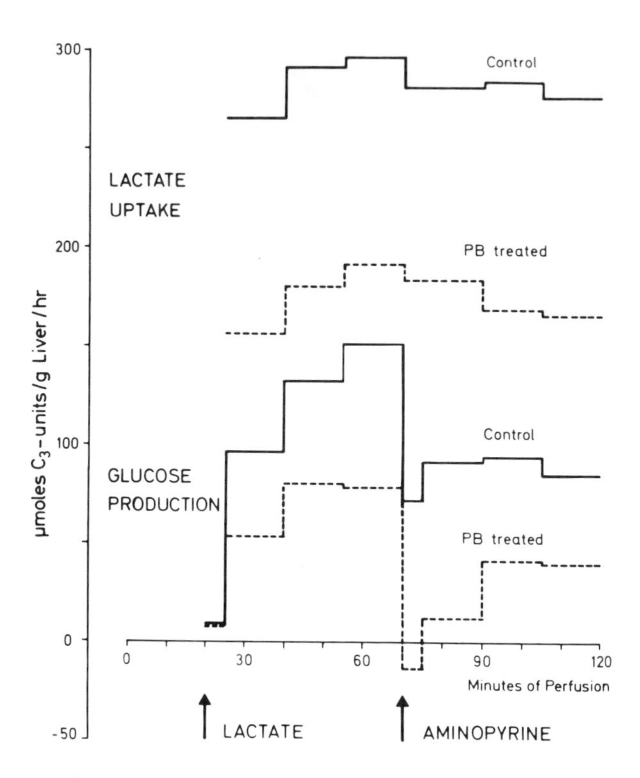

Fig. 4. Balance of three carbon units following lactate addition to control (——) and phenobarbital (----) treated rat livers and the effect of aminopyrine.

be obtained by the recording of flavin and pyridine nucleotide fluorescence from the surface of a hemoglobin free perfused liver. In previous studies it was observed that flavin fluorescence is derived predominantly from oxidized mitochondrial flavoproteins, whereas pyridine nucleotide fluorescence arises from protein bound cofactors in both extra- and intra-mitochondrial spaces (Scholz et al. 1969, Scholz 1968). It was concluded that changes in pyridine nucleotide fluorescence reflect the redox state of NADH with little contribution of NADPH since kinetic analyses showed close correlation between surface fluorescence, tissue contents of NADH (but not NADPH), and NAD^+ linked redox couples (Scholz and Bücher 1965, Scholz 1968, Bücher 1970 Following the addition of lactate, a rapid increase in pyridine nucleotide fluorescence and a decrease in flavin fluorescence was observed (Fig. 5) reflecting the reduction of extra- and intra-mitochondrial redox systems. The subsequent addition of aminopyrine caused a large decrease in the intensity of pyridine nucleotide fluorescence, but the flavin fluorescence was not changed significantly. In contrast, the ratios of NAD^+ linked redox couples in the perfusate (i.e. lactate/pyruvate and ß-hydroxybutyrate/acetoacetate) showed a slight reduction rather than an oxidation following the addition of aminopyrine (Fig. 6). Thus, ß-hydroxybutyrate/acetoacetate ratios conform to the fluorescence recording of flavoproteins which are in equilibrium with a mitochondrial NADH-NAD^+ pool. However, the decrease in pyridine nucleotide fluorescence goes in the opposite direction from the lactate/pyruvate ratios. A quenc ing artifact due to aminopyrine was excluded. It was concluded, therefore, that the fluorescence decrease following aminopyrine reflects a large and rapid oxidation of a cytosolic NADPH-$NADP^+$ pool due to the stimulated mixed function oxidation. However, this conclusion must be confirmed by tissue analyses of pyridine nucleotides.

Phenobarbital pretreatment affects the gluconeogenic capacity of the liver. In the absence of an exogenous substrate for mixed function oxidation, the rate of glucose production from both lactate and dihydroxyacetone was less in induced livers than in controls. Furthermore, phenobarbital caused an increase in liver weight of about 25% calculated on a dry weight basis (Table 1), Kunz et al. 1966. If this increase is exclusively due to cellular constituents not involved in gluconeogenesis, identical rates of glucose output should be obtained when referred to the rat weight. In fact, this is the case in experiments with dihydroxyacetone (Table 1). Only small differences between control and phenobarbital pretreated livers were observed when the rates were expressed per unit rat weight. However, when lactate was the gluconeogenic precursor, rates in induced livers were 30% lower than in controls, even with the correction applied. This suggests that the proliferation of the endoplasmic reticulum following pretreatment with phenobarbital dilutes the gluconeogenic "machinery" in the pathway between triose phosphates and glucose, whereas an additional interaction is expected in the sequence between lactate and triose phosphates.

Lactate uptake was also less in induced livers than in controls (Fig. 4), but the lactate uptake not accounted for as glucose or pyruvate was similar when referred to the rat weight (Table 1). Thus, it seems unlikely that an increased oxidation to acetyl-CoA can account for the lower rate of glucose synthesis in livers from phenobarbital pretreated rats. On the other hand, a decreased flux from lactate to glucose is consistent with a partial block in the gluconeogenic pathway, either by decreased activities of rate limiting enzymes or by increased turnover of intermediates in a "futile cycle" involving pyruvate carboxylase, malate dehydrogenase and malic enzyme. Of the enzymes unique to gluconeogenesis, the effect of phenobarbital pretreatment has only been studied on glucoe-6-phosphatase (Ernster and Orrenius 1965). It is less active in induced liver, but the remaining activity is still two orders of magnitude greater than the rate of glucose synthesis (Table 1). In contrast, the activity of malic enzyme is doubled in induced liver (Kunz et al. 1966) reaching values similar to that of pyruvate carboxylase in normal livers (Böttger et al. 1969). An increased flux through a NADPH generating "futile cycle," therefore, could limit the input of substrate into the gluconeogenic pathway.

Summary and Conclusions

The data presented above indicate that the glucose production from lactate and dihydroxyacetone in glycogen depleted livers is diminished by the addition of aminopyrine. The inhibition is enhanced by the induction of the mixed function oxygenase system following phenobarbital pretreatment and is abolished by an inhibitor of

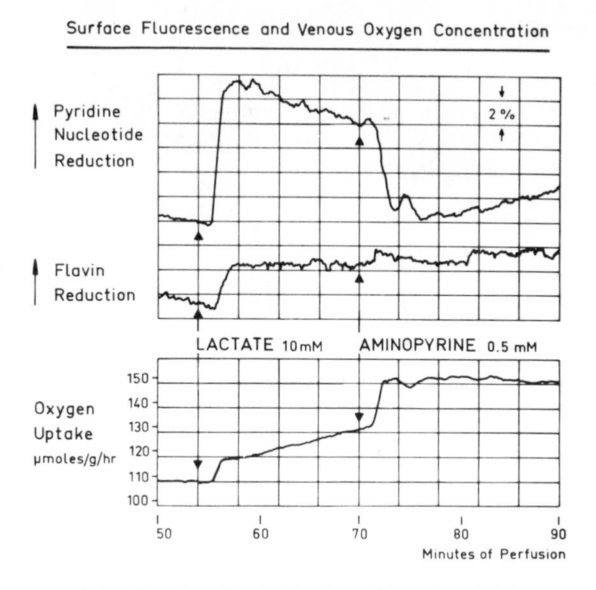

Fig. 5. Changes of surface fluorescence intensities and venous oxygen concentration following the addition of lactate and aminopyrine to perfused liver from a starved, phenobarbital pretreated rat. An upward deflection represents a reduction of pyridine nucleotides (upper trace, increase in fluorescence intensity excited at 366 mu) and flavoproteins (middle trace, decrease in fluorescence intensity excited at 436 mu) (Chance in press). Venous oxygen concentrations (lower trace) were converted into rates of oxygen uptake as in Fig. 3.

this system, metyrapone. Thus, it is concluded that it is not aminopyrine but consequences of its metabolism which affect gluconeogenesis. This observation demonstrates the close interaction of two important functions of the liver cell: synthesis of glucose and oxidation of compounds like steroids and drugs. If one wants to understand the control of gluconeogenesis and mixed function oxidation in vivo, their interactions have to be considered. Moreover, if one extrapolates these findings to man, a tendency toward hypoglycemia in barbiturate addicts is predicted, a question which may have clinical interest. In respect to clinical medicine, biguanides are active inhibitors of gluconeogenesis, but their hydroxylated products are not (Haeckel personal communication). In analogy with aminopyrine, the question is raised whether or not their metabolism by the mixed function oxidation system of liver is related to their ability to inhibit gluconeogenesis.

The mechanism by which enhanced mixed function oxidation inhibits gluconeogenesis, however, remains unclear. Since gluconeogenesis from both lactate and dihydroxyacetone was affected similarly, it is concluded that a site of interaction is between triose phosphates and glucose (Fig. 7). One attractive hypothesis is that enhance mixed function oxidation accelerates the flow of carbon through the pentose-

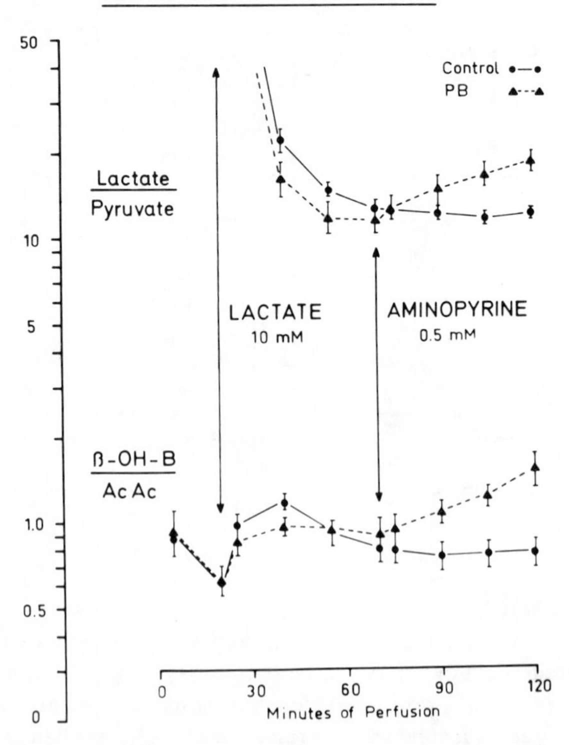

Fig. 6. Ratios of NAD^+ linked redox couples in the perfusate of livers from fasted control (——) and phenobarbital treated (----) rats. Conditions as in Fig. 1.

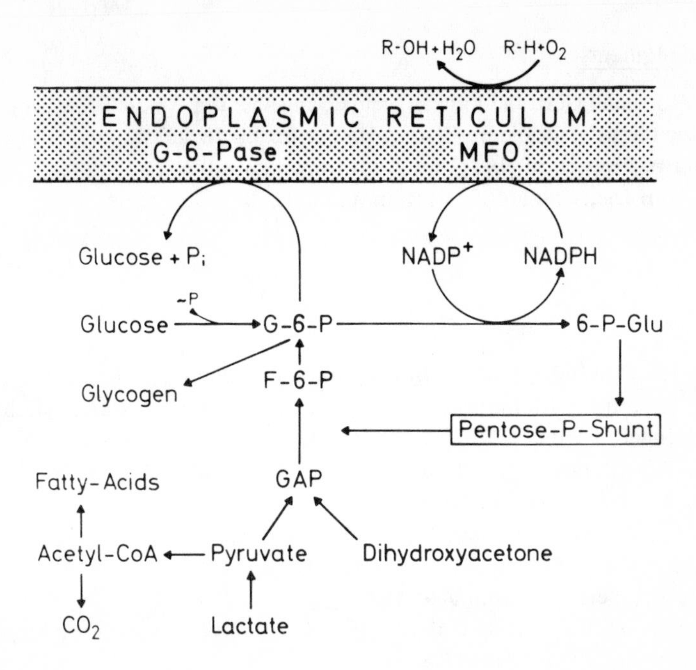

Fig. 7. Schematic representation of the interaction of mixed function oxidation and gluconeogenesis. R-H, aminopyrine or a suitable substrate for mixed function oxidation (MFO); R-OH, monomethylaminopyrine or other products of mixed function oxidation; G-6-P, glucose-6-phosphate; F-6-P, fructose-6-phosphate; GAP, glyceraldehyde phosphate; 6-P-Glu, 6-phospho-gluconolactone.

phosphate pathway due to increased oxidation of NADPH. This would in effect divert glucose-6-phosphate from its site of dephosphorylation, the endoplasmic reticulum. Although this mechanism probably contributes to the effect, the estimated rate of NADPH generation necessary for mixed function oxidation is too low to explain that glucose output is completely abolished. Another speculation is that activation of mixed function oxygenases in the endoplasmic reticulum inhibits glucose-6-phosphatase bound to the same membranes. Preliminary studies with isolated microsomes, however, did not support this view.

The key to the mechanism of the inhibitory action of mixed function oxidation on gluconeogenesis is probably the observation that lactate uptake was not diminished following the addition of aminopyrine (Fig. 4). Thus, large amounts of lactate are utilized which cannot be accounted for as glucose, glycogen, pyruvate or ketone bodies. Also, its disposal in tissue intermediates or its oxidation to carbon dioxide is remote. On the other hand, the diversion of these carbons into fatty acids is a possibility consistent with the present data. Whether a switch from gluco- to lipogenesis is a metabolic consequence of stimulated mixed function oxidation is a question for further studies.

Acknowledgments

We wish to thank Gertraud Lusch, Ursula Schwabe and Albert Schallweg for their expert technical assistance; Helmut Hofner for his technical advice and the Scientific Affairs Division of NATO for funds to purchase the fluorometer. Supported by a grant from Deutsche Forschungsgemeinschaft, Bad Godesberg, FRG.

References

Böttger, I., O. Wieland, D. Brdiczka, D. Pette: Europ. J. Biochem. 8 (1969) 113

Bücher, Th., in: Pyridine Nucleotide Dependent Dehydrogenase, Ed. by H. Sund, Springer, Berlin 1970, p. 439

Chance, B., G. Hollunger: J. biol. Chem. 238 (1963) 418

Chance, B., D. Mayer, V. Legallias, in press

Ernster, L., S. Orrenius: Fed. Proc. 24 (1965) 1190

Gilette, J.: Adv. Pharmacol. 2 (1964) 219

Haeckel, R., personal communication

Hildebrandt, A. G., K. C. Leibman, R. W. Estabrook: Biochem. biophys. Res. Commun. 37 (1969) 477

Klingenberg, M.: Arch. biochem. Biophys. 75 (1958) 376

Krebs, H. A., K. Henseleit: Hoppe-Seylers Z. physiol. Chem. 210 (1932) 33

Kunz, W., G. Schaude, H. Schimassek, W. Schmid, M. Siess, in: Proc. Europ. Soc. for Study of Drug Toxicity, Vol. VII, 1966, p. 138

Lu, A. Y. H., M. J. Coon: J. biol. Chem. 243 (1968) 1331

Netter, K. J., S. Jenner, K. Kajuschke: Naunyn-Schmiedeberg's Arch. Exp. Path. Pharmak. 259 (1967) 1

Omura, T., R. Sato: J. biol. Chem. 239 (1964) 2370

Scholz, R., Th. Bücher, in: Control of Energy Metabolism, Ed. by B. Chance, R. W. Estabrook and J. R. Williamson, Academic, New York 1965, p. 393

Scholz, R., F. Schwarz, Th. Bücher: Z. Klin. Chem. 4 (1966) 179

Scholz, R., in: Stoffwechsel der isoliert perfundierten Leber, Ed. by W. Staib und R. Scholz, Springer, Berlin 1968, p. 25

Scholz, R., R. G. Thurman, J. R. Williamson, B. Chance, Th. Bücher: J. biol. Chem. 244 (1969) 2317

Thurman, R. G., R. Scholz: Europ. J. Biochem. 10 (1969) 459

Thurman, R. G., R. Scholz, in press

Thurman, R. G., R. Scholz, unpublished data

Young, J., E. Shrago, H. A. Lardy, Biochemistry 3 (1964) 1687

Discussion to Thurman and Scholz

Söling: Can you imagine that a lot of the carbon you cannot account for is going into synthesis of nucleotides or RNA? At least under normal conditions the amount of G-6-P going into RNA is considerable. Hess and Brand (personal communication) have shown for yeast, meanwhile it has been shown also for liver and other tissues that the amount of carbon converted form G-6-P into RNA is tremendous, much more than going into CO_2.

Thurman: Certainly this is a possibility for the disposal of lactate carbon. However, the rate of lactate uptake following aminopyrine is 200 µmoles/g hr. I do not think that the removal of ribose-5-phosphate for RNA synthesis can reach such high rates.

Exton: Another possibility would be that the carbon which you cannot account for is going into α-glycerol phosphate. We are looking into this in normal livers perfused with ^{14}C-lactate. Under most conditions there is not much label going to glyceride-glycerol relative to glucose, but it may be that in your experiments, this pathway was significant. It would also produce a drain on NADH.

Schäfer: Any substance which can be oxidized by mixed function oxygenation should be an inhibitor for gluconeogenesis in Dr. Thurman's experiments, if his assumptions are correct. Referring to the metabolic scheme in Dr. Hanson's report, where he showed the glycerolneogenesis in adipose tissue, this type of control by mixed function oxygenation should be essentially absent, because we do not have a significant mixed function oxygenation in this tissue. I should like to ask Dr. Hanson whether he has any experience with inhibitors of gluconeogenesis in this system.

Hanson: I was asked previously whether quinolinic acid had an effect on glyceroneogenesis in adipose tissue. I do not know, but it is a good idea.

Thurman: Due to the generalized non-specificity of the mixed function oxidation system, it is possible that any inhibitor of gluconeogenesis is acting, at least in part, like aminopyrine. In experiments with new hypoglycemic agents, it might be worthwhile to rule out such possibilities from the onset with metyrapone, since the latter does not seem to effect gluconeogenesis itself.

Krebs: I was especially interested in the lack of balance between the amounts of lactate removed and products formed, because we also found this occasionally. We first came across this phenomenon when one of my coworkers, Federico Mayor, practiced the perfusion technique after his return to the University of Granada, and when he repeated experiments on gluconeogenesis from lactate. Although lactate disappeared no glucose was formed. A few times we have observed this also at Oxford. We could not correlate this phenomenon to special circumstances. There was no major accumulation of intermediary metabolites in the liver which could account for the removal of lactate. In particular there was no accumulation of α-glycerophosphate. Dr. Söling's suggestion that lactate may be converted into ribose phosphate seems to me attractive and could easily be checked by isotope experiments.

Exton: I would like to comment on Dr. Krebs'answer to point out that the α-glycerol phosphate could be esterified to triglycerides.

Krebs: This is not likely to be enough.

Exton: Well, it may be that the livers are also synthesizing fatty acids.

Thurman: That a switch from gluco-to lipogenesis has occurred following the addition of aminopyrine is the most plausible explanation for the recovery of lost carbon which is compatible with the changes in oxygen uptake observed.

Krebs: It does not seem probable to me that lactate is completely oxidized, though this cannot be ruled out. We have never noticed a major oxidation of lactate in the perfused liver.

D. H. Williamson: In the experiments where you find large changes in the redox state of the $NADP^+$-NADPH system, it would be interesting to measure total pyridine nucleotides because it might give an indication of the amounts of $NADP^+$ and NADPH in the cytoplasmic compartment.

Hanson: Did you consider cytosolic isocitrate dehydrogenase as a possible NADPH generating system?

Thurman: According to Cleland (Ann. Rev. Biochem. 36 (1967) 77), the cytosolic isocitrate dehydrogenase equilibrium favors isocitrate formation and would therefore be a NADPH utilizing system.

Krebs: There is definite evidence that the enzyme is responsible for the synthesis of citrate in the cytosol.

Some Aspects of Gluconeogenesis from Pyruvate in Hemoglobinfree Perfused Rat Liver

G. Müllhofer and O. Kuntzen
Institut für Physiologische Chemie und Physikalische Biochemie der Universität München, Munich, FRG

Summary

The addition of pyruvate-2-C^{14} 3.5 to 4.5 mM (specific activity=1.00) to perfused rat livers from animals starved for 48 hours produced an increase in oxygen uptake concurrent with the onset of gluconeogenesis and the appearance of lactate in the perfusate. Ketogenesis was nearly completely suppressed by substrate addition and urea levels, which were quite low, were diminished further.

After ten minutes of perfusion in an open system the liver was freeze clamped. Metabolites in the perfusate and tissue extract were measured and C^{14} specific activities are referred to the specific activity of the precursor pyruvate.

Although the activity of pyruvate leaving the liver was equal to 1.00, the value for lactate was 0.50. It was concluded that this lactate must arise from intracellular pyruvate and therefore the permeability of the liver cell membrane for pyruvate must be limited in our experiments. This postulate is supported by the detection of comparatively low levels of pyruvate in the tissue extract. To explain the low lactate activity, one must assume that a flux of inactive pyruvate into the intracellular pyruvate pool occurs.

Glucose was isolated from the perfusate and the C^{14} labeling pattern was determined. Relative specific activities of 0.32, 0.12, 0.12 were measured for carbon positions 4, 5 and 6. The corresponding labeling of glucose with lactate-2-C^{14} as substrate found in earlier experiments was 0.05, 0.30 and 0.30. One is tempted to explain the different labeling patterns of glucose in the experiments with lactate and pyruvate as glucogenic substrates by enhanced turnover of the citric acid cycle and a higher contribution for acetyl-CoA formation from pyruvate in the latter. But according to theoretical considerations, based on the generally assumed pathway for glucose formation from pyruvate, we would have to assume a C^{14} specific activity for acetyl-CoA of 1.5 to 2.0 times the intracellular pyruvate activity, to explain the glucose labeling pattern.

The extent of carbon-14 labeling of acetyl-CoA should also be reflected in the difference between citrate and malate activities, but for both these metabolites the same value of 0.95 was measured. Degradation of citrate with citrate lyase showed that the acetyl unit in citrate was not labeled at all, whereas for the ß-hydroxybutyrate, which is also connected with the acetyl-CoA pool, activities of 0.18 to 0.30 were found.

Intra-mitochondrial compartmentation of acetyl-CoA can only explain part of the shown discrepancies. The remaining difficulties in postulating pathways correlating all measured C^{14} activities are discussed.

Balances of substrates and products in the metabolism of pyruvate by rat liver and kidney cortex have been elucidated by (Ross et al. 1967, Krebs et al. 1967, Teufel et al. 1967) and the intracellular relationships in the pathway of gluconeogenesis from pyruvate have been discussed by Williamson et al. (1968, 1969a). In the studies presented here, pyruvate-2-^{14}C was used as gluconeogenic substrate in rat liver perfusions. The results indicate that there are still difficulties concerning the exact pathway of the carbon skeleton when glucose is formed from pyruvate.

Perfusion System

The hemoglobin-free perfused rat liver was used for these experiments. The experimental set up and analytical techniques were the same as described in former studies with lactate-2-^{14}C (Müllhofer et al. 1969). Livers of rats starved for 48 hours were perfused with Krebs Ringer bicarbonate buffer containing 7% dextran[+)] (Scholz 1968).

+ Kindly supplied by Knoll Co., Ludwigshafen, FRG.

After perfusing in a recirculating system at 33 ^{0}C without addition of exogenous substrate for 30 minutes, the perfusion system was opened and perfusion continued with fresh substrate free medium for ten minutes. The addition of pyruvate (4.6 mM) produced an increase in oxygen uptake concurrent with the onset of gluconeogenesis and the appearance of lactate in the perfusate (Fig. 1). Ketogenesis was nearly completely suppressed by substrate addition and urea production which was quite low, was diminished further. Twelve minutes after pyruvate addition the liver was freeze clamped. Metabolites were separated from the tissue extract and perfusate by ion-exchange chromatography and analyzed for their specific ^{14}C activities. All ^{14}C activities are related to the specific ^{14}C activity of the starting pyruvate-2-^{14}C.

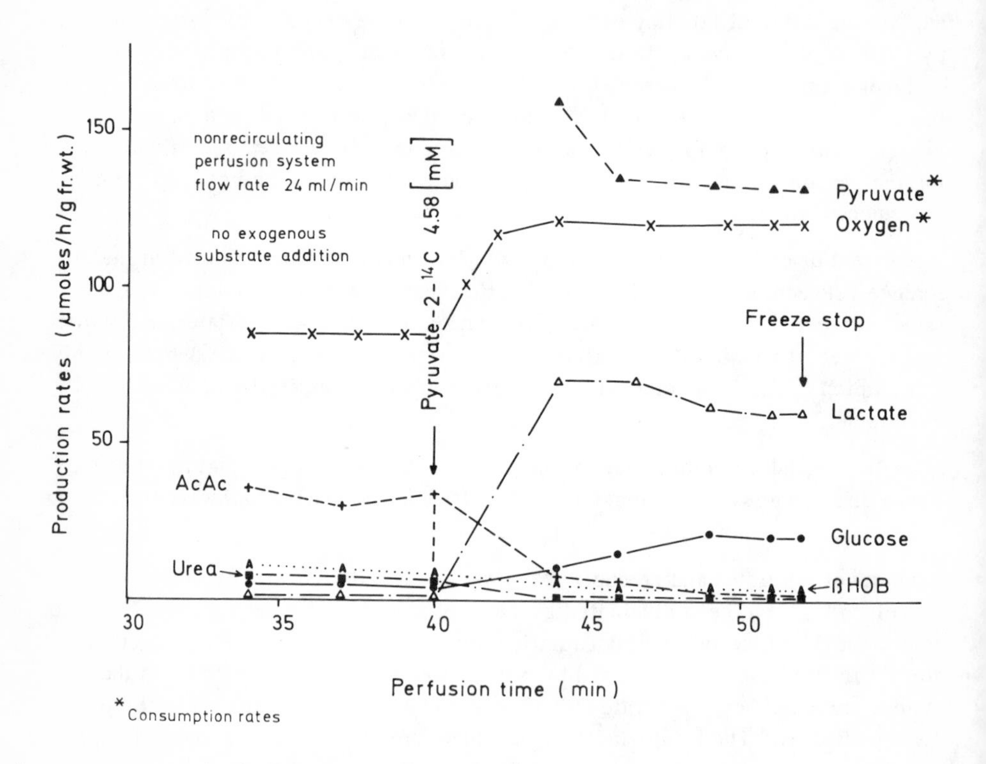

Fig. 1. Metabolic events on addition of pyruvate to the perfusion fluid. Wistar rat 183 g body weight. Freeze clamped liver 11.02 g (fresh weight) equivalent to 2.16 g liver dry weight. Rate of perfusion 24 ml/min; perfusion temperature 33^{0}C

The ^{14}C labeling pattern of the perfusate glucose from experiments with pyruvate-2-^{14}C as substrate is compared with the one found in earlier experiments with lactate-2-^{14}C (Müllhofer et al. 1969) Fig. 2. In both types of experiments glucose was symmetrically labeled with respect to the triose units. Detailed degradation and analysis of the ^{14}C activities in different carbon atoms was performed only for carbon atoms 4, 5 and 6. The different ^{14}C distributions for these positions were remarkable.

The labeling pattern of the glucose was undoubtably influenced by pathways interacting with intermediates of the gluconeogenic pathway such as the pentose-phosphate-and tricarboxylic acid cycle (TCA-cycle).

With pyruvate-2-^{14}C as substrate, a high turnover rate of the TCA-cycle would lead to dilution of ^{14}C activity in carbon atoms 5 and 6 of glucose by carbon atoms originating from the methyl-group of acetyl-CoA, while carbon atom 4 would gain activity by randomization processes.

Moreover, it is important for the labeling in the 4 position of glucose to what extent the carboxyl group of acetyl-CoA formed from pyruvate enters the TCA-cycle.

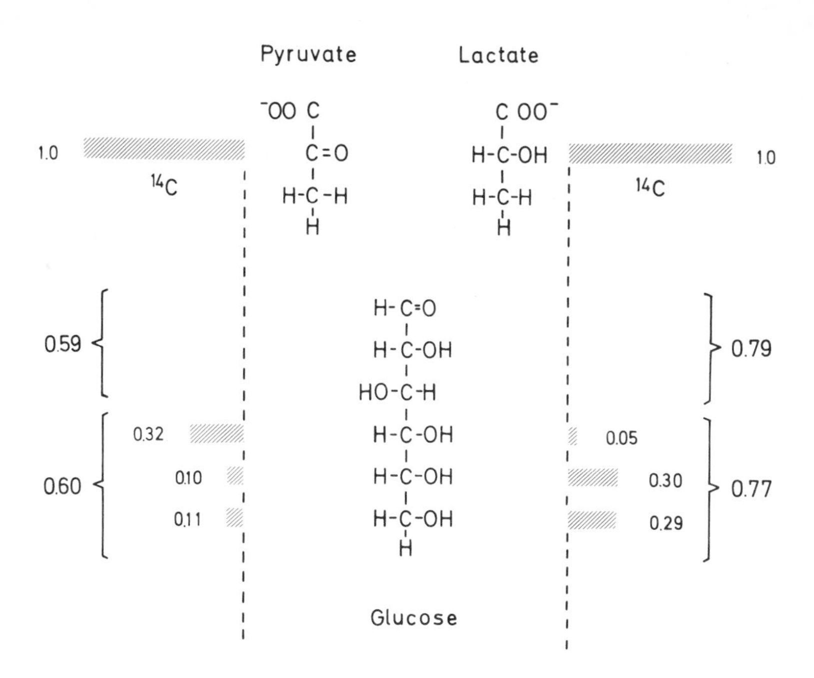

Fig. 2. The ^{14}C labeling patterns of glucose isolated from the perfusate with pyruvate-2-^{14}C and lactate-2-^{14}C as gluconeogenic substrates.

It is proposed that in the pyruvate relative to the lactate experiments a higher turnover rate of the TCA-cycle resulted in diluted ^{14}C activities of carbon positions 5 and 6. Furthermore, a high flux rate of pyruvate to acetyl-CoA which subsequently enters the TCA-cycle must be assumed in the experiments with pyruvate.

A high activity of the pentose-phosphate-cycle would primarily alter the ^{14}C pattern of positions 1, 2 and 3, resulting in an asymmetric ^{14}C distribution in the triose units. However, a high turnover rate of a "futile cycle" between fructose-6-phosphate and fructose-1, 6- di-phosphate in liver tissue (Newsholm and Gevers 1967) in connectio with the aldolase and triose-phosphate-isomerase reactions would re-equilibrate both triose units and complicate an exact evaluation of the pentose-phosphate-cycle.

From the data given in Table 1, it is evident that total ^{14}C activities of 3-phosphoglycerate and phosphoenolpyruvate correspond reasonably well with the ^{14}C activity of the triose units of glucose. Minimal dilution may originate from residual glycogenolysis. This may be used as an argument to exclude an important interaction

Metabolite	Rel. spec. ^{14}C Activity Pyruvate arterial = 1.00			
Glucose				
total	1.05			
4 position	0.32	0.53	C OO⁻	0.33
5 "	0.10		H-C-OH	0.10
6 "	0.11		H-C-H	0.11
3-Phosphoglycerate	0.59		C OO⁻	0.33
Phospoenolpyruvate	0.61			
Malate				
total	0.92			
4 position	0.33			

Table 1. Comparison of the ^{14}C labeling pattern of glucose 4,5 and 6 position with the ^{14}C activities of 3-phosphoglycerate, phosphoenolpyruvate and malate isolated from the tissue extract. The ^{14}C labeling of the malate 4 position was analyzed by conversion of malate with inactive acetyl-CoA and citrate synthetase (EC 4.1.3.7) to citrate, degradation of citrate with citrate lyase (EC 4.1.3.6) and decarboxylation of the oxaloacetate formed with oxaloacetate decarbonxylase (EC 4.1.1.3) to pyruvate. The ^{14}C activity of the 4 position of malate is equivalent to the ^{14}C activity difference between malate and the pyruvate, isolated as lactate. The ^{14}C activity pattern of malate is shown on the right hand side of the Table, assuming symmetrical labeling.

with the pentose-phosphate-cycle, but isotopic equilibration of intermediates between fructose-1, 6-di-phosphate and phosphoenolpyruvate, which is much faster than net flux through the reversed glycolytic chain has to be considered. On the other hand, the difference between total malate and phosphoenolpyruvate ^{14}C activity for the carboxyl group in position 4 of malate suggest that label in 4, 5 and 6 positions of glucose arise from malate without further rearrangements. This allows the assumption that the interaction of the gluconeogenic pathway with the pentose-phosphate -cycle is negligible under these conditions and that the ^{14}C activity pattern of glucose is mainly determined by the interaction of oxaloacetate formed by carboxylation of pyruvate with the TCA-cycle.

Mathematical Formulation of the Influence of the TCY-cycle on the ^{14}C-Labeling Pattern of Glucose

It is possible to derive equations describing the labeling pattern of glucose when pyruvate-2-^{14}C is substrate accounting for the turnover of oxaloacetate in the TCA-cycle. Similar calculations have been made by Exton and Park (1967). One must make the following assumptions: 1) The identical labeling of the 5 and 6 position of glucose indicates that oxaloacetate formed from pyruvate is rapidly equilibrated with malate and fumarate. 2) There should be no loss or inactive dilution of intermediates of the TCA-cycle. Inactive dilution by amino acid metabolism can be neglected because of the low urea production.

The oxaloacetate formed from pyruvate may be converted to phosphoenolpyruvate (with the probability 1-x) Fig. 3 or enter the TCA-cycle (with probability x). The main contribution to ^{14}C incorporation in 5 and 6 position of glucose is attributed to oxaloacetate which has not yet passed through the TCA-cycle ($\frac{P}{2}(1-x)$; rel. spec. activity of intracellular pyruvate = P). The following terms in equation Ia (Fig. 3) are contributions of oxaloacetate which has passed the cycle 1, 2, 3 . . . n times. The result is a geometric progression the threshold value for n against infinity is shown in equation I. With a known rate of gluconeogenesis it is possible to calculate the turnover rate of the TCA-cycle (rate of gluconeogenesis $\cdot x/1-x$). For the labeling of the 4 position of glucose similar considerations are valid. One must subtract the first term from equation I, because oxaloacetate which has not entered the cycle will not contribute to the label in 4 position of glucose. On the other hand, acetyl-CoA (re. spec. ^{14}C activity = A) originating from pyruvate-2-^{14}C will contribute to the label in 4 position ($\frac{A}{2}$ x).

If one calculates x from equation I and inserts it into equation II, the extent of labeling of acetyl-CoA entering the cycle can be determined.

To solve these equations the ^{14}C activity of intracellular pyruvate (P) has to be known. Pyruvate entering and leaving the liver had the same specific activity (1.00). A problem arises from the fact that the lactate leaving the liver had a ^{14}C activity of 0.46. This lactate must have originated from intracellular pyruvate and therefore it seems reasonable to assume that it represents the ^{14}C activity of the intracellular pyruvate pool.

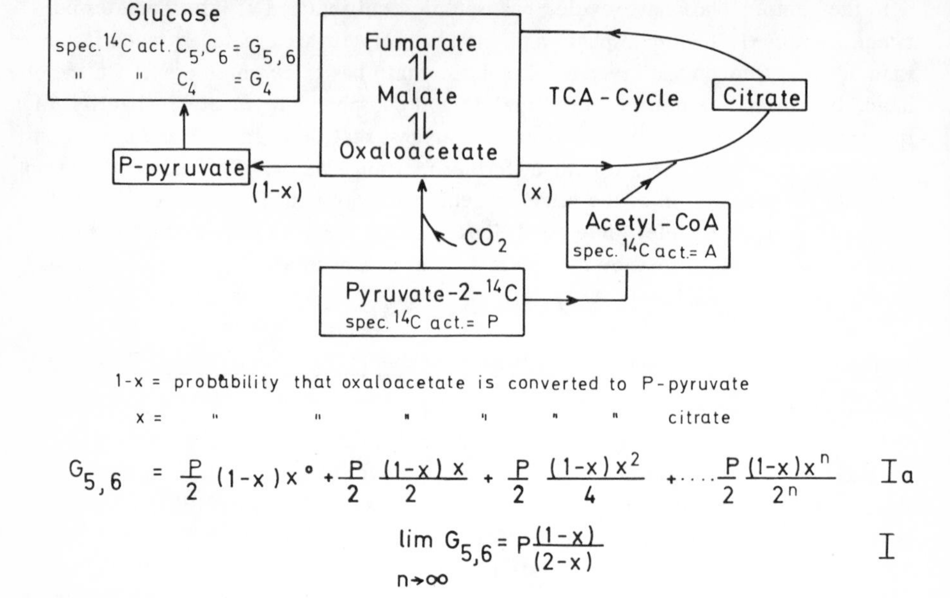

$$G_{5,6} = \frac{P}{2}(1-x)x^{\circ} + \frac{P}{2}\frac{(1-x)x}{2} + \frac{P}{2}\frac{(1-x)x^2}{4} + \cdots \frac{P}{2}\frac{(1-x)x^n}{2^n} \qquad \text{Ia}$$

$$\lim_{n \to \infty} G_{5,6} = P\frac{(1-x)}{(2-x)} \qquad \text{I}$$

$$\lim_{n \to \infty} G_4 = P\frac{(1-x)}{(2-x)} - \frac{P}{2}(1-x)x^{\circ} + \frac{A}{2}x \qquad \text{II}$$

Fig. 3. Theoretical evaluation of the interaction of the TCA cycle with the oxaloacetate formed from pyruvate-2-^{14}C.

From a comparison of the ^{14}C activity of glucose and perfusate pyruvate it was already clear that an inactive dilution must have occurred at some stage of gluconeogenesis. If the total dilution of the ^{14}C activity in glucose arose from an interaction of oxaloacetate with the TCA-cycle one would have to postulate a tremendous turnover rate of the TCA-cycle. This seems unlikely since the theoretical oxygen uptake for this high TCA-cycle activity would be much higher than the measured oxygen uptake of the liver.

When the intracellular pyruvate ^{14}C activity of 0.46 is inserted into equation I, together with the corrected ^{14}C activity of position 5 or 6 of glucose (0.12, corrected for small contribution of glycogenolysis), x will be 0.65 and the turnover rate of the TCA-cycle is calculated to be 75 μ moles/h/g liver fresh. This is somewhat higher than calculated on the basis of oxygen consumption (62 μ moles/h/g liver fresh), (10). With the corrected value for the glucose 4 position (0.35) we get from equation II (Fig. 3) a value for the ^{14}C activity of acetyl-CoA entering the TCA-cycle of 0.97. This would be two times the ^{14}C activity of the precursor pyruvate pool. These results indicate that interaction of the TCA-cycle with the gluconeogenic pathway may be the explanation for the diluted activity of position 5 and 6 of glucose, but cannot account for the high ^{14}C activity of the 4 position of glucose.

Pyruvate Permeation through the Liver Cell Membrane

From the low ^{14}C activity of lactate leaving the liver it was concluded that the ^{14}C activity of intracellular pyruvate was only half the ^{14}C activity of the extracellular pyruvate. The experimental data are summarized in Fig. 4.

The high ^{14}C activity found for the pyruvate isolated from the tissue extract together with the unchanged venous pyruvate ^{14}C activity is in this connection indicative of a low intracellular concentration of pyruvate and shows that the permeability of the liver cell membrane must be limited. This is supported further by the low pyruvate concentration found in the tissue extract. An exact calculation of the intracellular pyruvate concentration was not possible because the ratio of extra-to intracellular space has not yet been measured for our perfusion conditions.

From the difference between arterial and venous pyruvate concentrations an uptake of 132 μ moles/h/g liver fresh from the perfusate was calculated. To dilute the ^{14}C activity of the intracellular pyruvate pool to 0.46 one has to assume at the same time an inactive flux into this pool as high as 154 μ moles/h/g liver fresh. This proposed inactive flux also occurs in the experiments with lactate as gluconeogenic precursor. The origin of this inactive flux into the pyruvate pool is not yet identified.

Total influx into the intracellular pyruvate pool should be equal to the outflux, assuming steady state conditions. Only 35% of the outflux was found as glucose and lactate. Ketone bodies and alanine production were negligible. Compared with the starting pyruvate-2-^{14}C containing medium a decrease in ^{14}C activity of 8% was measured for the acidified medium leaving the liver. This activity difference represents carbon dioxide plus metabolites taken up by the liver. For the discrimination between both processes exact CO_2 determinations would be necessary. Calculated with the intracellular pyruvate ^{14}C activity this difference would be equal to a flux rate of 100 μ moles/h/g liver fresh. Seven and one-half percent of the ^{14}C activity in the perfusate leaving the liver was not identified and are equivalent to an outflow of 80 μmoles/h/g liver fresh from the intracellular pyruvate pool into the perfusate.

^{14}C- Activities of Other Intracellular Metabolites

In Table 2, ^{14}C activities of other metabolites isolated from the tissue extract are shown. Citrate had nearly the same ^{14}C activity as malate. Furthermore, degradation of citrate with citrate lyase (EC 4.1.3.6) showed that the acetate unit in citrate was hardly labeled and probably originated from an inactive acetyl-CoA pool. However, ß-hydroxybutyrate, which is also connected to the acetyl-CoA pool, incorporated ^{14}C activity. This may reflect the established compartmentation of the mitochondrial acetyl-CoA pool (Fritz 1968) but it is nevertheless, difficult to correlate these activities with the activity pattern found for glucose indicating the involvement of an acetyl-CoA pool of relatively high ^{14}C activity. The low ^{14}C activities measured for the amino acids aspartate and glutamate were also problematic if one assumes equilibration of these amino acids with the corresponding ketoacids of the TCA-cycle (Krebs and Veech 1969). The low ^{14}C activity of glycerol-3-phosphate in comparison with the triose units of glucose is indicative for a compartimentation of the triosephosphate pool as suggested by Müllhofer et al. (1969).

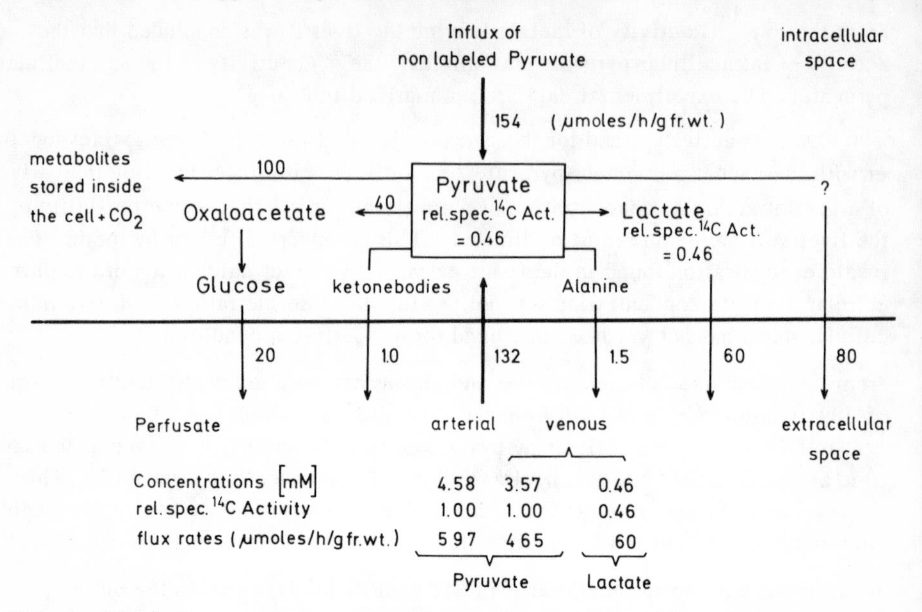

Fig. 4. Metabolic considerations concerning the intracellular pyruvate pool.

Tissue extract:		pyruvate	lactate
	Concentrations (mM)	1.97	0.34
	rel. spec. ^{14}C activities	0.96	0.45

Metabolite	Rel. spec. ^{14}C Activity Pyruvate arterial = 1.00
Malate	0.92
Citrate	
total	1.04
malate from citrate	0.95
ß-Hydroxybutyrate	0.34
Glutamate	0.46
Aspartate	0.40
Glycerol-3P	0.25

Table 2. The ^{14}C activities of different metabolites isolated from the tissue extract.

Conclusions

The results indicate that an interaction of the TCA-cycle with oxaloacetate formed by carboxylation of pyruvate-2-^{14}C as shown in Fig. 3 cannot explain the labeling pattern of glucose. More complicated interactions with other metabolites or unknown intracellular compartmentations have to be considered. The ^{14}C activity distribution especially in the citrate isolated from the tissue extract further supports this concept. For the oxaloacetate prepared from citrate by degradiation with citrate lyase a higher ^{14}C activity was measured as the one assumed for the intracellular pyruvate pool. This would indicate interaction with a TCA-cycle in which acetyl-CoA is highly ^{14}C labeled and is therefore incompatible with the non-labeled acetate formed in the citrate lyase reaction. A formation of citrate in the cytosol by reductive carboxylation of ketoglutarate cannot explain the measured ^{14}C activities. The main difficulties arise from the fact of the measured inactive dilution of the lactate pool. The production of lactate and glucose without pyruvate as exogenous substrate is very low. Therefore it must be concluded that this inactive dilution is initiated by the addition of pyruvate.

As shown in Fig. 4, glucose and lactate formation cannot account for the postulated high influx into the intracellular pyruvate pool. The identification of the remaining ^{14}C labeled compounds in the perfusate leaving the liver and an exact analysis of ^{14}C -carbondioxide production are required to understand the metabolic interactions.

Acknowledgments

We wish to thank Prof. Bücher for his many stimulating discussions. One of us (O.K.) is indebted to the Deutsche Forschungsgemeinschaft for a research grant. The technical assistance of Miss Ch. Miltz is gratefully acknowledged. This research was supported by the Sonderforschungsbereich 51 für Medizinische Biochemie und Molekularbiologie.

References

Exton, J. H., C. R. Park: J. biol. Chem. 242 (1967) 2622

Fritz, I. B., in: Cellular Compartmentalization and Control of Fatty Acid Metabolism, Ed. by F. C. Grain. Proc. IV Meeting of FEBS, Oslo 1967, Academic, New York 1968, p. 39

Krebs, H. A., R. L. Veech, in: The Energy Level and Metabolic Control in Mitochondria, Ed. by S. Papa, J. M. Tager, E. Ouagliariello and E. C. Slater. Adriatica Editrice, Bari 1969, p. 329

Krebs, H. A., T. Gascoyne, B. M. Notton: Biochem. J. 102 (1967) 275

Müllhofer, G., O. Kuntzen, S. Hesse, Th. Bücher: FEBS Letters 4 (1969) 33

Newsholm, E. A., W. Gevers: Vitam. and Horm. 25 (1967) 1

Ross, B. D., R. Hems, H. A. Krebs: Biochem. J. 102 (1967) 942

Scholz, R., in: Stoffwechsel der isoliert perfundierten Leber, Ed. by W. Staib and R. Scholz. Springer, Heidelberg 1968, p. 25

Teufel, H., L. A. Menahan, J. C. Shipp, S. Böning, O. Wieland: Europ. J. Biochem. 2 (1967) 182

Williamson, J. R., E. T. Browning, M. S. Olson, in: Advances in Enzyme Regulation, Ed. by G. Weber. Pergamon Press 1968, p. 67

Williamson, J. R., R. Scholz, R. G. Thurman, B. Chance, in: The Energy Level and Metabolic Control in Mitochondria, Ed. by S. Papa, J. M. Tager, E. Quagliariello and E. C. Slater. Adriatica Editrice, Bari 1969a, p. 411

Williamson, J. R., R. Scholz, E. T. Browning: J. biol. Chem. 244 (1969b) 4617

Discussion to Müllhofer and Kuntzen

Ballard: It occurs to me that you could explain these results if your isotope was 1-^{14}C-pyruvate rather than 2-^{14}C-pyruvate. Did you check the position of the label in pyruvate?

Müllhofer: We have checked this. The pyruvate was transferred to lactate and was oxidized with cer-IV-sulphate to acetaldehyde which had the same specific activity. We don't know whether the methyl group had ^{14}C activity.

Exton: I have at least 20 questions. As I remember, our calculation of the activity of the Krebs cycle was based on at least seven assumptions. One of these assumptions was that the citric acid cycle was intact, that is to say, essentially no carbon left the cycle besides CO_2 and oxaloacetate and nothing entered it besides acetyl-Coa and oxalacetate. As there is certainly some glutamate entering the cycle and some aspartate leaving, this assumption is certainly invalid. What really concerns me and what is difficult for me to understand in your studies is the tremendous difference between the specific activity of the pyruvate entering the liver and that of the lactate leaving it. This is very strange because in our experiments the two specific activities remained very close. If you had red blood cells in your system this would have been one source of dilution of the lactate. Another possibility would be lactate formation by non-parenchymal liver cells. These presumably perform glycolysis and could yield unlabeled lactate which would dilute the specific activity of lactate in the medium. In the cells carrying out gluconeogenesis, the specific activity of lactate would of course be much higher, because in these cells there would be no dilution. This would be a partial explanation. The other problem is the specific activity of acetyl-CoA which you computed as 0.97. Is this correct?

Müllhofer: All right.

Exton: But the ketone bodies which are directly derived from acetyl-CoA have only a specific activity of 0.34. That is very strange to me. Now, of course, it is understandable that there are different pools of acetyl-CoA, but I would assume that the ketone bodies are derived directly from the intra-mitochondrial acetyl-CoA pool.

I am concerned with this and find this troublesome.

In our initial studies we found that the bulk of oxaloacetate derived from lactate or pyruvate was used for gluconeogenesis, but this was with high substrate levels. Although I realize that we loaded the system so that oxaloacetate went preferentially to gluconeogenesis and not into the cycle, I am surprised that you get such a high amount going into the cycle. With your concentration of 5.5 mM pyruvate I would expect much more to go in direction to glucose formation.

Müllhofer: Let me first try to answer part of these questions. Concerning the dilution of the citric acid cycle intermediates by glutamate, I don't know where the inactive glutamate should come from.

Exton: Protein!

Müllhofer: When glutamate or aspartate from endogenous sources would dilute intermediates of the citric acid cycle there should be a corresponding urea production, which was not observed.

Exton: You can get transamination to alanine or aspartate. . .

Müllhofer: Almost no amino acids are leaving the liver using our experimental conditions. The intracellular levels of aspartate and glutamate were quite low (0.209 and 0.245 μ mol/g liver fresh). You think that intracellular lactate might have a higher activity than the lactate leaving the liver.

Exton: I mean the specific activity in the cells actually carrying out gluconeogenesis may be much higher than that in the medium leaving the liver. There may be a number of cell types in the liver which exhibit active glycolysis.

Müllhofer: We measured the same specific ^{14}C activity for lactate in the perfusate leaving the liver and lactate isolated from the tissue extract.

Exton: Above all I am thinking of the cells carrying out gluconeogenesis compared with these other cell types.

Müllhofer: This of course would make all results from tissue extracts doubtful. When I mentioned compartmentation of the acetyl-CoA pool, I thought of mitochondrial compartmentation as discussed by Fritz (Proc. IV Meeting of FEBS, Oslo, Academic Press, London 1967, p. 39).

I would say it might be possible that the ^{14}C activity of aspartate and glutamate really represent the ^{14}C activity of corresponding ketoacids as oxaloacetate and ketoglutarate, intermediates of the citric acid cycle. The only difficulty then is that we would have to assume a separate pathway for gluconeogenesis.

Hanson: Your findings are somewhat surprising because this type of experiment has been done previously with different results. Recently Dr. Sidney Weinhouse of the Fels Research Institute, has perfused livers from 24 hour starved rats with 5 mM pyruvate-2-^{14}C. He then degraded the lactate and glucose completely in order to determine the extent of pyruvate "recycling" through pyruvate kinase. His results indicate that both C-3 and C-4 of glucose were labeled to the same extent and contained a small percentage of the total ^{14}C found in carbons 1, 2 and 5, 6 of glucose. Lactate and pyruvate also had a similar specific activity. It might be well to point out, however, that he perfused for one hour as compared to your ten minute perfusion and he recycled his pyruvate.

Müllhofer: It is reasonable to find a difference in the ^{14}C labeling pattern of glucose when pyruvate or lactate are used as substrates. In the experiments with pyruvate as substrate relative large amounts of lactate are produced besides glucose in the non-recirculating perfusion system. The NADH necessary for these processes is presumably of mitochondrial origin. To furnish these hydrogen equivalents and also the ones necessary for respiration a higher turnover of the citric acid cycle is necessary as compared to the experiments with lactate as gluconeogenic substrate. The difference between the ^{14}C activities of malate and phosphoenolpyruvate or phosphoglycerate and the determined ^{14}C activity of the carboxyl group in position 4 of malate further support the measured ^{14}C labeling pattern of glucose.

D. H. Williamson: Your patterns of labeling in glucose do not seem to be consistent with those of many in vivo experiments using variously labeled pyruvate (e. g. Friedmann et al.: J. biol. Chem. 221 (1956) 665). Can you explain the discrepancy?

Müllhofer: But these experiments were done, I think , with liver slices.

D. H. Williamson: Yes, and also in vivo.

Müllhofer: In vivo experiments are very problematic because of contributions from other tissues.

D. H. Williamson: It is well established that there is a difference in the labeling pattern of carbons 1, 2 and 3 and 4, 5 and 6 of glucose. Did you also degrade the 1, 2 and 3 positions of glucose?

Müllhofer: We assume for both triose units of glucose the same ^{14}C activity distribution. This is quite reasonable with respect to the identical total specific activity. Nonsymmetrical labeling could only result from an interaction with the pentose phosphate cycle. Arguments against this interaction were presented. This has nothing to do with the theoretical considerations we made.

Seubert: How can the specific activity of malate be higher than that of lactate?

Müllhofer: Because the acetyl-CoA entering the citric acid cycle is labeled.

Seubert: And you said the incorporation into the 4 position of glucose is a measure of recycling of carbon via the citric acid cycle? Either by condensation with radioactive acetyl-CoA formed from the C_3-precursors or by an equilibration at the C_4 level.

Müllhofer: Carbon atoms originating from acetyl-CoA are incorporated into malate by interaction of the gluconeogenic pathway with the citric acid cycle.

Seubert: How then can the labeling of the C_4 position of glucose from 2-C^{14}-lactate and 2-C^{14} -pyruvate be different (0.05 and 0.32, respectively) if one assumes that oxaloacetate formed from lactate and pyruvate, and pyruvate formed from lactate are going through the same compartment, the mitochondrial compartment. Identical turnover rates of carbon from both lactate and pyruvate via the citric acid cycle and identical labeling of the 4 position of glucose should be expected in this case. From the lower contribution of carbon atom 2 from lactate- as compared with the 2 position of pyruvate- for the 4 position of glucose, an incomplete equilibration of pyruvate formed from lactate with the mitochondrial pyruvate pool has to be deduced in gluconeogenesis from lactate. This only can be achieved by an extra-mitochondrial PEP-synthesis from lactate.

Müllhofer: The ^{14}C labeling pattern of glucose is, as I already mentioned, determined by the turnover rate of the citric acid cycle relative to the rate of gluconeogenesis and the extent of acetyl-CoA formation from pyruvate.

Krebs: Is the following observation relevant? When liver is perfused with a medium containing little or no lactate, within a very short time lactate accumulates from endogenous sources to reach a concentration of 1 or 2 mM. Subsequently lactate production ceases. It probably originates from the glycogen store. Could this account for the difference in the labeling between pyruvate and lactate?

Müllhofer: We considered this possibility. Lactate, pyruvate and glucose production were very low without addition of pyruvate, this indicates that glycogen was really depleted. Of course we cannot exclude the fact that glycogenolysis was activated by pyruvate addition and that inactive pyruvate was produced by glycolysis independent of the pathway of gluconeogenesis as indicated by Threlfall and Heath (Biochem. J. 110 (1968) 303). We activated glycogenolysis in livers of 48 hour starved rats with cyclic AMP but there was no increased production of lactate, pyruvate and glucose; these metabolite levels remained very low.

General Discussion

Hanson: I would like to ask Dr. J. R. Williamson to comment on the difficulties and reliability of oxaloacetate measurements by existing techniques.

J. R. Williamson: The reproducibility of the oxaloacetate measurement is good if a good fluorometer is used. We can demonstrate excellent recoveries even with lyophylized tissue samples.

Hanson: Are you confident that you can measure changes in oxaloacetate concentration, say 20 or 30%, which make such large changes in crossover plots,

J. R. Williamson: Yes, but one need not rely entirely on changes of oxalacetate to illustrate control at pyruvate carboxylase, since aspartate levels often reflect the directional change of oxalacetate. I have a comment to Dr. Müllhofer. I suggest that a lot of your problems in interpreting the results arise from the use of ^{14}C-pyruvate and ^{14}C-lactate labeled in the 2 position. Malate will then be labeled by the reactions of the citric acid cycle in addition to pyruvate carboxylase and malic enzyme. For this reason we used 1-^{14}C-pyruvate and lactate in our experiments. My second comment is directed to Dr. Exton. I believe that you freeze your livers very shortly after the addition of isotopic substrate. Since substrate addition will alter the pool size of intermediates, spurious conclusions may be drawn from data obtained by measuring the count incorporation into the particular intermediates, particularly if the system is not in a steady state. I think it is very necessary for experiments with isotopes to be as short as possible, and to ensure that metabolism is proceeding under steady-state conditions.

Exton: Let me explain in greater detail how our experiments were carried out: There was an initial one-hour recirculation without substrate to produce a very steady state and, according to our experience, a low rate of gluconeogenesis. Then we changed to "flow-through" perfusion with substrate. This continued for 2.5, 8.5 or 15 minutes. The labeled material was usually infused for these time periods but in some cases it was infused for a shorter time. The intermediates in the control livers achieved stable levels as shown by the fact that the concentrations at 2.5 minutes were essentially the same as at 8.5 minutes. That means the livers were close to a steady-state with respect to the chemical levels of the intermediates. They were not, however, in a steady state with respect to radioactivity as indicated by changes in the specific activity from 2.5 to 8.5 minutes. We deliberately planned it this way since we wanted to observe early effects and to amplify these effects. We have also done long-term perfusions where a steady radioactive state is achieved but the changes are not so dramatic.

J. R. Williamson: Of course the incorporation of isotopes depends on the pool size of the particular intermediate. So it is better to express the results in terms of specific activity.

Exton: I agree with you. Every piece of information has to be interpreted in terms of precursor pool specific activity. One also has to take into consideration the possibility of different specific activities in different pools, and herein lies the great difficulty.

Walter: Dr. Müllhofer, have you done this type of experiment also as a function of time?

Müllhofer: We got exactly the same labeling pattern in all the metabolites whether we made the freeze stope after seven minutes or after 14 minutes. This means that isotopic steady state was already established.

Walter: Were the specific radioactivities the same too?

Müllhofer: Yes.

J. R. Williamson: I have been attacked about the accuracy of my oxaloacetate measurements, but I have a question for Dr. Exton: I can't see how you could isolate oxaloacetate and determine its specific activity!

Exton: Let me clarify this. We have measured oxaloacetate many times. In almost every experiment oxaloacetate was measured. However we didn't put any numbers in our tables because of the great variation and our lack of confidence in the measurements. Now, as far as radioactivity is concerned, as soon as the liver was frozen it was powdered and extracted with $HClO_4$. To the unneutralized extract we added unlabeled oxaloacetate and then 2, 4-dinitro-phenyl-hydrazine. The hydrazones of oxaloacetate, pyruvate and α- oxoglutarate were separated on cellulose chromatoplates.

J. R. Williamson: I fail to see how enough counts can be incorporated in oxaloacetate to measure them properly.

Exton: We add very large amounts of radioactivity in these experiments. I don't think there was any radioactivity measurement in the tables which was not hundreds of counts above background.

J. R. Williamson: How many microCurie did you use?

Exton: Usually in the range of ten or a hundred microCuries.

Söling: Dr. Exton, if I recall it right, you never observed this discrepancy between the specific activities of pyruvate inside and outside the liver cells seen by Dr. Müllhofer. Did you employ the same technical procedure ad Dr. Müllhofer, in that you compared the specific radioactivity of pyruvate with that of lactate. I ask this because Dr. Müllhofer had the same specific activity for pyruvate inside and outside the liver cells but found a lower specific activity in the lactate leaving the liver. Did you do this type of experiment, Dr. Exton? Because, apparently, to measure the specific activity of pyruvate inflowing into the liver and in the liver itself is not sufficient to measure the dilution of the endogenous pyruvate pool.

Exton: The measurements were done in frozen liver and we measured the specific activity of the pyruvate and of lactate.

Söling: And they were in the same range?

Exton: Yes.

J. R. Williamson: I would like to ask Prof. Krebs about his feelings concerning the use of isotopes.

Krebs: I emphasized some time ago that the distribution of isotopes can be very complicated (Krebs et al. : Biochem. J. 101 (1966) 242). Perhaps it was this that Dr. Müllhofer had in mind when he said that we still have a lot to learn about pathways. I think we really know the pathways in all their essentials, but there is still a lot to be learned about the kinetics, especially the kinetics of exchange reactions. In the paper quoted we showed that a large fraction of ^{14}C labeled acetoacetate appears in glucose when incubated with kidney slices, and a large fraction of ^{14}C labeled lactate appears as carbon dioxide, although lactate has not undergone net oxidation and acetoacetate cannot give rise to a net formation of glucose.

The transfer of label in this situation is connected with the fact that oxaloacetate is an intermediate in both gluconeogenesis and terminal oxidations. Other exchange reactions are brought about by transaminases. Because of exchange reactions and compartmentation the interpretation of isotope data is often very difficult.

Hanson: I wonder if Dr. Seubert would care to summarize what he feels is the current status of pyruvate carboxylase distribution in liver of various species . . .with emphasis on the rat?

Seubert: I think we should not discuss about feelings. I have feelings, that is right, but I would rather do more experiments. As an important result of the isotope experiments, I consider Dr. Müllhofer's finding that lactate and pyruvate behave so differently. With respect to the incorporation of carbon atom 2 into the 4 position of glucose, I cannot see how intermediates formed from both precursors should behave so differently if they have to go through the mitochondrial compartment.

Söling: One point which should be discussed more extensively, I think, was the finding of Dr. Schoner concerning the role and regulation of pyruvatekinase activity. I would like to ask the audience what it thinks about this type of regulation. Does alanine play a physiological role in regulation of pyruvatekinase and gluconeogenesis, and what is the physiological role of the ATP effect especially with respect to the point that ATP may rather act by complex formation with Mg^{++}?

J. R. Williamson: I was very happy to see these data, and I believe the effect is a very important one. I first learned of the alanine effect on pyruvate kinase from the talk Dr. Seubert gave in Indianapolis (Adv. Enzyme Regulat. 6 (1968) 153). Since then Dr. Brian Safer has done some work with the rat liver type L enzyme, and has shown that the alanine inhibition can be overcome by very low concentrations of FDP, in agreement with Dr. Schoner's work. Probably, the inhibitions by ATP and alanine are synergistic, and since they can both be overcome by FDP with complex kinetics dependent on the PEP concentration, there is the possibility of a very fine control of pyruvate kinase activity by ATP, ADP, alanine, PEP and FDP, with of course, a magnesium involvement as well. I think that an altered control of pyruvate kinase — or more properly, lack of control — accounts for abnormal gluconeogenesis in the adrenalectomized state. We followed up an observation from Dr. Park's laboratory, namely that livers from adrenalectomized rats failed to respond to glucagon by increased rates of gluconeogenesis. We confirmed this observation and showed in addition that gluconeogenesis in these livers was also not increased by oleate. However, the normal changes of acetyl-CoA, and increased oxygen consumption and mitochondrial oxidation-reduction state (i. e. increased ß-hydroxybutyrate/acetoacetate ratios) were observed. The crossover patterns of the gluconeogenic intermediates suggested an interaction between pyruvate and PEP (see proceedings of the symposium, "The Action of Hormones — Genes to Population," edited by P. P. Foa, in press). We suggested that the defect may lie in an inability to inhibit pyruvate kinase sufficiently, because of the low tissue alanine levels in livers from adrenalectomized rats. Activation of pyruvate carboxylase results in more PEP, but this is recycled back to pyruvate with a consequent waste of ATP, rather than converted to glucose. So I think that there is at least one metabolic situation where extensive recycling occurs. Diminished gluconeogenesis induced by ethanol may provide another example. However, a certain amount of recycling may always be occurring in vivo except under conditions of extreme starvation or diabetes, when enzyme adaptative changes have occurred fully.

D. H. Williamson : I would not disagree with Dr. John Williamson about this particular point, because I have no information on adrenalectomized animals. However, there are two points I would like to make concerning measurements of hepatic alanine in vivo: 1) alanine rises rapidly during ischemia, so that freeze-clamping of the tissue is essential (D. H. Williamson et al.: Biochem. J. 104 (1967) 497. 2) the hepatic alanine is about 1.2 mM in the fed state, but falls progressively in the starved rat, in the starved rat treated with phlorrizin and in the alloxan-diabetic rat. These are all situations where there is a high rate of gluconeogenesis and where pyruvate kinase might be expected to be inhibited. From this evidence one might reasonably conclude that alanine is not an important regular of pyruvate kinase in vivo.

Seubert: In all the situations you just referred to we never mentioned alanine as being an inhibitor (Seubert et al.: Adv. Enzyme Regul. 6 (1968), because in these situations the activity of pyruvate kinase is anyhow very low. Dr. Schoner has only referred to cortisol treatment (Schoner et al.: Hoppe Seylers Z. Physiol. Chem., in press), and there you certainly have a rise of alanine. You probably refer to the publication of Feigelson (Biochem. biophys. Acta 104 (1965) 92).

D. H. Williamson: No, I refer to our own in vivo measurements (see above).

Seubert: You mentioned the effects of ischemia on alanine levels. Dr. Schoner has done experiments, where he avoided ischemia, and he could confirm the data of Feigelson, not the absolute values, but with respect to the relation between the control and in the cortisol-treated animals. Let me stress once more: if we discuss the role of alanine, we only refer to situations after cortisol treatment.

Concerning your question, Dr. Söling and concerning the role of alanine, we are in the same situation now as with acetyl-CoA. In the case of pyruvate carboxylase so many effectors act together that you never can deduce from one metabolite to its physiological function. In the scheme of Dr. Schoner you have seen that three metabolites are put in direction of the H-form and two metabolites in the direction of the T-form.

Walter: I want to mention a publication of Dr. Kesner's group (G. Paleologos et al.: Proc. Soc. Exp. Biol. Med. 132 (1969) 270). They found that the level of alanine in liver was reduced under conditions of enhanced gluconeogenesis.

Seubert: As already mentioned, in this state you do not need this type of control. As has been shown by Prof. Krebs, the total activity of pyruvate kinase goes down. So another mechanism of control is involved.

Guder: Under the conditions where you found these inhibitions, Dr. Schoner, did you need to preincubate your enzyme with the inhibitor? And did you get this type of inhibition with a crystalized enzyme?

Schoner: In order to observe the inhibition of liver pyruvate kinase by L-alanine you do not need a preincubation. The inhibitory effect is immediate and was found with an enzyme preparation purified 100 times.

I would also like to mention that it seems to me very difficult to study the effect of alanine on pyruvate kinase in rat liver; since the function of the liver is to synthesize glucose and not to glycolyze (Glogner, P. and Schurek, H.: Z. Klin. Chem. u. Klin. Biochem. 7 (1969) 590). Since pyruvate kinase of yeast shows similar characteristics as hepatic pyruvate kinase, we looked for a possible inhibition of the enzyme by alanine in this organism. In the crude extract of yeast, pyruvate kinase was inhibited by about 50% by 13 mM alanine. We therefore hope that studies with alanine in this organism under different experimental conditions may give more information on the physiological significance of the alanine inhibition of pyruvate kinase.

Guder: Do you think that this has something to do with the enzyme constitution?

Schoner: Yes. If we compare the kinetics of the optical assay for pyruvate kinase in presence and absence of the negative effectors, ATP or alanine, we observe that in the presence of alanine or ATP only a part of PEP is hydrolized. We therefore suggest that hepatic pyruvate kinase exists in two conformational states.

Krebs: We discussed earlier the question of why it is difficult to demonstrate gluconeogenesis in homogenates. One of the factors is of course the high activity of pyruvate kinase which reconverts phosphopyruvate to pyruvate. We have tried to inhibit pyruvate kinase by addition of high concentrations of alanine, of ATP and of Ca ions but without success. In this context I would like to mention that G. Weber recently showed that phenylalanine is an effective inhibitor of pyruvate kinase, especially in the brain (Proc. Nat. Acad. Sci. 63 (1969) 1365). Weber suggests that the brain damage in phenylketonuria may be connected with an inhibition of pyruvate kinase. Finally, regarding Dr. Schoner's remark that the liver has to synthesize glucose this is true for the starved liver. When there is an excess of carbohydrate the liver can contribute to the synthesis of fat and this involves glycolysis to the stage of pyruvate. Thus, under some conditions, hepatic glycolysis and pyruvate kinase are important. In fact there are adaptive mechanisms which increase pyruvate kinase activity by a large factor when there is an excess of carbohydrate in the diet (Krebs and Eggleston: Biochem. J. 94 (1965) 3c).

Walter: I have a question to Dr. Seubert: I think it was your group which first proposed that acetyl-CoA was an inhibitor of pyruvate kinase, and then I think you showed that it was due to an impurity, but then later on Weber claimed that acetyl-CoA was an inhibitor. I would like to know whether he now also thinks that this was due to an impurity or whether he still sticks to his idea?

Seubert: Well, that is not my problem. I must admit we made a mistake when we stated this in a paper. We have then purified acetyl-CoA and then we could no longer observe this inhibition (W. Seubert et al.: Adv. Enzyme Regul. 6 (1968) 153). As far as I remember, G. Weber observed his inhibition only after a preincubation. This compares well with the fatty acyl-CoA inhibition.

Schoner: It is true that we made a mistake with respect of the acetyl-CoA inhibition of hepatic pyruvate kinase. The effects of Weber may be explained on the basis of a disulphate formation. As was recently shown by Dr. Kutzbach in Prof. Hess' laboratory (Zeitschr. Physiol. Chemie 351 (1970) 272) SH-groups of hepatic pyruvate kinase are involved in the activation or inhibition of the enzyme by allosteric effectors. It seems to me that Dr. Weber affected this SH-group by preincubating the enzyme with acetyl-CoA.

Concerning the remarks of Prof. Krebs, I would like to say that we discuss an inhibition of pyruvate kinase by alanine for the gluconeogenic state after cortisol treatment only (W. Seubert et al.: Adv. Enzyme Regul. 6 (1968) 153). During other gluconeogenic conditions hepatic pyruvate kinase may be regulated in a different way.

Söling: I have a question for Dr. J. R. Williamson. Do you believe that under conditions where you stimulate gluconeogenesis from lactate by fatty acids in rat liver the pyruvate kinase reaction is accelerated or slowed down? Because everything goes in the direction for acceleration; at least the effectors we know.

J. R. Williamson: The information we have on this point is that pyruvate kinase activity appears to be relatively unchanged upon addition of lactate. Thus, if six times the rate of glucose production obtained after lactate addition is divided by the increment of oxygen consumption to obtain an in vivo P/O ratio, values close to three are obtained (Williamson et al.: J. Biol. Chem. 244 (1969) 4617). A similar value was obtained after oleate addition, showing a lack of uncoupling effect of the fatty acid. However, the possibility remains that recycling between pyruvate and PEP occurs in substrate-free livers. We have recently obtained evidence in favor of this possibility by the use of xylitol in perfused livers, which increases glucose production (using about 1.2 moles of ATP per mole of glucose formed) but decreases the oxygen consumption. Xylitol is a strong reductant for the cytoplasmic pyridine nucleotides, and therefore causes large decreases in the levels of PEP, 2PGA and 3PGA. The fall of PEP presumably results in diminished pyruvate kinase activity.

Krebs: We do not know exactly why it is blocked and why "futile cycles" do not occur to a major extent.